Springer-Verlag Berlin Heidelberg GmbH

Robert J. White

RNA Polymerase III Transcription

Second Edition

Springer

Robert J. White

Institute of Biomedical and Life Sciences
Division of Biochemistry and Molecular Biology
University of Glasgow
Glasgow, Scotland, U.K.

ISBN 978-3-662-03520-7 ISBN 978-3-662-03518-4 (eBook)
DOI 10.1007/978-3-662-03518-4

Biotechnology Intelligence Unit

Library of Congress Cataloging-in-Publication data

White, Robert J., 1963-
RNA polymerase III transcription / Robert J. White. — 2nd ed.
p. cm.
Previously published: 1994.
Includes bibliographical references and index.
1. RNA polymerases. 2. Genetic transcription. I. Title.
QP606.R53W48 1998
572.8'845—dc21
for Library of Congress

98-10636
CIP

Typesetting: R.G. Landes Company, Georgetown, TX, U.S.A.

SPIN 10676413 31/3111 - 5 4 3 2 1 0 - Printed on acid-free paper

Dedication

To Clare

PREFACE

Eukaryotic organisms contain three nuclear RNA polymerases, each of which is responsible for synthesizing a distinct set of essential products.[1] RNA polymerase I (pol I) synthesizes the 5.8S, 18S and 28S ribosomal RNAs (rRNAs), RNA polymerase II (pol II) synthesizes messenger RNA (mRNA) and most small nuclear RNAs (snRNA), and RNA polymerase III (pol III) synthesizes transfer RNA (tRNA), 5S rRNA, and a variety of other small cellular and viral RNAs.[1-7] The aim of this book is to review the substantial body of work that has been published concerning transcription by pol III—the largest and most complex of the eukaryotic RNA polymerases.

Chapter 1 begins by describing the genes that serve as templates for pol III transcription. These are often referred to as class III genes. Chapter 2 then examines the promoter structures of these genes and, in so doing, tries to ascertain what features of their sequence and organization mark them for transcription by pol III. Chapter 3 deals in detail with what is known about the pol III enzyme. Specific transcription by pol III requires the involvement of accessory factors in addition to the polymerase itself.[5,8,9] Chapter 4 describes the complex array of transcription factors that are involved in recruiting pol III to the appropriate sites on the appropriate sets of genes. Chapter 5 then examines how these various factors interact with one another and with the pol III enzyme in order to assemble a correctly positioned initiation complex. Chapter 6 describes what is known about the process of transcription itself. It is divided into sections dealing with initiation, elongation, termination and reinitiation. Most transcriptional studies have used naked DNA as template. However, genes in the living cell are packaged into chromatin and this can have a powerful influence upon their level of expression. Chapter 7 considers the chromatin structure of class III genes. Chapter 8 deals with the regulatory proteins that have been shown to modulate the level of pol III transcription. Chapter 9 then describes the mechanisms which cells employ to regulate expression of class III genes in response to changes in environmental conditions. Finally, chapter 10 provides a brief overview of the current state of the field and the challenges that still face it.

Naturally, it is not possible to provide a detailed analysis of every aspect of pol III research in a book of this size. As a consequence, I have concentrated upon those topics that I consider to be of the greatest interest and importance. I have, however, tried to at least mention all the areas of class III transcription that have been the subject of study and to provide references which will allow the reader to pursue each topic further. I can only apologize to those of my colleagues whose work has been omitted owing to oversight or lack of space. I would, however, like to draw the reader's attention to other review articles dealing with aspects of pol III transcription that have been published over the years; these may help to fill in some of the gaps left by this particular publication.[10-50]

References

1. Chambon P. Eukaryotic nuclear RNA polymerases. Annu Rev Biochem 1975; 44:613-635.
2. Roeder RG, Rutter WJ. Multiple forms of DNA-dependent RNA polymerase in eukaryotic organisms. Nature 1969; 224:234-237.
3. Weinmann R, Raskas HJ, Roeder RG. Role of DNA-dependent RNA polymerases II and III in transcription of the adenovirus genome late in productive infection. Proc Natl Acad Sci USA 1974; 71:3426-3430.
4. Weinmann R, Brendler TG, Raskas HJ et al. Low molecular weight viral RNAs transcribed by RNA polymerase III during Ad2-infection. Cell 1976; 7:557-566.
5. Parker CS, Roeder RG. Selective and accurate transcription of the *Xenopus laevis* 5S RNA genes in isolated chromatin by purified RNA polymerase III. Proc Natl Acad Sci USA 1977; 74:44-48.
6. Rosa MD, Gottlieb E, Lerner M et al. Striking similarities are exhibited by two small Epstein-Barr virus-encoded ribonucleic acids and the adenovirus-associated ribonucleic acids VAI and VAII. Mol Cell Biol 1981; 1:785-796.
7. Zieve GV. Two groups of small stable RNAs. Cell 1981; 25:296-297.
8. Ng SY, Parker CS, Roeder RG. Transcription of cloned *Xenopus* 5S RNA genes by *X. laevis* polymerase III in reconstituted systems. Proc Natl Acad Sci USA 1979; 76:136-140.
9. Weil PA, Segall J, Harris B et al. Faithful transcription of eukaryotic genes by RNA polymerase III in systems reconstituted with purified DNA templates. J Biol Chem 1979; 254:6163-6173.
10. Hall BD, Clarkson SG, Tocchini-Valentini G. Transcription initiation of eukaryotic transfer RNA genes. Cell 1982; 29:3-5.
11. Korn LJ. Transcription of *Xenopus* 5S ribosomal RNA genes. Nature 1982; 295:101-105.
12. Ciliberto G, Castagnoli L, Cortese R. Transcription by RNA polymerase III. Curr Topics Dev Biol 1983; 18:59-88.
13. Brown DD. The role of stable complexes that repress and activate eukaryotic genes. Cell 1984; 37:359-365.
14. Sharp SJ, Schaack J, Cooley L et al. Structure and transcription of eukaryotic tRNA genes. CRC Crit Rev Biochem 1984; 19:107-144.
15. Sentenac A. Eukaryotic RNA polymerases. CRC Crit Rev Biochem 1985; 18:31-90.
16. Geiduschek EP, Tocchini-Valentini GP. Transcription by RNA polymerase III. Annu Rev Biochem 1988; 57:873-914.
17. Sollner-Webb B. Surprises in polymerase III transcription. Cell 1988; 52:153-154.
18. Wolffe AP, Brown DD. Developmental regulation of two 5S ribosomal RNA genes. Science 1988; 241:1626-1632.
19. Millstein LS, Gottesfeld JM. Control of gene expression in eukaryotic cells: lessons from class III genes. Curr Opin Cell Biol 1989; 1:497-502.
20. Murphy S, Moorefield B, Pieler T. Common mechanisms of promoter recognition by RNA polymerases II and III. Trends Genet 1989; 5:122-126.
21. Palmer JM, Folk WR. Unraveling the complexities of transcription by RNA polymerase III. Trends Biochem Sci 1990; 15:300-304.
22. Paule MR. In search of the single factor. Nature 1990; 344:819-820.
23. Dahlberg JE, Lund E. How does III x II make U6? Science 1991; 254:1462-1463.
24. Gabrielsen OS, Sentenac A. RNA polymerase III (C) and its transcription factors. Trends Biochem Sci 1991; 16:412-416.

25. Kunkel GR. RNA polymerase III transcription of genes that lack internal control regions. Biochim Biophys Acta 1991; 1088:1-9.
26. Shastry BS. Xenopus transcription factor IIIA (XTFIIIA): after a decade of research. Prog Biophys Mol Biol 1991; 56:135-144.
27. Wolffe AP. RNA polymerase III transcription. Curr Opin Cell Biol 1991; 3:461-466.
28. Wolffe AP. Developmental regulation of chromatin structure and function. Trends Cell Biol 1991; 1:61-66.
29. Geiduschek EP, Kassavetis GA. RNA polymerase III transcription complexes. In: McKnight SL, Yamamoto KR, eds. Transcriptional Regulation. Cold Spring Harbor, NY: Cold Spring Harbor Laboratory 1992:247-280.
30. Gill G. Complexes with a common core. Curr Biol 1992; 2:565-567.
31. Green MR. Transcriptional transgressions. Nature 1992; 357:364- 365.
32. Hernandez N. Transcription of vertebrate snRNA genes and related genes. In: McKnight SL, Yamamoto KR, eds. Transcriptional Regulation. Cold Spring Harbor, NY: Cold Spring Harbor Laboratory 1992:281-313.
33. Sentenac A, Riva M, Thuriaux P et al. Yeast RNA polymerase subunits and genes. In: McKnight SL, Yamamoto KR, eds. Transcriptional Regulation. Cold Spring Harbor, NY: Cold Spring Harbor Laboratory 1992:27-53.
34. Sharp PA. TATA-binding protein is a classless factor. Cell 1992; 68:819-821.
35. Sprague KU. New twists in class III transcription. Curr Opin Cell Biol 1992; 4:475-479.
36. Thuriaux P, Sentenac A. Yeast nuclear RNA polymerases. In: The Molecular and Cellular Biology of the Yeast *Saccharomyces*: Gene Expression. Cold Spring Harbor, NY: Cold Spring Harbor Laboratory 1992; II:1-48.
37. White RJ, Jackson SP. The TATA-binding protein: a central role in transcription by RNA polymerases I, II and III. Trends Genet 1992; 8:284-288.
38. White RJ, Rigby PWJ, Jackson SP. The TATA-binding protein is a general transcription factor for RNA polymerase III. J Cell Science 1992; 16 (Supp): 1-7.
39. Hernandez N. TBP, a universal eukaryotic transcription factor? Genes Dev 1993; 7:1291-1308.
40. Pieler T, Theunissen O. TFIIIA: nine fingers-three hands? Trends Biochem Sci 1993; 18:226-230.
41. Rigby PWJ. Three in one and one in three: it all depends on TBP. Cell 1993; 72:7-10.
42. Tafuri SR, Wolffe AP. Dual roles for transcription and translation factors in the RNA storage particles of *Xenopus* oocytes. Trends Cell Biol 1993; 3:94-98.
43. Willis IM. RNA polymerase III. Genes, factors and transcriptional specificity. Eur J Biochem 1993; 212:1-11.
44. Struhl K. Duality of TBP, the universal transcription factor. Science 1994; 263:1103-1104.
45. Geiduschek EP, Kassavetis GA. Comparing transcriptional initiation by RNA polymerases I and III. Curr Opin Cell Biol 1995; 7:344-351.
46. White RJ. Coordination of nuclear RNA polymerase activity. J NIH Research 1995; 7:48-49.
47. Zawel L, Reinberg D. Common themes in assembly and function of eukaryotic transcription complexes. Annu Rev Biochem 1995; 64:533-561.
48. Gottesfeld JM, Forbes DJ. Mitotic repression of the transcriptional machinery. Trends Biochem Sci 1997; 22:197-202.
49. White RJ. Regulation of RNA polymerases I and III by the retinoblastoma protein: a mechanism for growth control? Trends Biochem Sci 1997; 22:77-80.
50. Larminie CGC, Alzuherri HM, Cairns CA et al. Transcription by RNA polymerases I and III: a potential link between cell growth, protein synthesis and the retinoblastoma protein. J Mol Med 1998; 76:94-103.

CONTENTS

1. **Class III Genes** 1
   - 5S rRNA GENES 1
   - tRNA Genes 2
   - Viral Class III Genes 4
   - U6 snRNA Genes 4
   - Other Class III Genes Encoding RNP Components 5
   - SINEs 7
   - Class II Genes 11

2. **Promoter Structure of Class III Genes** 23
   - 5S rRNA Genes 24
   - tRNA Genes 27
   - The VA_I Gene 31
   - The Vault RNA Gene 32
   - The EBER2 Gene 32
   - BC1 Genes 34
   - 7SL Genes 34
   - Alu SINEs 34
   - U6 Genes 36
   - H1 Genes 40
   - MRP Genes 40
   - Y RNA Genes 41
   - 7SK Genes 41
   - Class II Genes 41

3. **RNA Polymerase III** 57
   - Biochemical Characterization 57
   - Genetic Characterization 61

4. **Transcription Factors Utilized by RNA Polymerase III** 77
   - General Factors 77
   - Gene-Specific Factors 102

5. **Transcription Complex Formation on Class III Genes** 131
   - Transcription Complex Assembly on Type II Promoters 133
   - Transcription Complex Assembly on Type I Promoters 141
   - Transcription Complex Assembly on Type III Promoters 146
   - Complex Formation in the Absence of DNA 148
   - Stability of Complexes 149

6. **Transcription** 163
   - Initiation 163
   - Elongation 164
   - Termination 166
   - Reinitiation 168

7. Chromatin Structure of Class III Genes 173
5S Genes 176
tRNA Genes 180
U6 Genes 181
SINEs 182
Methylation 183

8. Proteins that Modulate the Rate of RNA Polymerase III Transcription 191
Activities that Reduce Pol III Transcription 191
Activities that Stimulate Pol III Transcription 198
Kinases and Phosphatases 203

9. Regulation of RNA Polymerase III Transcription 213
Developmental Regulation 213
Tissue-Specific Regulation 224
Regulation in Response to Growth Conditions 225
Regulation in Response to Viruses 232
Regulation in Response to Transformation 237

10. Perspective 253

Index 267

CHAPTER 1

Class III Genes

The genes transcribed by pol III encode a variety of small RNA molecules. (Table 1.1) Many of these have essential functions in cellular metabolism, such as tRNA and 5S rRNA, which are required for protein synthesis, 7SL RNA, which is involved in intracellular protein transport, and the U6, H1 and MRP RNAs, which are involved in post-transcriptional processing. The VA RNAs encoded by adenovirus are also synthesized by pol III, and these serve to divert the translational machinery of an infected cell towards the more effective production of viral proteins. Other class III genes encode transcripts with no known function. This category includes the 7SK genes and the short interspersed repeat (SINE) gene families which constitute the majority of class III templates in mammals. The aim of this chapter is to describe these various gene families, which are the DNA templates for pol III transcription.

5S rRNA GENES

5S rRNA is approximately 120 nucleotides long and is found associated with the large subunit of ribosomes in all eukaryotic organisms. The genes for 5S rRNA are often situated in repetitive clusters, but this is not invariably the case. In *Saccharomyces cerevisiae* and in *Dictyostelium discoideum* the genes for each of the four rRNAs are located in a shared repeat, despite the fact that 5S rRNA is made by pol III and the 28S, 18S and 5.8S rRNAs are synthesized by pol I.[1,2] There are 140 5S genes in the haploid genome of *S. cerevisiae*.[3] Each lies within a ribosomal repeat between the major promoter element and the initiation site of the large rRNA gene, but is transcribed in the opposite orientation.[3] Expression of the rRNAs is not interdependent, since synthesis of large rRNA is not affected by the inhibition of pol III transcription.[4] In *Schizosaccharomyces pombe* and *Neurospora crassa* the 5S genes are dispersed instead of tandemly repeated and are separate from the other rRNA genes.[5,6]

The 5S genes are highly reiterated in the *Xenopus laevis* genome.[7] Several distinct classes of this gene exist and each is organized into tandemly repeated clusters situated at unique chromosomal locations.[8,9] (Table 1.2) The most abundant classes lie in an A/T-rich repeat of 650 to 850 bp, which is present in 20,000 copies per haploid genome.[10,11] Each contains a major oocyte gene and an 80% homologous pseudogene that was generated by duplication.[10,11] Overall, these repeats display a high degree of sequence uniformity.[12] However, the major oocyte 5S rRNAs contain ~1% heterogeneity at any particular position.[11,13] Over a thousand copies of these repeats are found clustered at the telomeres of most *X. laevis* chromosomes.[8] A trace oocyte class of 5S gene occurs in 1300 copies of a 310 bp A/T-rich

RNA Polymerase III Transcription, Second Edition, by Robert J. White.

Table 1.1. Pol III products

Product	Function	Size
5S rRNA	Ribosomal component	120 nt
tRNA	Translational adaptor	70-90 nt
U6 snRNA	mRNA splicing	106 nt
VA	Translational control	160 nt
EBER	Translational control?	165 nt
7SL	SRP component	300 nt
7SK	Unknown	330 nt
H1	RNase P component	369 nt
MRP	rRNA splicing	265 nt
vRNA	Vault component	89-141 nt
Y RNA	Unknown	69-112 nt
SINEs	Unknown	87-300 nt
BC200	Unknown	200 nt
BC1	Unknown	152 nt

repeat.[14] A major cluster of these repeats has been localized to the distal end of chromosome 13.[9] The fourth class is the somatic 5S gene, which is found in a single cluster of 400 copies of an 880 bp repeat with a G/C-rich spacer.[9,14] The coding regions of the various 5S classes differ at only a few positions, but their flanking regions display little homology. The amounts of transcripts produced by the different 5S gene classes do not reflect the relative abundance of the various DNA templates. For example, the 20,000 pseudogenes produce no detectable RNA in vivo.[15-17] The major and trace oocytic genes are active in oocytes but inactive in somatic cells, whereas the somatic genes are expressed in both oocytes and somatic cells.[15,16,18,19] The molecular basis for this differential gene expression has been the subject of intensive study (reviewed by Korn[20] and by Wolffe and Brown;[21] see chapter 9).

The haploid human genome contains 300 to 400 5S genes and approximately 1500 5S pseudogenes and gene variants.[22] Many of these occur in clusters of tandem repeats, although some are probably dispersed as single copies.[22]

tRNA Genes

tRNA molecules are 70 to 90 nucleotides in length. They function as adaptors which serve to translate the genetic information carried by messenger RNA into a particular order of amino acid residues in a protein. They are able to do this because any given tRNA will only recognize a particular amino acid and match this

Table 1.2. 5S Gene Organization in *Xenopus laevis*

Gene Type	Copy Number	Repeat Size	Spacer
Somatic	400	880 bp	G/C-rich
Major Oocyte	20,000	650-850 bp	A/T-rich
Trace Oocyte	1,300	310 bp	A/T-rich

to a specific codon in the message. Each eukaryotic cell contains 50 to 100 distinct tRNA species.[23] The relative amounts of the different tRNAs vary considerably from one cell type to another, but correlate well with codon usage.[24] As well as serving as the donor of selenocysteine during translation, $tRNA^{Sec}$ also functions as the carrier molecule upon which selenocysteine is synthesized.[25]

There is considerable redundancy among the tRNA genes, which often occur in complex multigene families dispersed around the genome (reviewed by Sharp et al[23]). Members of the individual families maintain a high degree of homology, apparently as a result of intergenic conversion.[26] In *S. cerevisiae* there are approximately 350 tRNA genes, which constitute over 0.1% of the genome.[27] Most are dispersed throughout the genome without clustering.[27] However, a few yeast tRNA genes are arranged in tandem pairs, with the promoter of the upstream gene directing synthesis of a dimeric readthrough precursor from which the individual tRNA species are processed.[28-30] In *S. pombe*, three out of the four initiator $tRNA^{Met}$ genes are positioned 7 bp downstream from a $tRNA^{Ser}$ gene.[28] These $tRNA^{Met}$ genes are dependent on their upstream partners and are not expressed following inactivation of the linked $tRNA^{Ser}$ promoter.[31] This appears to be because the $tRNA^{Met}$ genes have inherently weak promoters.[32] In *S. cerevisiae*, several sets of $tRNA^{Arg}$-$tRNA^{Asp}$ pairs are found separated by 10 bp.[29] The downstream $tRNA^{Asp}$ gene is silenced following deletion or point mutation of the $tRNA^{Arg}$ gene.[33] However, deletion of $tRNA^{Arg}$ and the spacer allows the $tRNA^{Asp}$ to be expressed.[33] Thus, in this case a potentially active gene is repressed by sequences present in the upstream gene and spacer region.

An irregular clustering of tRNAs is the rule in higher eukaryotes.[34,35] For example, 18 tRNA genes (eight $tRNA^{Asn}$, five $tRNA^{Lys}$, four $tRNA^{Arg}$ and one $tRNA^{Ile}$) lie within a 46 kb cluster on chromosome 2 of *D. melanogaster*.[36] In *X. laevis*, eight different tRNA genes lie within a 3.2 kb cluster that is tandemly reiterated approximately 150 times at a single locus.[37] This cluster includes an oocyte-specific $tRNA^{Tyr}$ gene,[37] whereas the $tRNA^{Tyr}$ gene that is expressed in somatic cells is present in only one to three copies.[38] Overall, *X. laevis* contains around 8000 tRNA genes per haploid genome,[39] many of which are situated in large multigene families.[35,40]

The haploid human genome contains approximately 1300 tRNA genes and pseudogenes encoding 60-90 tRNA isoacceptors.[41] This gives an average copy number of 10-20 genes per isoacceptor. Most tRNAs appear to conform to this average, although 60 genes code for $tRNA^{Asn}$ and only one encodes the selenocysteyl-$tRNA^{Sec}$.[42] The sequences immediately flanking the coding regions are generally unrelated, although most of the 13 $tRNA^{Glu}$ genes are flanked by DNA of very similar sequence for at least 5 kb.[43] Some human tRNA genes, such as initiator $tRNA^{Met}$, are scattered individually throughout the genome,[44] but clustering of tRNA genes also occurs in man.[45] The largest cluster identified in humans contains 5 tRNA genes within a 4.2 kb fragment.[42] This cluster is itself repeated three times within a small region of chromosome 17.[42]

Viral Class III Genes

Several viruses contain short class III transcription units within their genomes. The best characterized example is that of adenovirus, which encodes two small pol III transcripts, called VA_I and VA_{II}, that are synthesized at high levels during the late stages of viral infection.[46,47] The VA_I and VA_{II} genes are each approximately 160 bp long and are separated from one another by only 98 bp.[48] Whereas VA_I is required for efficient expression of the adenovirus genome, deletion of the VA_{II} gene does not have a major effect upon the viral life-cycle.[49] The VA RNAs act by stimulating the translation of adenoviral mRNA at late times after infection.[49]

The genome of Epstein-Barr virus (EBV) also contains two small adjacent genes that are transcribed by pol III; these are approximately 165 bp long and are called EBER1 and EBER2.[50] They are the most highly expressed viral genes during the infection and immortalization of human B lymphocytes by EBV, accumulating in 10^7 copies per cell.[51] However, EBER1 and EBER2 are not required for EBV to infect or immortalize B lymphocytes in culture, and a normal replication cycle can occur in their absence.[52] Multiple copies of the EBERs can substitute for VA_I during adenovirus infection[53] and a role for the EBER gene products in translational control seems likely (see Schwemmle et al[54] and references therein). The EBER RNAs associate with a 14.8 kD cellular protein called EAP (EBER-associated protein).[55]

U6 snRNA Genes

Small nuclear ribonucleoproteins (snRNPs) are a group of structurally related RNA-protein complexes that are found in the nuclei of eukaryotic cells. The most abundant snRNPs are the spliceosomes, which are present at $\sim10^6$ copies per mammalian cell and are involved in the splicing of pre-mRNA (reviewed by Maniatis and Reed[56] and Mattaj et al[57]). Spliceosomes contain five snRNA species, four of which are made by pol II whereas the smallest, U6, is made by pol III.[58-60] An exception to this is provided by trypanosomes, which appear to use pol III to transcribe U2 as well as U6 genes.[61] U3 snRNA, which is involved in pre-rRNA splicing, is also synthesized by pol III in plants, although it is a pol II product in most other eukaryotes.[62] Oligonucleotide-directed RNase cleavage was used to demonstrate the involvement of U6 snRNA in the splicing of pre-mRNA.[63] The 106 nucleotide U6 transcript is the most highly conserved of the spliceosomal RNAs, with 95% identity between human and *Drosophila* and 60% identity between human and yeast.[64,65] It is encoded by an essential single-copy gene in both *S. cerevisiae* and *S. pombe*.[65,66] In *S. cerevisiae*, a 4-fold drop in the steady-state level of U6 RNA makes no difference to the growth rate, whereas a 10-fold drop is lethal.[67] This suggests that yeast contain at least four times more U6 RNA than is required. This is consistent with the fact that U4 RNA is 4- to 5-fold less abundant than the U6 RNA with which it interacts.[67] The abundance of U snRNAs is even greater in higher organisms than it is in yeasts, which may reflect a considerable increase in gene copy number. The *Drosophila* genome contains three U6 genes and these are clustered together in a head to tail orientation within 2 kb of one another.[64] The coding regions of the three genes are identical, but the flanking sequences display significant divergence.[64] In *Xenopus*, U6 genes are arranged in two distinct tandemly repeated families.[68] One repeat is 1 kb long and is present in about 500 copies per haploid genome, whereas the other is 1.6 kb long and is about half as abundant.[68] The human genome is estimated to contain approximately 200 U6 loci.[69]

Other Class III Genes Encoding RNP Components

A variety of other pol III transcripts are components of ribonucleoprotein (RNP) complexes, such as 7SL, 7SK, H1, and MRP. The 300 nucleotide 7SL RNA forms the scaffold of the signal recognition particle, an RNP involved in the cotranslational insertion of nascent polypeptides into the endoplasmic reticulum.[70] 7SL RNA shows a high degree of evolutionary stability, with 87% homology between human and *Xenopus* and 64% homology between human and *Drosophila*.[71] The human genome contains four 7SL genes and approximately 500 7SL pseudogenes which are truncated at one or both ends.[72]

7SK is an abundant 330 nucleotide RNA that forms part of a 12S RNP together with eight proteins.[57,73] Although its function has yet to be determined, its ubiquity in higher eukaryotes and its substantial evolutionary conservation suggest an important role.[74] 7SK displays limited similarity to U6 at its 5' end and extensive sequence complimentarity to both U4 and U11 RNAs.[75] The human genome contains a large family of truncated 7SK pseudogenes, but only one full-length 7SK gene.[73,76,77]

H1 is the 369 nucleotide RNA component of RNase P, an endoribonuclease that processes the 5' termini of pre-tRNA.[78-81] The RNase P of *Schizosaccharomyces pombe* has been purified to homogeneity and shown to contain a single polypeptide in addition to H1 RNA.[81] Although there is very little sequence homology between H1 RNA from different organisms, the tertiary structure is thought to be conserved.[81] H1 is encoded by a single copy gene in both yeast and humans.[82,83]

RNase MRP is an endoribonuclease that was identified because of its ability to cleave the mitochondrial transcript to generate an RNA primer for replication of mitochondrial DNA.[84-86] However, most RNase MRP is located in the nucleolus where it performs an important role in the endonucleolytic processing of pre-rRNA.[81,87-89] RNase MRP contains a 265 nucleotide RNA which shows several blocks of sequence homology to H1 RNA.[90] The MRP and H1 RNAs can also be folded into similar secondary structures.[81] In addition, the RNases P and MRP share a common protein component.[91] It is therefore clear that RNase P and RNase MRP are closely related. Since RNase MRP has only been found in eukaryotes, whereas RNase P also exists in bacteria and archaea, it is likely that RNase MRP is derived from RNase P.[81] In all probability the MRP RNA arose through duplication of the H1 gene in an early eukaryote. In yeast, the MRP RNA is encoded by the single copy nuclear gene NME1, which is essential for viability.[87] It is also encoded by a single copy nuclear gene in mice and humans, although several pseudogenes exist as well.[85,86,92] The mouse and human genes are 84% identical within their coding regions and are also 70% homologous over 715 bp of upstream sequence.[86] This unusual degree of flanking homology may reflect an overlap with another gene. Although MRP RNA is synthesized by pol III in mammals, this appears not to be the case in *S. cerevisiae*, where the MRP coding sequence contains a run of 8 Ts and would therefore be expected to terminate pol III transcription.[93] Inactivation of pol III is reported to have no effect on the level of MRP RNA in *S. cerevisiae*.[93]

Vaults are the largest known cytoplasmic RNPs, with a mass of 13 MD and dimensions of 35 x 65 nm.[94] Although most of the mass is due to protein, they also contain multiple copies of a pol III transcript called vault RNA.[94] Vaults are found in species as divergent as humans and amoeba, but their function is unknown.[94] It has been suggested that they form part of the nuclear pore complex, although 95% of a cell's vaults are located in the cytoplasm.[94] A role for vaults in nucleocytoplasmic transport is consistent with the demonstration that overexpression of the major vault protein is associated with multidrug resistance in some human tumors.[95] Vault RNA (vRNA) from rats is 141 bases long and is encoded by a single copy gene that is expressed ubiquitously.[96] Bullfrogs have more than one vRNA gene and produce shorter vRNAs (89-94 nucleotides) which are 65% identical to those of rat.[96] Both rat and bullfrog vRNAs are predicted to form secondary structures similar to those of VA and EBER transcripts.[96]

Y RNAs are pol III products of 69 to 112 nucleotides that are found associated with the Ro autoantigen.[97] Humans produce four discrete Y RNAs, which are designated hY1, hY3, hY4, and hY5 (hY2 is a truncated version of hY1).[13] However, many vertebrates, including rodents, do not contain Y4 or Y5. The Y1 and Y3 transcripts are highly conserved through evolution.[98] For example, murine Y1 and Y3 RNAs are 95-97% identical to their human counterparts.[99] All the Y RNAs are predicted to adopt a structure containing a large internal loop and a long stem with the 5' and 3' ends base-paired.[98] The 60 kD Ro autoantigen binds to the stem, as revealed by ribonuclease protection.[100] There are ~10^5 copies of each Ro RNP in mammalian cells, which is ~1% of the number of ribosomes.[13] Despite such high levels of expression, the human and mouse Y RNAs appear to be encoded by single copy genes.[97,99] All four human Y genes are tightly linked on chromosome 7.[101] This would suggest that Y4 and Y5 arose by duplication in primates. Many Y RNA pseudogenes with inactive promoters are dispersed through mammalian genomes and these are presumed to have arisen by retrotransposition.[101] The high degree of evolutionary conservation of Y RNAs implies an important function. However, their role has proved elusive. Ro was found to associate with mutant 5S rRNAs that contain internal substitutions and 3' terminal extensions.[13] Because these abnormal rRNAs are processed inefficiently, it was suggested that the Ro protein may function as part of a quality control pathway for discarding defective 5S rRNA precursors.[13] The Y RNAs could be involved in such a pathway, as part of the Ro RNP, although this remains to be demonstrated.

BC1 RNA is a 152 nt pol III transcript that is assembled into a cytoplasmic RNP in rodents.[102] It evolved from a $tRNA^{Ala}$ and is encoded by a single copy gene.[102] Its expression is developmentally regulated and is restricted to the somatic and/or dendritic domains of a specific subset of neurons in the central and peripheral nervous systems.[102,103] Primates have a pol III transcript called BC200 that is expressed in the equivalent subset of human neurons to those that produce BC1 in rat.[103,104] BC200 has evolved from Alu or 7SL RNA and bears little homology to BC1.[105] Although BC1 and BC200 have different origins, the striking similarity in their neuronal expression patterns, both at the regional and subcellular levels, has led to the idea that they might perform similar functions.[103,104] Postulated roles include the transport and/or translation of dendritic mRNAs.[103,104]

The RNA component of the telomerase RNP serves as the template for synthesizing chromosomal telomeres.[106] This RNA has been shown to be a pol III product in the ciliated protozoan *Tetrahymena thermophila*.[106] This is also the case in *Euplotes* and *Oxytricha* species, but pol II appears to carry out this role in humans.[107]

SINEs

A variety of repetitive short interspersed elements (SINEs) constitute quantitatively important classes of pol III template in higher organisms (reviewed in refs. 108-110). (Table 1.3) For instance, the genome of *Bombyx mori* contains ~20,000 copies of a repetitive element called BmX that is actively transcribed by pol III.[111] *Xenopus* has a multicopy gene family, called OAX or satellite I, that is expressed in oocytes and gastrula stage embryos but is largely silent at later stages of development.[112-115] The 741 bp satellite I repeat is highly reiterated and comprises ~1% of the total genome of *X. laevis*.[113] The major SINE in primates is the Alu family, of which there are about 500,000 to one million copies in the haploid human genome.[116-120] Alu genes consist of two imperfect repeats separated by an 18 bp spacer.[116,118] The upstream repeat contains a functional pol III promoter whereas the promoter in the downstream repeat is disrupted by a 31 bp insertion.[121] The Alu consensus sequence is 282 nucleotides long, but transcription continues into a downstream A-rich region to produce poly$(A)^+$ RNAs with an average length of ~400 bases.[122] Truncated poly$(A)^-$ transcripts of 118 nucleotides are also found in the cytoplasm; these are called scAlu (small cytoplasmic Alu) and correspond to just the left monomer of the Alu element.[123] ScAlu RNA appears to be a stable processing product of primary Alu transcripts and has a half-life at least five times longer than the ~30 minutes displayed by full-length Alu RNA.[124] Different Alu members display an average of 14% deviation from the familial consensus sequence.[118] Rodent species contain a variety of SINE families, of which the B1 and B2 genes are the most abundant.[125] B1 genes are approximately 80% homologous to human Alu genes, but with only one of the two repeats. They are ~130 bp long and are present in about 100,000 copies per haploid mouse genome.[125-128] The B2 family is specific to rodents,[128] although B2-like SINEs are widespread in metazoa.[129] Each B2 gene is ~180 bp long, with ~80,000 copies per haploid mouse genome.[125,128] The rodent-specific ID family is represented in the mouse genome by about 12,000 copies, but is present in 135,000 copies in the rat.[130] Additional SINE families of lower copy number also occur in rodents.[125,131]

Clearly, SINE DNA accounts for a significant proportion of mammalian genomes. B2 genes alone constitute ~0.7% of total mouse genomic DNA and Alu repeats contribute ~5% of all human genetic material. If randomly dispersed, this frequency would result in an Alu element every 4-6 kb. In fact, clustering of SINEs is common. An extreme example is provided by the growth hormone gene of the Sprague-Dawley rat, where a 5.8 kb genomic fragment contains 8 copies of B2 and 2 copies of ID genes.[132] The frequent occurrence of such genes in tandem may reflect a preferential location of SINEs in A/T-rich areas. Alu or B1 genes are also found immediately downstream of 7SK, H1 and MRP genes.[73,77,82,85] A more general regional preference is suggested by cases where homologous loci have different SINEs in similar, but not identical, positions. For example, different B1 and B2 genes are found in the untranslated regions of many murine H-2 genes.[128,133,134] In situ hybridization to metaphase spreads showed that B1 and B2 sequences are

Table 1.3. Properties of Representative SINE Families

SINE	Organism	Size	Copy Number
Alu	Human	300 bp	500,000
B1	Mouse	130 bp	100,000
B2	Mouse	180 bp	80,000
ID	Rat	87 bp	135,000
Satellite 1	Frog	180 bp	20,000
BmX	Silkworm	91 bp	20,000

present on all chromosomes rather than being clustered at a small number of sites.[125] A more detailed study using human chromosomes revealed strong regional variations in the localization of Alu genes, although hybridization occurred along every chromosome.[135] The positions of Alu elements correspond quite precisely with the Reverse (R) bands of metaphase chromosomes.[135,136] R bands replicate their DNA early in S-phase, condense late in mitotic prophase, and are thought to be the chromosomal regions in which active genes are concentrated.[137]

The dispersion and amplification of SINEs is believed to occur by retrotransposition, in which pol III transcripts are reverse transcribed into DNA and then integrated into new genomic sites (reviewed by Weiner et al[122]). Polymorphisms due to the de novo insertion of Alu elements suggest that this process is currently active and is capable of causing genetic variation, sometimes with detrimental effects.[138-140] Heritable retroposition events must occur in the germ cells or in early embryonic development, before the separation of the germ cell and somatic cell lineages. SINE families are strongly transcribed in embryonic stem cells, oocytes and early embryonic stages.[125,141-145] Because the promoters of many class III genes are internal and therefore included in the transcript, they will be duplicated during retroposition. This means that each new gene copy could potentially be transcribed and generate further copies. This feature may be responsible for the high rate of transposition of SINEs relative to other retroposons.

Several SINE families, such as B2 and ID, appear to have evolved from tRNA genes, and so can be regarded as amplified tRNA pseudogenes.[146] The mouse B2 consensus shows 64% homology to a rat $tRNA^{Ser}$ throughout its length.[146] Structures resembling tRNA stems and loops can be drawn for several SINEs[128,146] and tRNA methyl transferase recognizes the secondary structure of ID RNA,[147] providing further support for the postulated evolutionary relationship. ID elements arose from the BC1 gene, which in turn arose from a $tRNA^{Ala}$.[148] RNase fingerprinting and primer extension analyses of SINE transcripts synthesized both in vitro and in vivo demonstrate that the 5' terminus of the RNA corresponds precisely to that of the SINE sequence.[123,127,149,150] This is strong evidence that the internal promoter defines the 5'-boundary of the SINE and that transposition of the element occurs through an RNA intermediate. In contrast, the 3'-terminal A-rich tracts of SINEs lie far downstream of the tRNA homology, suggesting that these units arose by retroposition of a readthrough transcript.[122] Satellite I genes appear to be derived from a fusion of two tRNA genes in their anticodon regions.[114] In contrast, the B1 and Alu families are thought to have evolved from the 7SL gene.[71] The Alu consensus sequence shows ~90% homology to 100 bp at the 5' end of the 7SL gene and to 45 bp at its 3' end.[74] The Alu progenitor therefore probably arose by the deletion of ~155 bp from the center of a 7SL gene, followed by duplication and then inactivation of the promoter in the downstream repeat.[71] One of the polypeptides from the signal recognition particle has been shown to bind to Alu transcripts in vivo and influence their metabolism.[151] Britten et al[152] demonstrated four distinct human Alu subfamilies which become successively closer to 7SL with increasing age. The oldest Alu repeats display greater than 10% divergence from the Alu consensus sequence and are calculated to have arisen about 65 million years ago.[153] In contrast, the newest Alu subfamily shows 0.1% divergence from its consensus and may be only 660,000 years old.[153]

There is substantial evidence that SINEs arise by the amplification of a small number of "founder elements" or "master genes" that give rise to a succession of distinct subfamilies.[154-156] The four human Alu subfamilies have expanded simultaneously since man diverged from the apes.[153] Although multiple dispersed Alu source genes are currently capable of retroposition in humans, the recent amplifications are responsible for only 0.4% of the total Alu repeats present in the genome.[153] These are all derived from a single specific subset of older Alu elements, which suggests that earlier in primate evolution there was one, or very few, much more prolific master gene(s).[153] This may have arisen through the fortuitous combination of a powerful internal promoter and an optimal chromosomal environment. In general, the more ancient subfamilies are transcribed less actively than the newer subfamilies, since promoters of the older genes have become inactivated by accumulated mutations.[157,158] However, the overall level of Alu expression is extremely low, with only one-hundred to one-thousand transcripts being detected per HeLa cell (c.f. one million 7SL RNA molecules).[158] In one case, the fortuitous integration of an Alu gene alongside stimulatory cis-acting sequences appears to have generated a founder element with high transcriptional activity.[159]

No functional role has been demonstrated unequivocally for a SINE family, despite a broad range of speculations (reviewed by Howard and Sakamoto[110]). According to the "selfish DNA" theory, nonviral retroposons constitute molecular parasites that infest the genome but rarely confer a selective advantage, with evolutionary pressures serving merely to optimize their ability to amplify.[160,161] Since the major SINE families appear to be derived from class III genes of known physiological significance, and since inactivation of these progenitors was probably a prerequisite of their amplification, it is quite possible that SINEs represent large numbers of pseudogenes that are entirely bereft of function. However, certain SINE transcripts may have acquired roles during the course of evolution. Furthermore, the integration of repetitive elements near or within functional genes will inevitably have physiological effects in some cases. Given the opportunism of natural selection, such effects might have been utilized to confer a selective advantage.

One of the first functions proposed for SINEs was in regulating the expression of adjacent genes.[162] The model held that coordinate expression of a number of genes could be achieved by means of common DNA regulatory elements near each gene, and that SINEs might serve in this way. The mobility of SINEs would provide relatively frequent opportunities to form new integrative combinations of preexisting genes. The original model suggested that regulation would occur at the transcriptional level, but it was later extended to include the possibility of post-transcriptional regulation mediated by cotranscribed repetitive sequences.[163] Correlations have sometimes been observed between levels of pol III SINE transcripts and levels of particular pol II transcripts.[133,141,164-168] Furthermore, several studies have provided evidence that class III genes can influence the expression of adjacent class II genes.[169-178] For example, Glaichenhaus and Cuzin[170] cloned a set of rodent mRNAs of unknown function which are coordinately upregulated in response to serum or transformation. Nuclear run-on assays showed that the response occurs at the post-transcriptional level.[170,176] The 3'-untranslated regions of these mRNAs share several short blocks of sequence homology and also contain a SINE.[176] Whereas the mRNAs cloned from rats contain an ID gene, their homologues from mice carry a B1 gene instead.[170,176] When cloned into the corresponding region of the rabbit β-globin gene, fragments from the 3'-untranslated regions of these genes conferred serum- and transformation-responsiveness upon the globin transcript, again at the post-transcriptional level.[170,176] Both the SINE itself and its flanking sequence were required for this response.[176] ID elements have also been proposed to be involved in controlling neuronal gene expression.[166] Many Alu genes contain binding sites for retinoic acid receptors, and these have been shown to confer retinoic acid responsiveness to a class II reporter gene in transiently transfected cells.[178] The conservation in Alu and B1 transcripts of a sequence motif that occurs throughout evolution in the translational control domain of 7SL RNA has lead to the suggestion that Alu and B1 RNAs play a role in the cytoplasmic regulation of gene expression.[179] Other functions that have been suggested for particular SINEs include a role in splicing,[126,180] translation,[151] DNA replication,[117,181-184] the cell stress response,[185,186] and in regulating growth[187] or the turnover of specific mRNAs.[188]

The most serious objection to hypothesized roles for SINE transcripts in processes such as gene regulation or RNA processing is that these phenomena clearly antedate the relatively recent multiplication of repetitive families, so that any function performed by SINEs must be either subsidiary or very evolutionarily adaptable. However, a subset of family members could have taken over some previous cellular function or added a new layer of complexity to a preexisting process. The rate of divergence of a SINE family can indicate whether it is subject to selective pressure, as would be predicted for a functionally important sequence. A neutral divergence rate is observed for seven Alu members located in the globin clusters of chimpanzee and human.[189] However, analyses of the recently transposed Alu elements suggest a moderate degree of selection.[152,190] Whether the selective pressure on active Alu genes reflects a function for Alu or merely the restraints imposed by retroposition remains to be determined.

Even if SINE transcripts have no function, the insertion of multiple repetitive elements into new genomic locations will inevitably have had major effects upon the structure and evolution of the genome. The most obvious impact of SINEs is their mutagenic potential due to disruption of sequences at the site of integration. The apparently preferential insertion of SINEs in and around other genes will accentuate their mutagenic impact. Cases have been documented of de novo Alu retroposition into the coding regions of genes encoding Factor IX and cholinesterase; in both instances the Alu introduced stop codons and caused premature translational termination.[140] The presence of so many short, homologous sequences scattered throughout the genome is also likely to increase the level of recombination. Alu elements are frequently involved in unequal, homologous, and nonhomologous exchange processes. For example, five different hereditary defects in the low density lipoprotein receptor gene, causing familial hypercholesterolemia, were found to result from deletions or duplications with Alu sequences at the rearrangement break points.[191] SINEs clearly make a major contribution to the fluidity of mammalian genomes.

Although SINEs possess internal promoters which direct accurately initiated pol III transcription,[127,149,150] they are also subject to extensive readthrough transcription by pol II.[125,144,192-194] For example, pol II is responsible for 69% of B2 transcription in isolated liver nuclei.[125] Indeed, B1 and B2 sequences were originally identified as major components of heterogeneous nuclear RNA (hnRNA)[195] where they may account for up to 2.5% of the total.[196] In humans, Alu sequences comprise 10% of hnRNA and are found in the introns of most class II genes.[140,158] Northern blot analyses reveal the extreme heterogeneity of hnRNAs carrying such sequences.[119,141,144,164,194,196,197] Most of the SINE sequences found in hnRNA are due to the integration of these elements into the untranslated regions of class II genes and their consequent inclusion in primary transcripts. As such, the bulk of these sequences do not survive processing. However, a few proteins have been found to contain Alu sequences.[140] The most striking example is the RMSA-1 gene (regulator of mitotic spindle assembly 1), which has two Alu elements located within its coding region.[198] These Alus contribute ~40% of the RMSA-1 translated sequence and encode 111 amino acid residues of the protein product.[198] This example is highly unusual, since most Alus contain numerous translation termination signals and therefore cause premature termination when incorporated into an exon.[140]

Class II Genes

A variety of class II genes can be transcribed by pol III under particular conditions. For example, following injection into *Xenopus* oocytes, c-myc is transcribed by pol II at low template concentrations and by pol III at higher concentrations.[199,200] Pols II and III can both initiate transcription in vitro from the same start sites of the c-myc gene.[199,200] The same is true of the human T-lymphotropic virus type I (HTLV-I) promoter.[201] The human L1 retrotransposon is transcribed by pol III in vitro.[202] The brain creatine kinase gene can be transcribed by pol III if it is preincubated with class III transcription factors.[203,204] The early E2 (E2E) promoter of adenovirus directs transcription by pol II in high salt nuclear extracts and by pol III in more dilute extracts prepared at lower salt concentrations.[205] These observations suggest that the choice of RNA polymerase can be dictated by a particular set of circumstances. In some of these cases, pol III transcription of class II templates may be an artifactual response to abnormal in vitro conditions. However, the E2E promoter has been shown to direct transcription by both pols II and III in vivo.[206] In the same infected cells, the adenovirus major late promoter is exclusively active for pol II.[206] Although pols II and III initiate at similar rates at the E2E promoter, the pol III transcripts are degraded preferentially and fewer than ten copies per cell are found in the cytoplasm.[206] The physiological significance of these findings is unclear. Pol III transcription from the E2E and c-myc promoters is restricted to 5' regions and terminates prematurely at T-rich sequences.[199,200,205,206] The transcripts are therefore unlikely to be functional. It has been suggested that pol III may serve to compete with pol II and downregulate it at these transcription units.[199,206]

Evidence for competition between pols II and III has been provided in the case of the *Xenopus* TFIIIA gene. Three different transcriptional start sites have been mapped for this gene.[207] Pol II initiates at +1 and -284, with the former predominating in oocytes and the latter predominating in somatic cells.[207] Transcripts that begin at -284 encode a TFIIIA protein that has 22 additional amino acids at its N-terminus.[208] The gene is also transcribed by pol III, with initiation occurring at -70 and +1.[207] Although the +1 pol III transcripts may not occur in vivo, the -70 transcripts are clearly detectable in somatic cell RNA.[207] Pol III reads into the TFIIIA coding region, but the transcripts are not polyadenylated and probably not spliced, and their functional significance is unclear.[207] If tagetitoxin is used to inhibit pol III specifically, pol II transcription of the TFIIIA gene increases.[207] This suggests that the two polymerases may compete on this promoter.

References

1. Maizels N. *Dictyostelium* 17S, 25S, and 5S rDNAs lie within a 38,000 base pair repeated unit. Cell 1976; 9:431-438.
2. Nath K, Bollen AP. Organization of the yeast ribosomal RNA gene cluster via cloning and restriction analysis. J Biol Chem 1977; 252:6562-6571.
3. Elion EA, Warner JR. The major promoter element of rRNA transcription in yeast lies 2 kb upstream. Cell 1984; 39:663-673.
4. Gudenus R, Mariotte S, Moenne A et al. Conditional mutants of *RPC160*, the gene encoding the largest subunit of RNA polymerase C in *Saccharomyces cerevisiae*. Genetics 1988; 119:517-526.
5. Selker EU, Yanofsky C, Driftmier K et al. Dispersed 5S RNA genes in *N. crassa*: structure, expression, and evolution. Cell 1981; 24:819-828.
6. Mao J-I, Appel B, Schaack J et al. The 5S RNA genes of *Schizosaccharomyces pombe*. Nucleic Acids Res 1982; 10:487-500.
7. Brown DD, Wensink PC, Jordan E. Purification and some characteristics of 5S DNA from *Xenopus laevis*. Proc Natl Acad Sci USA 1971; 68:3175-3179.
8. Pardue ML, Brown DD, Birnstiel ML. Location of the genes for 5S ribosomal RNA in *Xenopus laevis*. Chromosoma 1973; 42:191-203.
9. Harper ME, Price J, Korn LJ. Chromosomal mapping of *Xenopus* 5S genes: somatic-type versus oocyte-type. Nucleic Acids Res 1983; 11:2313-2333.
10. Fedoroff NV, Brown DD. The nucleotide sequence of oocyte 5S DNA in *Xenopus laevis*. I. The AT-rich spacer. Cell 1978; 13:701-716.
11. Miller JR, Cartwright EM, Brownlee GG et al. The nucleotide sequence of oocyte 5S DNA in *Xenopus laevis*. II. The GC-rich region. Cell 1978; 13:717-725.
12. Engelke DR, Gottesfeld JM. Chromosomal footprinting of transcriptionally active and inactive oocyte-type 5S RNA genes of *Xenopus laevis*. Nucleic Acids Res 1990; 18:6031-6037.
13. O'Brien CA, Wolin SL. A possible role for the 60-kD Ro autoantigen in a discard pathway for defective 5S rRNA precursors. Genes Dev 1994; 8:2891-2903.
14. Peterson RC, Doering JL, Brown DD. Characterization of two *Xenopus* somatic 5S DNA and one minor oocyte-specific DNA. Cell 1980; 20:131-141.
15. Wegnez M, Monier R, Denis H. Sequence heterogeneity of 5S RNA in *Xenopus laevis*. FEBS Lett 1972; 25:13-20.
16. Ford PJ, Southern EM. Different sequences for 5S RNA in kidney cells and ovaries of *Xenopus laevis*. Nature New Biol 1973; 241:7-12.

17. Jacq C, Miller JR, Brownlee GG. A pseudogene structure in 5S DNA of *Xenopus laevis*. Cell 1977; 12:109-120.
18. Brown DD, Carroll D, Brown RD. The isolation and characterization of a second oocyte 5S DNA from *Xenopus laevis*. Cell 1977; 12:1045-1056.
19. Korn LJ, Gurdon JB. The reactivation of developmentally inert 5S genes in somatic nuclei injected into *Xenopus* oocytes. Nature 1981; 289:461-465.
20. Korn LJ. Transcription of *Xenopus* 5S ribosomal RNA genes. Nature 1982; 295:101-105.
21. Wolffe AP, Brown DD. Developmental regulation of two 5S ribosomal RNA genes. Science 1988; 241:1626-1632.
22. Sorensen PD, Frederiksen S. Characterization of human 5S rRNA genes. Nucleic Acids Res 1991; 19:4147-4151.
23. Sharp SJ, Schaack J, Cooley L et al. Structure and transcription of eukaryotic tRNA genes. CRC Crit Rev Biochem 1984; 19:107-144.
24. Garel JP. Quantitative adaptation of isoacceptor tRNAs to mRNA codons of alanine, glycine and serine. Nature 1976; 260:805-806.
25. Low SC, Berry MJ. Knowing when to stop: selenocysteine incorporation in eukaryotes. Trends Biochem Sci 1996; 21:203-208.
26. Munz P, Amstutz H, Kohli J et al. Recombination between dispersed serine tRNA genes in *Schizosaccharomyces pombe*. Nature 1982; 300:225-231.
27. Guthrie C, Abelson J. Organization and expression of tRNA genes in *Saccharomyces cerevisiae*. In: Strathern JN, Jones EW, Broach JR, eds. The Molecular Biology of the Yeast *Saccharomyces*, Metabolism and Gene Expression. Cold Spring Harbor, NY: Cold Spring Harbor Laboratory, 1982:487.
28. Mao J, Schmidt O, Soll D. Dimeric transfer RNA precursors in *S. pombe*. Cell 1980; 21:509-516.
29. Schmidt O, Mao J, Ogden R et al. Dimeric tRNA precursors in yeast. Nature 1980; 287:750-752.
30. Willis I, Hottinger H, Pearson D et al. Mutations affecting excision of the intron from a eukaryotic dimeric tRNA precursor. EMBO J 1984; 3:1573-1580.
31. Nichols M, Bell J, Klekamp MS et al. Multiple mutations of the first gene of a dimeric tRNA gene abolish in vitro tRNA gene transcription. J Biol Chem 1989; 264:17084-17090.
32. Johnson DL, Nichols M, Bolger MB et al. Interaction of yeast transcription factor IIIC with dimeric *Schizosaccharomyces pombe* $tRNA^{Ser}$-$tRNA^{Met}$ genes. J Biol Chem 1989; 264:19221-19227.
33. Reyes VM, Newman A, Abelson J. Mutational analysis of the coordinate expression of the yeast $tRNA^{Arg}$-$tRNA^{Asp}$ gene tandem. Mol Cell Biol 1986; 6:2436-2442.
34. Kubli E. The genetics of transfer RNA in *Drosophila*. Adv Genet 1982; 21:123.
35. Rosenthal DS, Doering JL. The genomic organization of dispersed tRNA and 5S RNA genes in *Xenopus laevis*. J Biol Chem 1983; 258:7402-7410.
36. Dingermann T, Burke DJ, Sharp S et al. The 5' flanking sequences of *Drosophila* $tRNA^{Arg}$ genes control their in vitro transcription in a *Drosophila* cell extract. J Biol Chem 1982; 257:14738-14744.
37. Muller F, Clarkson SG, Galas DJ. Sequence of a 3.18 kb tandem repeat of *Xenopus laevis* DNA containing 8 tRNA genes. Nucleic Acids Res 1987; 15:7191.
38. Stutz F, Gouilloud E, Clarkson SG. Oocyte and somatic tyrosine tRNA genes in *Xenopus laevis*. Genes Dev 1989; 3:1190-1198.
39. Clarkson SG, Birnstiel ML, Serra V. Reiterated transfer RNA genes of *Xenopus laevis*. J Mol Biol 1973; 79:391-410.
40. Muller F, Clarkson SG. Nucleotide sequence of genes coding for $tRNA^{Phe}$ and $tRNA^{Tyr}$ from a repeating unit of *X. laevis* DNA. Cell 1980; 19:345-353.

41. Hatlen L, Attardi G. Proportion of the HeLa cell genome complementary to transfer RNA and 5S RNA. J Mol Biol 1971; 56:535.
42. Bourn D, Carr T, Livingstone D et al. An intron-containing $tRNA^{Arg}$ gene within a large cluster of human tRNA genes. DNA Sequence 1994; 5:83-92.
43. Gonos ES, Goddard JP. Human $tRNA^{Glu}$ genes: their copy number and organization. FEBS Lett 1990; 276:138-142.
44. Santos T, Zasloff M. Comparative analysis of human chromosomal segments bearing nonallelic dispersed $tRNA^{Met}$ genes. Cell 1981; 23:699-709.
45. Roy KL, Cooke H, Buckland R. Nucleotide sequence of a segment of human DNA containing the three tRNA genes. Nucleic Acids Res 1982; 10:7313-7322.
46. Soderlund H, Pettersson U, Vennstom B et al. A new species of virus-coded low molecular weight RNA from cells infected with Adenovirus type 2. Cell 1976; 7:585-593.
47. Weinmann R, Raskas HJ, Roeder RG. Role of DNA-dependent RNA polymerases II and III in transcription of the adenovirus genome late in productive infection. Proc Natl Acad Sci USA 1974; 71:3426-3430.
48. Akusjarvi G, Mathews MB, Andersson P et al. Structure of genes for virus associated RNA_I and RNA_{II} of adenovirus type 2. Proc Natl Acad Sci USA 1980; 77:2424-2428.
49. Thimmappaya B, Weinberger C, Schneider RJ et al. Adenovirus VAI RNA is required for efficient translation of viral mRNA at late times after infection. Cell 1982; 31:543-551.
50. Rosa MD, Gottlieb E, Lerner M et al. Striking similarities are exhibited by two small Epstein-Barr virus-encoded ribonucleic acids and the adenovirus-associated ribonucleic acids VAI and VAII. Mol Cell Biol 1981; 1:785-796.
51. Arrand JR, Rymo L. Characterization of the major Epstein-Barr virus-specific RNA in Burkitt lymphoma-derived cells. J Virol 1982; 41:376-389.
52. Swaminathan S, Tomkinson B, Kieff E. Recombinant Epstein-Barr virus with small RNA (EBER) genes deleted transforms lymphocytes and replicates in vitro. Proc Natl Acad Sci USA 1991; 88:1546-1550.
53. Bhat RA, Thimmappaya B. Construction and analysis of additional adenovirus substitution mutants confirm the complementation of VA_I RNA function by two small RNAs encoded by Epstein-Barr virus. J Virol 1985; 56:750-756.
54. Schwemmle M, Clemens MJ, Hilse K et al. Localization of Epstein-Barr virus-encoded RNAs EBER-1 and EBER-2 in interphase and mitotic Burkitt lymphoma cells. Proc Natl Acad Sci USA 1992; 89:10292-10296.
55. Toczyski DPW, Steitz JA. EAP, a highly conserved cellular protein associated with Epstein-Barr virus small RNAs (EBERs). EMBO J 1991; 10:459-466.
56. Maniatis T, Reed R. The role of small nuclear ribonucleoprotein particles in pre-mRNA splicing. Nature 1987; 325:673-678.
57. Mattaj IW, Tollervey D, Seraphin B. Small nuclear RNAs in messenger RNA and ribosomal RNA processing. FASEB J 1993; 7:47-53.
58. Kunkel GR, Maser RL, Calvet JP et al. U6 small nuclear RNA is transcribed by RNA polymerase III. Proc Natl Acad Sci USA 1986; 83:8575-8579.
59. Reddy R, Henning D, Das G et al. The capped U6 small nuclear RNA is transcribed by RNA polymerase III. J Biol Chem 1987; 262:75-81.
60. Moenne A, Camier S, Anderson G et al. The U6 gene of *Saccharomyces cerevisiae* is transcribed by RNA polymerase C (III) in vivo and in vitro. EMBO J 1990; 9:271-277.
61. Fantoni A, Dare AO, Tschudi C. RNA polymerase III-mediated transcription of the trypanosome U2 small nuclear RNA gene is controlled by both intragenic and extragenic regulatory elements. Mol Cell Biol 1994; 14:2021-2028.

62. Kiss T, Marshallsay C, Filipowicz W. Alteration of the RNA polymerase specificity of U3 snRNA genes during evolution and in vitro. Cell 1991; 65:517-526.
63. Black DL, Steitz JA. Pre-mRNA splicing in vitro requires intact U4/U6 small nuclear ribonucleoprotein. Cell 1986; 46:697-704.
64. Das G, Henning D, Reddy R. Structure, organization, and transcription of *Drosophila* U6 small nuclear RNA genes. J Biol Chem 1987; 262:1187-1193.
65. Brow DA, Guthrie C. Spliceosomal RNA U6 is remarkably conserved from yeast to mammals. Nature 1988; 334:213-218.
66. Tani T, Ohshima Y. The gene for the U6 small nuclear RNA in fission yeast has an intron. Nature 1989; 337:87-90.
67. Kaiser MW, Brow DA. Lethal mutations in a yeast U6 RNA gene B block promoter element identify essential contacts with transcription factor-IIIC. J Biol Chem 1995; 270:11398-11405.
68. Krol A, Carbon P, Ebel J-P et al. *Xenopus tropicalis* U6 snRNA genes transcribed by pol III contain the upstream promoter elements used by pol II dependent U snRNA genes. Nucleic Acids Res 1987; 15:2463-2478.
69. Hayashi K. Organization of sequences related to U6 RNA in the human genome. Nucleic Acids Res 1981; 9:3379-3388.
70. Walter P, Blobel G. Signal recognition particle contains a 7S RNA essential for protein translocation across the endoplasmic reticulum. Nature 1982; 299:691-698.
71. Ullu E, Tschudi C. Alu sequences are processed 7SL RNA genes. Nature 1984; 312:171-172.
72. Ullu E, Weiner AM. Human genes and pseudogenes for the 7SL RNA component of signal recognition particle. EMBO J 1984; 3:3303-3310.
73. Murphy S, Tripodi M, Melli M. A sequence upstream from the coding region is required for the transcription of the 7SK RNA genes. Nucleic Acids Res 1986; 14:9243-9260.
74. Ullu E, Esposito V, Melli M. Evolutionary conservation of the human 7 S RNA sequences. J Mol Biol 1982; 161:195-201.
75. Wassarman DA, Steitz JA. Structural analyses of the 7SK ribonucleoprotein (RNP), the most abundant human small RNP of unknown function. Mol Cell Biol 1991; 11:3432-3445.
76. Murphy S, Altruda F, Ullu E et al. DNA sequences complementary to human 7 SK RNA show structural similarities to the short mobile elements of the mammalian genome. J Mol Biol 1984; 177:575-590.
77. Kruger W, Benecke B-J. Structural and functional analysis of a human 7S K RNA gene. J Mol Biol 1987; 195:31-41.
78. Bartkiewicz M, Gold H, Altman S. Identification and characterization of an RNA molecule that copurifies with RNase P activity in HeLa cells. Genes Dev 1989; 3:488-499.
79. Lee J-Y, Engelke DR. Partial characterization of an RNA component that copurifies with *Saccharomyces cerevisiae* RNase P. Mol Cell Biol 1989; 9:2536-2543.
80. Lee J-Y, Rohlman CE, Molony LE et al. Characterization of *RPR1*, an essential gene encoding the RNA component of *Saccharomyces cerevisiae* nuclear RNase P. Mol Cell Biol 1991; 11:721-730.
81. Morrissey JP, Tollervey D. Birth of the snoRNPs: the evolution of RNase MRP and the eukaryotic pre-rRNA-processing system. Trends Biochem Sci 1995; 20:78-82.
82. Baer M, Nilsen TW, Costigan C et al. Structure and transcription of a human gene for H1 RNA, the RNA component of human RNase P. Nucleic Acids Res 1990; 18:97-103.

83. Lee JY, Evans CF, Engelke DR. Expression of RNase P RNA in *Saccharomyces cerevisiae* is controlled by an unusual RNA polymerase III promoter. Proc Natl Acad Sci USA 1991; 88:6986-6990.
84. Chang DD, Clayton DA. A mammalian mitochondrial RNA processing activity contains nucleus-encoded RNA. Science 1987; 235:1178-1184.
85. Chang DD, Clayton DA. Mouse RNAase MRP RNA is encoded by a nuclear gene and contains a decamer sequence complementary to a conserved region of mitochondrial RNA substrate. Cell 1989; 56:131-139.
86. Topper JN, Clayton DA. Characterization of human MRP/Th RNA and its nuclear gene: full length MRP/Th RNA is an active endoribonuclease when assembled as an RNP. Nucleic Acids Res 1990; 18:793-799.
87. Schmitt ME, Clayton DA. Nuclear RNase MRP is required for correct processing of pre-5.8S rRNA in *Saccharomyces cerevisiae*. Mol Cell Biol 1993; 13:7935-7941.
88. Clayton DA. A nuclear function for RNase MRP. Proc Natl Acad Sci USA 1994; 91:4615-4617.
89. Lygerou Z, Allmang C, Tollervey D et al. Accurate processing of a eukaryotic precursor ribosomal RNA by ribonuclease MRP in vitro. Science 1996; 272:268-270.
90. Gold HA, Topper JN, Clayton DA et al. The RNA processing enzyme RNase MRP is identical to the Th RNP and related to RNase P. Science 1989; 245:1377-1380.
91. Lygerou Z, Pluk H, van Venrooij WJ et al. hPop1: an autoantigenic protein subunit shared by the human RNase P and RNase MRP ribonucleoproteins. EMBO J 1996; 15:5936-5948.
92. Yuan Y, Reddy R. 5' flanking sequences of human MRP/7-2 RNA gene are required and sufficient for the transcription by RNA polymerase III. Biochim Biophys Acta 1991; 1089:33-39.
93. Hermann-Le Denmat S, Werner M, Sentenac A et al. Suppression of yeast RNA polymerase III mutations by *FHL1*, a gene coding for a *fork head* protein involved in rRNA processing. Mol Cell Biol 1994; 14:2905-2913.
94. Rome LH, Kedersha NL, Chugani DC. Unlocking vaults: organelles in search of a function. Trends Cell Biol 1991; 1:47-50.
95. Scheffer GL, Wijngaard PLJ, Flens MJ et al. The drug resistance-related protein LRP is the human major vault protein. Nature Medicine 1995; 1:578-582.
96. Kickhoefer VA, Searles RP, Kedersha NL et al. Vault ribonucleoprotein particles from rat and bullfrog contain a related small RNA that is transcribed by RNA polymerase III. J Biol Chem 1993; 268:7868-7873.
97. Wolin SL, Steitz JA. Genes for two small cytoplasmic Ro RNAs are adjacent and appear to be single-copy in the human genome. Cell 1983; 32:735-744.
98. O'Brien CA, Margelot K, Wolin SL. *Xenopus* Ro ribonucleoproteins: members of an evolutionarily conserved class of cytoplasmic ribonucleoproteins. Proc Natl Acad Sci USA 1993; 90:7250-7254.
99. Farris AD, Gross JK, Hanas JS et al. Genes for murine Y1 and Y3 Ro RNAs have class 3 RNA polymerase III promoter structures and are unlinked on mouse chromosome 6. Gene 1996; 174:35-42.
100. Wolin SL, Steitz JA. The Ro small cytoplasmic ribonucleoproteins: identification of the antigenic protein and its binding site on the Ro RNAs. Proc Natl Acad Sci USA 1984; 81:1996-2000.
101. Maraia RJ, Sasaki-Tozawa N, Driscoll CT et al. The human Y4 small cytoplasmic RNA gene is controlled by upstream elements and resides on chromosome 7 with all other hY scRNA genes. Nucleic Acids Res 1994; 22:3045-3052.
102. DeChiara TM, Brosius J. Neural BC1 RNA: cDNA clones reveal nonrepetitive sequence content. Proc Natl Acad Sci USA 1987; 84:2624-2628.

103. Tiedge H, Fremeau RT, Weinstock PH et al. Dendritic location of neural BC1 RNA. Proc Natl Acad Sci USA 1991; 88:2093-2097.
104. Tiedge H, Chen W, Brosius J. Primary structure, neural-specific expression, and dendritic location of human BC200 RNA. J Neurosci 1993; 13:2382-2390.
105. Martignetti JA, Brosius J. BC200 RNA: a neural RNA polymerase III product encoded by a monomeric Alu element. Proc Natl Acad Sci USA 1993; 90:11563-11567.
106. Yu G-L, Bradley JD, Attardi LD et al. In vivo alteration of telomere sequences and senescence caused by mutated *Tetrahymena* telomerase RNAs. Nature 1990; 344:126-132.
107. Feng J, Funk WD, Wang S-S et al. The RNA component of human telomerase. Science 1995; 269:1236-1241.
108. Jelinek WR, Schmid CW. Repetitive sequences in eukaryotic DNA and their expression. Annu Rev Biochem 1982; 51:813-844.
109. Singer MF. SINEs and LINEs: highly repeated short and long interspersed sequences in mammalian genomes. Cell 1982; 28:433-434.
110. Howard BH, Sakamoto K. Alu interspersed repeats: selfish DNA or a functional gene family. New Biol 1990; 2:759-770.
111. Wilson ET, Condliffe DP, Sprague KU. Transcriptional properties of BmX, a moderately repetitive silkworm gene that is an RNA polymerase III template. Mol Cell Biol 1988; 8:624-631.
112. Ackerman EJ. Molecular cloning and sequencing of OAX DNA: an abundant gene family transcribed and activated in *Xenopus* oocytes. EMBO J 1983; 2:1417-1422.
113. Lam BS, Carroll D. Tandemly repeated DNA sequences from *Xenopus laevis* I. Studies on sequence organization and variation in satellite 1 DNA (741 base-pair repeat). J Mol Biol 1983; 165:567-585.
114. Andrews MT, Loo S, Wilson LR. Coordinate inactivation of class III genes during the Gastrula-Neurula Transition in *Xenopus*. Dev Biol 1991; 146:250-254.
115. Cohen I, Reynolds WF. The *Xenopus* YB3 protein binds the B box element of the class III promoter. Nucleic Acids Res 1991; 19:4753-4759.
116. Rubin CM, Houck CM, Deininger PL et al. Partial nucleotide sequence of the 300-nucleotide interspersed repeated human DNA sequences. Nature 1980; 284:372-374.
117. Jelinek WR, Toomey TP, Leinwand L et al. Ubiquitous, interspersed repeated sequences in mammalian genomes. Proc Natl Acad Sci USA 1980; 77:1398-1402.
118. Deininger PL, Jolly DJ, Rubin CM et al. Base sequence studies of 300 nucleotide renatured repeated human DNA clones. J Mol Biol 1981; 151:17-33.
119. Pan J, Elder JT, Duncan CH et al. Structural analysis of interspersed repetitive polymerase III transcription units in human DNA. Nucleic Acids Res 1981; 9:1151-1170.
120. Britten RJ. Evidence that most human *Alu* sequences were inserted in a process that ceased about 30 million years ago. Proc Natl Acad Sci USA 1994; 91:6148-6150.
121. Paolella G, Lucero MA, Murphy MH et al. The *Alu* family repeat promoter has a tRNA-like bipartite structure. EMBO J 1983; 2:691-696.
122. Weiner AM, Deininger PL, Efstratiadis A. Nonviral retroposons: genes, pseudogenes, and transposable elements generated by the reverse flow of genetic information. Annu Rev Biochem 1986; 55:631-661.
123. Maraia RJ, Driscoll CT, Bilyeu T et al. Multiple dispersed loci produce small cytoplasmic *Alu* RNA. Mol Cell Biol 1993; 13:4233-4241.
124. Chu WM, Liu WM, Schmid CW. RNA polymerase III promoter and terminator elements affect *Alu* RNA expression. Nucleic Acids Res 1995; 23:1750-1757.
125. Bennett KL, Hill RE, Pietras DF et al. Most highly repeated dispersed DNA families in the mouse genome. Mol Cell Biol 1984; 4:1561-1571.

126. Krayev AS, Kramerov DA, Skryabin KG et al. The nucleotide sequence of the ubiquitous repetitive DNA sequence B1 complementary to the most abundant class of mouse fold-back RNA. Nucleic Acids Res 1980; 8:1201-1215.
127. Haynes SR, Jelinek WR. Low molecular weight RNAs transcribed in vitro by RNA polymerase III from Alu-type dispersed repeats in Chinese hamster DNA are also found in vivo. Proc Natl Acad Sci USA 1981; 78:6130-6134.
128. Rogers JH. The origin and evolution of retroposons. Int Rev Cytol 1985; 93:187-279.
129. Okada N. SINEs. Curr Opin Genet Dev 1991; 1:498-504.
130. Sapienza C, St-Jacques B. "Brain-specific" transcription and evolution of the identifier sequence. Nature 1986; 319:418-420.
131. Saba JA, Busch H, Reddy R. A new moderately repetitive rat DNA sequence detected by a cloned 4.5 S_I DNA. J Biol Chem 1985; 260:1354-1357.
132. Sutcliffe JG, Milner RJ, Bloom FE et al. Common 82-nucleotide sequence unique to brain RNA. Proc Natl Acad Sci USA 1982; 79:4942-4946.
133. Brickell PM, Latchman DS, Murphy D et al. Activation of a *Qa/Tla* class I major histocompatibility antigen gene is a general feature of oncogenesis in the mouse. Nature 1983; 306:756-760.
134. Kress M, Barra Y, Seidman JG et al. Functional insertion of an Alu type 2 (B2 SINE) repetitive sequence in murine class I genes. Science 1984; 226:974-977.
135. Korenberg JR, Rykowski MC. Human genome organization: Alu, Lines, and the molecular structure of metaphase chromosome bands. Cell 1988; 53:391-400.
136. Chen TL, Manuelidis L. SINEs and LINEs cluster in distinct DNA fragments of Giemsa band size. Chromosoma 1989; 98:309-316.
137. Korenberg JR, Thermann E, Denniston C. Hotspots and functional organization of human chromosomes. Hum Genet 1978; 43:13-22.
138. Wallace MR, Andersen LB, Saulino AM et al. A de novo *Alu* insertion results in neurofibromatosis type 1. Nature 1991; 353:864-866.
139. Goldberg YP, Rommens JM, Andrew SE et al. Identification of an Alu retrotransposition event in close proximity to a strong candidate gene for Huntington's disease. Nature 1993; 362:370-373.
140. Makalowski W, Mitchell GA, Labuda D. Alu sequences in the coding regions of mRNA: a source of protein variability. Trends Genet 1994; 10:188-193.
141. Murphy D, Brickell PM, Latchman DS et al. Transcripts regulated during normal embryonic development and oncogenic transformation share a repetitive element. Cell 1983; 35:865-871.
142. Kaplan G, Jelinek WR, Bachvarova R. Repetitive sequence transcripts and U1 RNA in mouse oocytes and eggs. Dev Biol 1985; 109:15-24.
143. Vasseur M, Condamine H, Duprey P. RNAs containing B2 repeated sequences are transcribed in the early stages of mouse embryogenesis. EMBO J 1985; 4:1749-1753.
144. White RJ, Stott D, Rigby PWJ. Regulation of RNA polymerase III transcription in response to F9 embryonal carcinoma stem cell differentiation. Cell 1989; 59:1081-1092.
145. Maraia RJ. The subset of mouse B1 (Alu-equivalent) sequences expressed as small processed cytoplasmic transcripts. Nucleic Acids Res 1991; 19:5695-5702.
146. Daniels GR, Deininger PL. Repeat sequence families derived from mammalian tRNA genes. Nature 1985; 317:819-822.
147. Sakamoto K, Okada N. Methylcytidylic modification of in vitro transcript from the rat identifier sequence: evidence that the transcript forms a tRNA-like structure. Nucleic Acids Res 1985; 13:7195-7206.
148. Kim J, Martignetti JA, Shen MR et al. Rodent BC1 RNA gene as a master gene for ID element amplification. Proc Natl Acad Sci USA 1994; 91:3607-3611.

149. Singh K, Carey M, Saragosti S et al. Expression of enhanced levels of small RNA polymerase III transcripts encoded by the B2 repeats in simian virus 40-transformed mouse cells. Nature 1985; 314:553-556.
150. Carey MF, Singh K, Botchan M et al. Induction of specific transcription by RNA polymerase III in transformed cells. Mol Cell Biol 1986; 6:3068-3076.
151. Chang D-Y, Nelson B, Bilyeu T et al. A human *Alu* RNA-binding protein whose expression is associated with accumulation of small cytoplasmic *Alu* RNA. Mol Cell Biol 1994; 14:3949-3959.
152. Britten RJ, Baron WF, Stout DB et al. Sources and evolution of human *Alu* repeated sequences. Proc Natl Acad Sci USA 1988; 85:4770-4774.
153. Batzer MA, Rubin CM, Hellmann-Blumberg U et al. Dispersion and insertion polymorphism in two small subfamilies of recently amplified human *Alu* repeats. J Mol Biol 1995; 247:418-427.
154. Schmid C, Maraia R. Transcriptional regulation and transpositional selection of active SINE sequences. Curr Opin Genet Dev 1992; 2:874-882.
155. Deininger PL, Batzer MA, Hutchison CA et al. Master genes in mammalian repetitive DNA amplification. Trends Genet 1992; 8:307-311.
156. Brookfield JFY. The human Alu SINE sequences—is there a role for selection in their evolution? Bioessays 1994; 16:793-795.
157. Liu W-M, Schmid CW. Proposed roles for DNA methylation in *Alu* transcriptional repression and mutational inactivation. Nucleic Acids Res 1993; 21:1351-1359.
158. Liu W-M, Maraia RJ, Rubin CM et al. Alu transcripts: cytoplasmic localization and regulation by DNA methylation. Nucleic Acids Res 1994; 22:1087-1095.
159. Chesnokov I, Schmid CW. Flanking sequences of an Alu source stimulate transcription in vitro by interacting with sequence-specific transcription factors. J Mol Evol 1996; 42:30-36.
160. Doolittle WF, Sapienza C. Selfish genes, the phenotype paradigm and genome evolution. Nature 1980; 284:601-603.
161. Orgel LE, Crick FHC. Selfish DNA: the ultimate parasite. Nature 1980; 284:604-607.
162. Britten RJ, Davidson EH. Gene regulation for higher cells: a theory. Science 1969; 165:349-357.
163. Davidson EH, Britten RJ. Regulation of gene expression: possible role of repetitive sequences. Science 1979; 204:1052-1059.
164. Scott MRD, Westphal K-H, Rigby PWJ. Activation of mouse genes in transformed cells. Cell 1983; 34:557-567.
165. Sutcliffe JG, Milner RJ, Gottesfeld JM et al. Identifier sequences are transcribed specifically in brain. Nature 1984; 308:237-241.
166. Sutcliffe JG, Milner RJ, Gottesfeld JM et al. Control of neuronal gene expression. Science 1984; 225:1308-1315.
167. Majello B, La Mantia G, Simeone A et al. Activation of major histocompatibility complex class I mRNA containing an *Alu*-like repeat in polyoma virus-transformed rat cells. Nature 1985; 314:457-459.
168. Vasseur M, Duprey P, Brulet P et al. One gene and one pseudogene for the cytokeratin endo A. Proc Natl Acad Sci USA 1985; 82:1155-1159.
169. McKinnon RD, Shinnick TM, Sutcliffe JG. The neuronal identifier element is a *cis*-acting positive regulator of gene expression. Proc Natl Acad Sci USA 1986; 83:3751-3755.
170. Glaichenhaus N, Cuzin F. A role for ID repetitive sequences in growth- and transformation-dependent regulation of gene expression in rat fibroblasts. Cell 1987; 50:1081-1089.
171. Oliviero S, Monaci P. RNA polymerase III promoter elements enhance transcription of RNA polymerase II genes. Nucleic Acids Res 1988; 16:1285-1293.

172. Saffer JD, Thurston SJ. A negative regulatory element with properties similar to those of enhancers is contained within an Alu sequence. Mol Cell Biol 1989; 9:355-364.
173. Brini AT, Lee GM, Kinet J-P. Involvement of Alu sequences in the cell-specific regulation of transcription of the γ chain of Fc and T cell receptors. J Biol Chem 1993; 268:1355-1361.
174. Hambor JE, Mennone J, Coon ME et al. Identification and characterization of an *Alu*-containing, T-cell-specific enhancer located in the last intron of the human CD8α gene. Mol Cell Biol 1993; 13:7056-7070.
175. Thorey IS, Cecna G, Reynolds W et al. *Alu* sequence involvement in transcriptional insulation of the keratin 18 gene in transgenic mice. Mol Cell Biol 1993; 13:6742-6751.
176. Vidal F, Mougneau E, Glaichenhaus N et al. Coordinated post-transcriptional control of gene expression by modular elements including *Alu*-like repetitive sequences. Proc Natl Acad Sci USA 1993; 90:208-212.
177. Hull MW, Erickson J, Johnston M et al. tRNA genes as transcriptional repressor elements. Mol Cell Biol 1994; 14:1266-1277.
178. Vansant G, Reynolds WF. The consensus sequence of a major *Alu* subfamily contains a functional retinoic acid response element. Proc Natl Acad Sci USA 1995; 92:8229-8233.
179. Strub K, Moss JB, Walter P. Binding sites of the 9 and 14 kD heterodimeric protein subunit of the signal recognition particle (SRP) are contained exclusively in the Alu domain of SRP DNA and contain a sequence motif that is conserved in evolution. Mol Cell Biol 1991; 11:3949-3959.
180. Krayev AS, Markusheva TV, Kramerov DA et al. Ubiquitous transposon-like repeats B1 and B2 of the mouse genome: B2 sequencing. Nucleic Acids Res 1982; 10:7461-7475.
181. Anachkova B, Todorova M, Vassilev L et al. Isolation of short interspersed repetitive DNA sequences present in the regions of initiation of mammalian DNA replication. Eur J Biochem 1984; 141:105-106.
182. Ariga H. Replication of cloned DNA containing the Alu family sequence during cell extract-promoting simian virus 40 DNA synthesis. Mol Cell Biol 1984; 4:1476-1482.
183. Anachkova B, Russev M, Altmann H. Identification of the short dispersed repetitive DNA sequences isolated from the zones of initiation of DNA synthesis in human cells as Alu-elements. Biochem Biophys Res Commun 1985; 128:101-106.
184. Johnson EM, Jelinek WR. Replication of a plasmid bearing a human Alu-family repeat in monkey COS-7 cells. Proc Natl Acad Sci USA 1986; 83:4660-4664.
185. Fornace AJ, Mitchell JB. Induction of B2 RNA polymerase III transcription by heat shock: enrichment for heat shock induced sequences in rodent cells by hybridization subtraction. Nucleic Acids Res 1986; 14:5793-5811.
186. Liu W-M, Chu W-M, Choudary PV et al. Cell stress and translational inhibitors transiently increase the abundance of mammalian SINE transcripts. Nucleic Acids Res 1995; 23:1758-1765.
187. Sakamoto K, Fordis CM, Corsico CD et al. Modulation of HeLa cell growth by transfected 7SL RNA and Alu gene sequences. J Biol Chem 1991; 266:3031-3038.
188. Clemens MJ. A potential role for RNA transcribed from B2 repeats in the regulation of mRNA stability. Cell 1987; 49:157-158.
189. Deininger PL, Daniels GR. The recent evolution of mammalian repetitive DNA elements. Trends Genet 1986; 2:76-80.
190. Deininger PL, Slagel VK. Recently amplified *Alu* family members share a common parental *Alu* sequence. Mol Cell Biol 1988; 8:4566-4569.

191. Lehrman MA, Goldstein JL, Russell DW et al. Duplication of seven exons in LDL receptor gene caused by *Alu-Alu* recombination in a subject with familial hypercholesterolemia. Cell 1987; 48:827-835.
192. Kramerov DA, Tillib SV, Lekakh IV et al. Biosynthesis and cytoplasmic distribution of small poly(A)-containing B2 RNA. Biochim Biophys Acta 1985; 824:85-98.
193. Kramerov DA, Tillib SV, Shumyatsky GP et al. The most abundant nascent poly(A)$^+$ RNAs are transcribed by RNA polymerase III in murine tumor cells. Nucleic Acids Res 1990; 18:4499-4506.
194. White RJ, Stott D, Rigby PWJ. Regulation of RNA polymerase III transcription in response to Simian virus 40 transformation. EMBO J 1990; 9:3713-3721.
195. Kramerov DA, Grigoryan AA, Ryskov AP et al. Long double-stranded sequences (dsRNA-B) of nuclear pre-mRNA consist of a few highly abundant classes of sequences: evidence from DNA cloning experiments. Nucleic Acids Res 1979; 6:697-713.
196. Ryskov AP, Ivanov PL, Kramerov DA et al. Mouse ubiquitous B2 repeat in polysomal and cytoplasmic poly(A)$^+$ RNAs: unidirectional orientation and 3'-end localization. Nucleic Acids Res 1983; 11:6541-6558.
197. Kramerov DA, Lekakh IV, Samarina OP et al. The sequences homologous to major interspersed repeats B1 and B2 of mouse genome are present in mRNA and cytoplasmic poly(A)$^+$ RNA. Nucleic Acids Res 1982; 10:7477-7491.
198. Margalit H, Nadir E, Ben-Sasson SA. A complete Alu element within the coding sequence of a central gene. Cell 1994; 78:173-174.
199. Chung J, Sussman DJ, Zeller R et al. The *c-myc* gene encodes superimposed RNA polymerase II and III promoters. Cell 1987; 51:1001-1008.
200. Bentley DL, Brown WL, Groudine M. Accurate, TATA box-dependent polymerase III transcription from promoters of the *c-myc* gene in injected *Xenopus* oocytes. Genes Dev 1989; 3:1179-1189.
201. Piras G, Kashanchi F, Radonovich MF et al. Transcription of the human T-cell lymphotropic virus type I promoter by an α-amanitin-resistant polymerase. J Virol 1994; 68:6170-6179.
202. Kurose K, Hata K, Hattori M et al. RNA polymerase III dependence of the human L1 promoter and possible participation of the RNA polymerase II factor YY1 in the RNA polymerase III transcription system. Nucleic Acids Res 1995; 23:3704-3709.
203. Mitchell MT, Hobson GM, Benfield PA. TATA box-mediated polymerase III transcription in vitro. J Biol Chem 1992; 267:1995-2005.
204. Mitchell MT, Benfield PA. TATA box-mediated in vitro transcription by RNA polymerase III. J Biol Chem 1993; 268:1141-1150.
205. Pruzan R, Chatterjee PK, Flint SJ. Specific transcription from the adenovirus E2E promoter by RNA polymerase III requires a subpopulation of TFIID. Nucleic Acids Res 1992; 20:5705-5712.
206. Huang W, Pruzan R, Flint SJ. In vivo transcription from the adenovirus E2 early promoter by RNA polymerase III. Proc Natl Acad Sci USA 1994; 91:1265-1269.
207. Martinez E, Lagna G, Roeder RG. Overlapping transcription by RNA polymerases II and III of the *Xenopus* TFIIIA gene in somatic cells. J Biol Chem 1994; 269:25692-25698.
208. Kim SH, Darby MK, Joho KE et al. The characterization of the TFIIIA synthesized in somatic cells of *Xenopus laevis*. Genes Dev 1990; 4:1602-1610.

CHAPTER 2

Promoter Structure of Class III Genes

The promoters of most class III genes include discontinuous intragenic structures, termed internal control regions (ICRs), that are composed of essential sequence blocks separated by nonessential nucleotides. The ICRs of 5S rRNA genes are sometimes referred to as type I. These comprise two functional domains: an A-block and a second domain consisting of an intermediate element and a C-block. Most class III genes, including tRNA, VA, Alu, EBER, 7SL, 4.5S, B1, and B2 genes, have type II ICRs: these again have two domains, an A-block and a B-block. The A-blocks of types I and II are homologous and can substitute for one another in *Xenopus*,[1] although not in *Neurospora*.[2] The A-block is located much further from the start site in type I than it is in type II promoters. As well as the ICR, extragenic sequences can also affect the strength of type I and II promoters. However, substitutions in the extragenic regions are generally well tolerated, unlike mutations in the ICR. In contrast, with type III promoters, such as those of the vertebrate U6 and 7SK genes, transcription is independent of intragenic elements and is dictated solely by 5' flanking regions.[3-8] A schematic illustration of the three types of class III promoter is provided in Figure 2.1.

ICR sequences are highly conserved between different genes and different species. In contrast, the flanking sequences of type I and II promoters frequently show little or no conservation, although they can often have powerful modulatory effects. This suggests that the flanking sequences are more likely to be recognized by gene- or species-specific factors, or that their cognate factors have very flexible DNA-binding specificities.

Since studies on promoter structure involve the replacement of one sequence by another, it is always important to ensure that observed effects are due to changes in the wild-type sequence rather than fortuitous responses to an introduced sequence that is assumed to be neutral. An extreme example in which substituted sequences had a major effect upon expression was provided by an in vivo study of the yeast SUP4-o $tRNA^{Tyr}$ promoter, in which the introduction of three BamHI linkers at the -18 junction of a 5' deletion abolished expression, whereas a single BamHI linker left the gene active.[9] Since many pol III promoter elements are within transcribed regions, it is important to determine whether changes in expression levels following mutagenesis result from altered transcription or reduced RNA stability. The chosen assay conditions can also have a major influence upon the apparent extent of promoter elements. For instance, distal upstream sequences sometimes have an effect in vivo but not in vitro.[10-12] The result of a mutation can be

RNA Polymerase III Transcription, Second Edition, by Robert J. White.

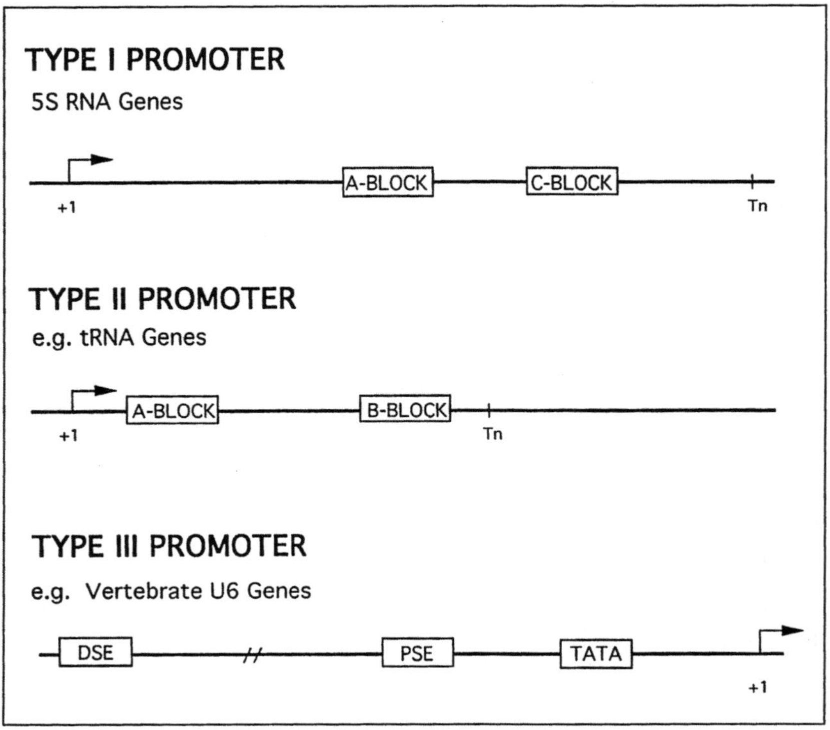

Fig. 2.1. Diagram depicting the three types of promoter arrangement that are utilized by class III genes. The site of transcription initiation is indicated by +1 and the site of termination is indicated by Tn. Promoter elements are shown as open boxes.

tissue- or species-dependent.[13-16] Examples have also been reported in which the requirement for particular sequences in vitro depends upon the template, extract, or salt concentration chosen for the assay.[17-21]

A number of review articles have been published that deal with aspects of class III promoter structure.[22-28]

5S rRNA Genes

The promoter structures of the *Xenopus* 5S rRNA genes have been subject to intensive study (Fig. 2.2). Deletion analyses of a somatic 5S gene first demonstrated that internal sequences are necessary and sufficient to promote transcription in a *Xenopus* oocyte nuclear extract.[29,30] This ICR also plays the dominant role in determining expression levels in living *Xenopus* eggs.[31] Linker scanning and point mutational analyses showed that the ICR lies between +50 and +97, and consists of three separate elements, viz the A-block (+50 to +64), the intermediate element (+67 to +72), and the C-block (+80 to +97).[32-34] The bases in between these regions serve as spacers and do not influence transcription efficiency.[35] Addition of up to 10 bp or removal of up to 3 bp from between the A-block and the intermediate

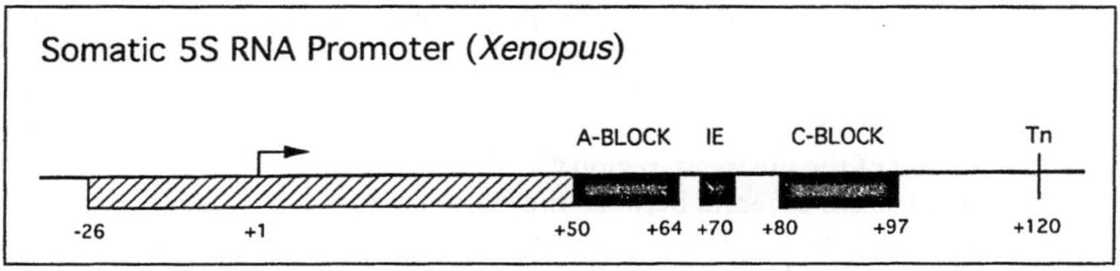

Fig. 2.2. Diagram depicting the promoter arrangement of the *Xenopus* somatic 5S rRNA genes. The site of transcription initiation is indicated by +1 and the site of termination is indicated by Tn. Essential promoter elements are shown as closed boxes and modulatory promoter elements are shown as cross-hatched boxes. IE is an abbreviation of intermediate element.

element was tolerated, but reduced transcription and prevented formation of a stable transcription complex.[35] Mutations in the A- and C-blocks of a major oocyte 5S gene also abolished transcription.[36]

Deletion of sequences upstream of the ICR reduced competitive strength and the accuracy of initiation of the somatic 5S gene in a *Xenopus* oocyte nuclear extract.[37] For example, 5' deletion to +28 decreased competitive strength to 35-40% of the wild-type level.[21,37] The same deletion reduced transcription to 5-25% when assayed in a whole oocyte S-150 extract.[21,36] The different results obtained using different types of extract are likely to reflect variations in the relative concentrations of transcription factors, although qualitative changes in factor populations cannot be ruled out. Rates of 5S transcription can be 100-fold greater in an optimized oocyte nuclear extract than they are in an S-150.[21] Whereas substitution of nucleotides between +18 and +37 had little or no deleterious effect upon transcription of the somatic 5S gene under conditions of factor excess,[38,39] a linker scanning mutation between +33 and +39 of the major oocyte 5S gene reduced expression in an S-150 to 18% of the wild-type level.[36] Severe loss of activity was also associated with mutation of residues +10 to +13 of the major oocyte gene when assayed in an oocyte S-150.[36] In contrast, linker substitution of nucleotides +8 to +15 of the somatic 5S gene reduced transcription by only 25% in a *Xenopus* oocyte extract,[38] but by 85% in a HeLa extract.[39] It is clear from these studies that the region between the A-block and the start site can have a major effect upon expression of the *Xenopus* 5S genes. This effect is greatest under conditions that are suboptimal for transcription, whereas the minimal ICR (+50 to +97) can suffice under optimal conditions. Initiation occurs approximately 50 bp upstream of the A-block, even in 5' deleted constructs.[1,29,35,37]

Deletion of the 5' flanking sequence of the *Xenopus* somatic 5S gene to -26 had no effect upon transcription or competitive strength in an oocyte nuclear extract, but removal of all 5' sequence reduced competition by 60% and transcription by about 40%.[37] However, the influence of 5' and 3' flanking sequences was diminished and eventually lost altogether as the ratio of nuclear extract to template was raised in order to place transcription factors in large excess over 5S genes.[21] Deletion of the flanking sequences produced a 2- to 3-fold decrease in transcription of the somatic 5S gene in an oocyte S-150, but had no detrimental effect upon the major oocyte 5S gene when assayed under the same conditions.[36] Substitution of the region from -34 to +5 of the somatic gene caused a 4-fold reduction in transcription using a reconstituted HeLa system.[40] Several transition mutations between -10 and -23 resulted in a modest (12-38%) increase in transcription of the somatic 5S gene in a *Xenopus* ovary S100 extract; each of these mutations increased the A/T content of the upstream region.[41]

There are 8 bases different between the major oocyte 5S gene and the somatic 5S gene of *X. laevis*, and 6 of these occur within the ICR.[42] The sequences preceding these genes are highly dissimilar, except for some homology between -29 and -14.[43,44] The somatic gene is transcribed 4- to 10-fold more efficiently than the major oocyte gene in microinjected oocyte nuclei or in HeLa or oocyte nuclear extracts and is transcribed 20- to 200-fold more efficiently in microinjected two-cell embryos or oocyte S-150 extracts.[37,42,45-50] These differences are exaggerated when transcription is conducted under competitive conditions.[42,46-49] The base differences at +47, +53, +55 and +56 can account for part of this differential, since changing some or all of these bases in the major oocyte gene to match those of the somatic gene stimulates transcription in an S-150 by 5- to 10-fold.[42] The somatic sequences between -32 and +37 can also contribute an advantage of up to 10-fold relative to the corresponding major oocyte 5S sequences; this effect is observed in oocyte S-150 extracts and in microinjected embryos, but not in HeLa extracts, oocyte nuclear extracts or microinjected oocyte nuclei.[49,50] The differences in sequence between -32 and +37 and those at +47 to +56 have been shown to affect the competitive strength of the two types of 5S gene.[37,42,48-50]

A synthetic 5S gene consisting solely of the coding region for the major human 5S rRNA can be transcribed in a HeLa extract, indicating that for human, as for *Xenopus* 5S genes, expression is possible in the absence of specific flanking sequences.[51] However, deletion analysis suggests that the 5'-flanking region of a complete human 5S gene can stimulate transcription in a HeLa S-100 extract by 10-fold or more.[52] Although the upstream promoter elements of mammalian 5S genes have not been mapped, GC boxes that resemble Sp1 binding sites are found at -40 of a human 5S gene and at -80 of a hamster 5S gene.[52]

The sequence from -2 to +20 is very highly conserved in 5S genes from a wide variety of organisms. Thus, these bases in the *X. borealis* somatic 5S gene show a 21/22 bp match with the corresponding sequence from a hamster 5S gene, a 19/22 bp match with a *Drosophila* 5S gene, and a 17/22 bp match with a yeast 5S gene.[39] A linker scanning mutation at +3 to +14 of the *Drosophila* 5S gene compromised transcription in *Drosophila* cell extracts.[53] The internal promoters of *Drosophila* and *Neurospora* 5S genes resemble those of *Xenopus* in comprising A- and C-blocks as well as important sequences lying between the A-block and the start site.[53,54] In addition, the 5S genes from *Drosophila*, *Neurospora* and *Bombyx* all require 5' flanking promoter sequences.[53-58] In each of these cases, the upstream modulatory sequences are located between -39 and the start site and include a TATA motif situated around -30.[53-57] Expression of the *Bombyx* 5S gene is also affected by 3' flanking sequences.[58]

Mutagenesis of the 5S gene of *S. cerevisiae* identified only two promoter elements that are essential for transcription in vitro.[59] Deletion analysis localized the essential C-block region as lying between +81 and +94.[59] A second promoter element extends from -14 to +8.[59] This region, which has been named sse, strongly affects the efficiency of transcription, but does not dictate start site selection.[59,60] None of the sequences between +9 and +80, including an A-block homology, are absolutely required in vitro, although they do contribute to the efficiency of transcription.[59] In vivo, a substantial decrease in expression was observed following point mutation of the A-block, the intermediate element or the C-block.[61] Sequences between -40 and -14 also have a net positive influence.[59] Changing the distance between the sse and the C-block by more than a few base pairs has a strongly deleterious effect upon expression.[59]

tRNA Genes

Type II promoters are split into two essential and highly conserved regions of about 10 bp each, the A- and B-blocks, that are generally separated by 30-40 bp. Chemically synthesized oligonucleotides corresponding to these two sequence blocks are sufficient to direct efficient transcription in a HeLa cell extract when separated by a 51 bp spacer.[62] The A- and B-blocks constitute the essential promoter elements of a tRNA gene, whereas additional internal or flanking sequences can often have modulatory effects (Fig. 2.3). Deletion analyses demonstrated that sequences 5' to +13 and 3' to +64 are not required for transcription of a *Xenopus* $tRNA^{Leu}$ gene following oocyte injection, and internal sequences between +21 and +50 could be replaced by polylinkers without abolishing expression.[63] Comparable ICR arrangements have been defined for a *Xenopus* $tRNA^{Met}$ gene,[64] a *Drosophila* $tRNA^{Arg}$ gene,[65] a nematode $tRNA^{Pro}$ gene[66] and a yeast $tRNA^{Tyr}$ gene.[67] The split promoter sequences mapped to nucleotides +13 to +20 and +51 to +64 of the *Xenopus* $tRNA^{Leu}$ gene coincide closely with two highly conserved sequence blocks present in all eukaryotic tRNA genes, as well as VA, EBER, Alu, 4.5S, B1 and B2 genes.[63,67,68] It was therefore possible to derive a consensus of TGGCNNAGTGG for the A-block and GGTTCGANNCC for the B-block.[63] Chimeric genes containing the 5' half of $tRNA^{Leu}$ and the 3' half of $tRNA^{Met}$ or vice versa were active, thereby demonstrating the functional compatibility of A- and B-blocks from different genes.[63] These sequences are extremely well conserved and even occur in certain bacterial and chloroplast tRNA genes which, as a consequence, can serve as templates for pol III.[69,70] To a large extent, the conservation of the A- and B-blocks may result from the fact that they encode the D- and T-loops which are required for the function of the tRNA gene product. The consensus is therefore likely to reflect selection for both tRNA and promoter functions. However, it is clear that point mutations in the A- and B-blocks can have a substantial effect upon transcription efficiency.[14,67,71-75] Indeed, 10 out of 12 substitution mutations affecting the transcription or competitive strength of the yeast SUP4 $tRNA^{Tyr}$ gene were found to lie within the A- or B-blocks; in general, promoter down-mutations reduced the homology with the consensus and up-mutations increased it, indicating that the consensus sequences coincide well, although not perfectly, with the optimal sequences for promoter activity.[67] Similarly, most of the major promoter determinants of the *Xenopus* $tRNA^{Met}$ gene lie within the A- and B-blocks.[76] Gaeta et al[14] carried out saturation mutagenesis of the B-block of a *Drosophila* $tRNA^{Arg}$ gene.[14] The mutants were assayed in both *Drosophila* and HeLa extracts for their ability to support transcription and stable complex formation.[14] Although the two systems provided similar results, *Drosophila* extracts were more tolerant of substitutions.[14] The wild-type sequence is a perfect match to the B-block consensus and this gave optimal activity.[14] The majority of substitutions were deleterious, although a few were neutral.[14] Positions 1, 3 and 4 of the B-block would not tolerate changes and are therefore likely to be the primary contact points for protein binding.[14] In general,

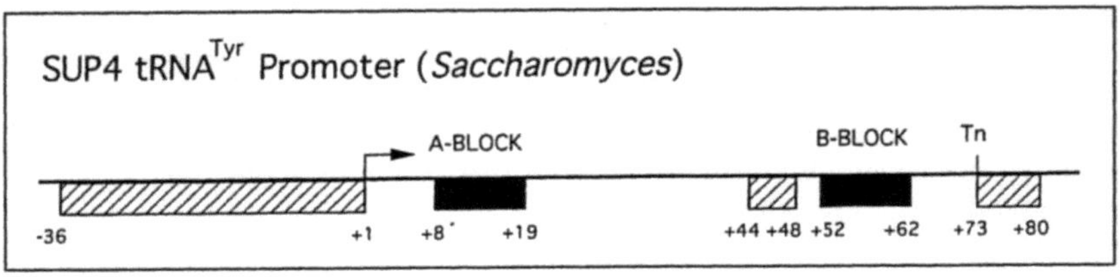

Fig. 2.3. Diagram depicting the promoter arrangement of the SUP4 tRNA^{Tyr} gene of *Saccharomyces cerevisiae*. The site of transcription initiation is indicated by +1 and the site of termination is indicated by Tn. Essential promoter elements are shown as closed boxes and modulatory promoter elements are shown as cross-hatched boxes.

mutations that reduced template activity also impaired factor binding.[14] However, several exceptions to this correlation[14] illustrate the point that stable complex formation is not a prerequisite for transcription. The effects of several point mutations could not be explained in terms of major or minor groove interactions, which raised the possibility that local DNA geometry contributes to B-block function.[14]

A partially conserved region between +45 and +51 encodes the extra loop of tRNAs and can also have an effect upon transcription efficiency.[67,72,77,78] This element may be regarded as a 5' extension of the B-block and the entire region has been referred to as box B+.[25] A functional difference between the A- and B-blocks was indicated by the more severe effects of tRNA^{Tyr} mutations in the latter relative to the former, especially with regard to competitive strength.[67,74] In contrast to the results with most tRNA genes, Wilson et al[18] found that deletion of the B-block of a silkworm tRNA^{AlaC} gene could be tolerated at high template concentrations; however, this result is fairly exceptional and probably reflects the fact that this gene shows only poor homology to the B-block consensus. For instance, *Bombyx* tRNA^{AlaC} contains an A at position 54 instead of the T that is found in almost all other tRNA genes.[24]

A wide variety of separations between the A- and B-blocks are compatible with transcription, as is necessitated by the varying lengths of extra arms and the presence of introns in some tRNA genes.[17,23,79,80] For instance, about 10% of yeast tRNA genes have introns, and the interblock distance can vary from 27 to 93 bp.[24,25] The relative helical orientation of the A and B blocks does not seem to be important in determining transcriptional efficiency.[80] Separations of approximately 30 to 60 bp are optimal, although distances of 400 bp can be tolerated.[79,80] Suboptimal spacings diminish both transcription efficiency and the ability to form stable transcription complexes.[79] A tRNA^{Leu3} construct with an interblock separation of 365 bp uses its normal A- and B-blocks, but is transcribed less efficiently than the wild-type gene.[81] However, it is more common for alternative regions with partial fortuitous A-block homology to be exploited in constructs with especially unfavorable ICR configurations.[80,81] For example, an upstream pseudoA-block is used by tRNA^{Leu3} if the interblock distance is reduced below 19 bp.[80]

The position of the start site is fixed primarily relative to the A-block rather than the B-block.[1,80,81] However, the sequence of the 5' flanking region can also influence the position of initiation.[23,64,82-84] Introduction of a TATA box upstream of the yeast SUP2 tRNA^{Tyr} gene had a minor effect upon start site selection.[85] The sequence of the initiation region is important. Pol III has a strong preference for initiating at a purine, and this is optimally preceded by a pyrimidine.[23,86] Sequence homologies have been noted around the start site of yeast tRNA genes and substitutions in this region can often be strongly detrimental to transcription,[17,86-89] although one study observed little effect.[90] Several workers have deduced consensus sequences for the start sites of *S. cerevisiae* tRNA genes: $CA_{+1}ACAA$,[88] $THTCA_{+1}W$-AAAWW,[91] and $WWWCA_{+1}AnA$,[86] where H signifies A, T or C, W signifies A or T, and n signifies a nonconserved base. The obvious features that these have in common are that transcription initiates at an A that is preceded by a C, with A/T-rich flanking sequences. A looser consensus of $W_{(1-3)}$PyPu was found in 104 out of 115 tRNA genes surveyed, where Py signifies pyrimidine and Pu signifies purine.[86] Replacing the TTT at -4 to -2 of the $tRNA^{Leu3}$ gene with GGC reduced transcription to 15% of the wild-type level.[86] In contrast, substituting the AAA at +4 to +6 of the same gene with CGG made little difference.[86] Substitutions at the start site can result in minor shifts in the position of initiation so that a PyPu motif is selected.[86] Such shifts are accompanied by reduced rates of transcription and can only be tolerated within a narrow range.[86] Overall, it appears that the A-block dictates the area in which initiation can occur, but the precise start site within that area is determined by local sequence and can also be influenced by the upstream flanking region.

Differential regulation of tRNA genes is necessary for the tRNA population to adapt to different codon frequencies and amino acid utilization in different cell types. Since ICRs are highly conserved whereas flanking regions show little homology, the latter are obvious candidates for mediating selective transcriptional control. Variations in 5' flanking sequences result in the differential expression of members of the *Xenopus* $tRNA^{Tyr}$ gene family,[92] which displays strong developmental regulation.[93] Two short A/T-rich sequences between -30 and -11 are required for transcription of the silkworm $tRNA^{AlaC}$ gene in silkworm extracts, but are dispensable for transcription in a *Xenopus* system.[94,95] Homologous sequences which include a TATA motif occur at corresponding regions of silkworm $tRNA^{Gly}$ and 5S genes, where they are also required for transcription.[96] A silkgland-specific $tRNA^{AlaSG}$ gene lacks these sequences and is less actively transcribed; the use of chimera generated by swapping upstream regions between the constitutive and silkgland-specific $tRNA^{Ala}$ genes demonstrated that the distinctive transcriptional properties are conferred by the 5'-flanking sequences.[96] Variations in flanking sequences are also responsible for substantial differences in the activities of individual members of the silkworm $tRNA^{Gly1}$ family, which have identical coding regions.[97]

5' flanking sequences are likely to have some influence upon the transcription of most, if not all, tRNA genes. Numerous examples of this have been reported for genes from yeast,[9,17,73,84,87,88,98] fruitflies,[99-105] silkworms,[13,18,94-97,106,107] frogs,[83,92,108,109] and humans.[110-115] The 5' elements that affect tRNA transcription are generally located within about 80 bp of the initiation site, and much shorter separations are most commonly found. For instance, an element situated between -1 and -15 of yeast $tRNA^{Leu3}$ stimulates expression both in vivo and in vitro.[17,87,88] However, one regulatory element has been reported to lie 800 bp upstream of a silkworm $tRNA^{Gly1}$ gene.[97] In certain cases, the modulatory effects can be extremely position-dependent. For example, the inhibitory influence of an upstream negative element can be diminished by short deletions or insertions that alter its distance from the coding sequence of a *Drosophila* $tRNA^{Lys2}$ gene.[99] One upstream regulatory element was found to function in either a positive or a negative fashion according to its position.[107] In most cases the 5' flanking sequences have an overall stimulatory influence upon transcription, although repressive effects can also occur.[73,83,97,99,100,107]

In general, there is little or no extensive sequence homology conserved between the 5' flanking regions of different tRNA genes. This is often true even of different genes that encode the same tRNA isoacceptor species. Although there are some instances in which limited homologies are shared between the upstream regions of several tRNA genes of a particular species (e.g., a partially conserved pentanucleotide in yeast[88]), such sequences are not well conserved between different species. One exception to this is the presence of short TATA motifs, which occur upstream of some tRNA genes in many organisms.[95-97,107,108,116] In fact, general A/T-richness is found in the 50 bp upstream of tRNA genes from fungi, protozoa, insects and plants, but not vertebrates.[95] Regions of especially high A/T content are centered around -20 and -30.[95] Huibregste and Engelke[116] compared the upstream regions of 12 yeast tRNA genes and found an average of 68% A/T content between -1 and -40; many, but not all of these, contain TATA elements. Joazeiro et al[84] mutated the A/T-rich 5'-flanking region of the yeast SUP4 $tRNA^{Tyr}$ gene so that it became highly G/C enriched. This decreased the level of expression and increased the proportion of aberrantly initiated transcripts.[84] When A/T bases were reintroduced into this G/C-rich background, start site utilization was strongly influenced by their location.[84] In this context, initiation occurred 28-30 bp downstream of the 5' end of the engineered A/T sequence.[84] As mentioned above, the A/T-rich blocks upstream of the silkworm $tRNA^{AlaC}$ gene can influence its transcription level strongly.[94-96] Whereas raising the G/C content of this region can abolish expression, many mutations that maintain the A/T level are tolerated.[95] Despite this clear preference in lower organisms, the 50 bp upstream of vertebrate tRNA genes have below average A/T content.[95] However, three human $tRNA^{Val}$ genes are located in extensive regions (~600 bp) of 70-90% A/T-richness.[110] The transcription efficiencies of the major and minor $tRNA^{Val}$ genes vary by an order of magnitude in a HeLa cell extract, thereby mimicking the relative concentrations of their transcripts in vivo. Deletion analysis and domain swapping experiments demonstrated that both the 5'- and the 3'-flanking regions contribute to this effect.[112,113] Neither region influenced the stability of transcription complexes, but both increased the level of expression. Sequences between -53 and -31 account for the 5-fold greater transcriptional activity of murine $tRNA^{Asp2}$ relative to $tRNA^{Asp1}$; this region is 78% A/T in $tRNA^{Asp2}$ and only 45% A/T in $tRNA^{Asp1}$.[117] Whereas these distal sequences do not affect competitive strength, proximal sequences between -9 and -1 enhance factor binding as well as overall transcription.[117] Thus, binding of factors and later steps in transcription can be modulated by distinct flanking regions in tRNA genes.

An extreme example of dependence upon upstream flanking sequences, that is unusual for a vertebrate tRNA gene, is provided by the *Xenopus* selenocysteine $tRNA^{Sec}$ promoter.[108,118,119] $tRNA^{Sec}$ is the carrier upon which selenocysteine is synthesized for incorporation into nascent selenoproteins in response to specific UGA codons. The 5'-flanking region of the $tRNA^{Sec}$ gene is sufficient to direct transcription, whereas 5' deletion to -4 abolishes expression in microinjected oocytes.[108,118] A consensus TATA box is located at -30 to -25 and mutation of this virtually abolishes transcription by pol III, but allows a low level of pol II transcription that is not seen with the wild-type gene.[108,118,119] Bases flanking the TATA sequence also contribute to expression.[120] A region that closely resembles the proximal sequence elements of U snRNA genes in both its sequence and its position is centered around -60, and mutation of this motif is also severely deleterious to transcription.[108,119] An activator element containing a single SPH motif is located between -209 and -195 and stimulates expression by 10- to 20-fold in microinjected oocytes.[119,121] This element is also found in human, mouse and bovine $tRNA^{Sec}$ promoters and has been shown to bind to the transcription factor Staf.[122] In addition to its upstream control region, the $tRNA^{Sec}$ promoter contains an A-block at +8 to +20 and a B-block at +65 to +75. Whereas the B-block stimulates expression by approximately 5-fold, the A-block is inactive due to a 2 bp insertion.[108] The inactivity of the A-block in this unusual tRNA gene may explain its extreme dependence upon upstream regulatory elements.

The 3' flanking regions of tRNA genes can also influence transcription efficiency. For example, stable complex formation on a *Drosophila* $tRNA^{Arg}$ gene is enhanced by sequences that extend approximately 35 bp 3' to the termination site.[101] The termination sequence itself can influence factor binding to the yeast SUP4-o $tRNA^{Tyr}$ gene.[123] Deletion of the termination region reduces transcription of the *Xenopus* $tRNA^{Met1}$ gene.[124] Sequences downstream of the silkworm $tRNA^{AlaC}$ gene influence both the level of transcription and the ability to compete for factors, in contrast to the upstream region, which affects transcription efficiency without altering competitive strength.[18] The extent of the apparent 3' control region depends upon the concentration of $tRNA^{AlaC}$ gene assayed, extending to +44 with 6 nM template, but to +146 with 0.5 nM template, even though transcription terminates at +98.[18] Genes in which the ICR is weak, such as silkworm $tRNA^{AlaC}$, may show exaggerated dependence upon flanking sequences. For example, a 3 bp substitution at -8 to -6 does not affect the wild-type yeast $tRNA^{Leu3}$ gene, but the same substitution abolishes the residual expression of this gene when it carries a point-mutation in its B-block.[25]

The principles that have been determined concerning the promoter structures of tRNA genes are likely to also apply to the various SINE families that have evolved from tRNA genes. For example, B2 genes contain two areas of strong homology to the A- and B-blocks at comparable internal positions to those found in tRNA genes.[68] Furthermore, B2 promoter sequences cross-compete with those of tRNA or VA_I genes.[125,126] Cross-competition with tRNA genes has also been demonstrated for the *Xenopus* OAX repeat.[45] A silkworm BmX gene has been shown to have very similar promoter requirements to tRNA genes.[20]

The VA_I Gene

Deletion analyses revealed that the adenovirus VA_I gene contains an internal promoter, with outer limits at +10 and +69, that contains all of the information required for transcription.[127-129] The only sequences that are essential for VA_I transcription in vitro correspond to the A-block at +13 to +24 and the B-block at +58 to +69, although interblock and external sequences have modulatory effects.[124,129-131] (Fig. 2.4) As with tRNA genes, the B-block is the major quantitative determinant of VA_I promoter activity; mutation of the B-block severely diminishes both transcription and competitive strength, whereas mutation of the A-block reduces expression but has little effect upon competition.[129-131] Although transcription normally initiates 11-18 bp upstream from the A-block of type II promoters, new start sites may be dictated by the VA_I B-block when its interaction with the A-block is

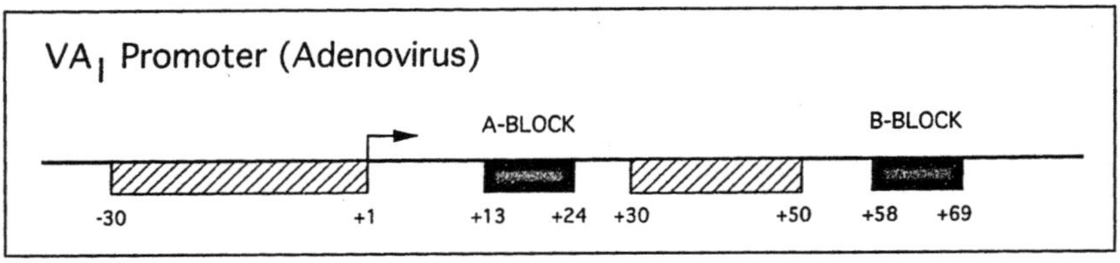

Fig. 2.4. Diagram depicting the promoter arrangement of the adenovirus VA$_I$ gene. The site of transcription initiation is indicated by +1. Essential promoter elements are shown as closed boxes and modulatory promoter elements are shown as cross-hatched boxes.

weakened by abnormal spacing; for example, the wild-type start of VA$_I$ is employed when the interblock spacing is less than 105 bp, but when it is greater, several new initiation sites closer to the B-block are used preferentially.[132]

Sequences upstream of the VA$_I$ start site also influence its level of transcription.[127,129,131,133] Linker scanning mutations identified both positive and negative putative control regions upstream of the VA$_I$ gene.[131] Substitution of the region between -36 and -25 reduced transcription in a human KB cell extract by 30%; substitutions between -25 and -17 raised transcription by up to 54%; and substitutions between -17 and +2 lowered transcription by up to 45%, as well as altering the start site.[131] None of these mutations affected the competitive strength of VA$_I$.[131] Deletion analysis suggests that sequences around the termination site can also affect transcriptional efficiency.[124,129]

The Vault RNA Gene

The vault RNA (vRNA) gene in rats has an A-block and two B-blocks, as well as upstream regulatory elements.[134] (Fig. 2.5) Both the ICR and the 5'-flanking sequences are required for efficient expression.[134] Each B-block can function, although they appear not to operate simultaneously and the upstream B-block is used preferentially.[134] Thus, mutation of the upstream B-block reduces transcription by 10-fold in vitro, mutation of the downstream B-block halves the level of expression, and inactivation of both decreases transcription by 100-fold.[134] A TATA box is centered at -25 and a putative proximal sequence element is centered at -70; mutation of either can diminish transcription significantly.[134] Deletion of the 5'-flanking sequences reduces expression by over 30-fold.[134] The proximity of the two B-blocks appears to preclude stable complex formation in the absence of upstream sequences.[134] However, mutating the downstream B-block or increasing the separation between the two can overcome the requirement for the 5' flanking region.[134] In the wild-type promoter, the upstream region appears to interact synergistically with the internal elements to facilitate the function of an unusual ICR arrangement.

The EBER2 Gene

The EBER genes of Epstein-Barr virus have type II ICRs homologous to the A- and B-blocks of tRNA genes.[135] The A- and B-blocks are both essential for EBER2 transcription, but upstream sequences also play a major role.[136] (Fig. 2.6) Whereas 5' deletion to -80 has no effect, deletion to -46 reduces EBER2 expression to 7% of the wild-type level in transfected BJAB or HeLa cells.[136] As well as a TATA box at

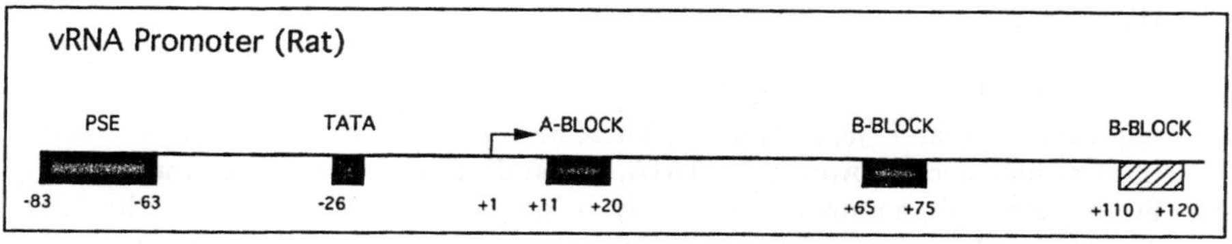

Fig. 2.5. Diagram depicting the promoter arrangement of the rat vault RNA gene. The site of transcription initiation is indicated by +1. Essential promoter elements are shown as closed boxes and modulatory promoter elements are shown as cross-hatched boxes.

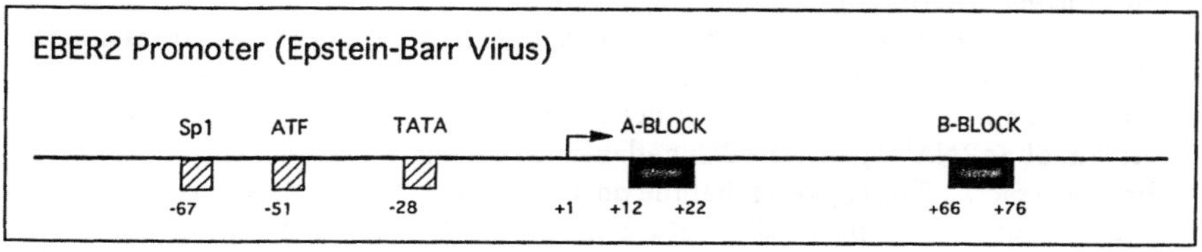

Fig. 2.6. Diagram depicting the promoter arrangement of the EBER2 gene of Epstein-Barr virus. The site of transcription initiation is indicated by +1. Essential promoter elements are shown as closed boxes and modulatory promoter elements are shown as cross-hatched boxes.

-28 to -23, the upstream region contains potential binding sites for ATF at -51 to -44, and for Sp1 at -67 to -61.[136,137] However, the 5'-flanking region (-120 to -1) alone was unable to serve as a promoter in its own right when fused directly to a CAT reporter gene and transfected into Raji cells.[137] EBER2 expression was reduced about 5-, 2-, and 5-fold by triple point mutations in the TATA, ATF, and Sp1 motifs, respectively.[136] In competition experiments using BJAB cells, cotransfection of a plasmid carrying multiple copies of either an Sp1 site or an ATF site reduced EBER2 expression by approximately 5-fold.[136]

The EBER2 TATA motif has the sequence TATAGAG, which differs at two positions from the TATAAAA sequence found in the well-characterized class II promoter of the adenovirus major late (AdML) gene. Substitution of the two G bases in the EBER2 motif in order to match the AdML sequence reduces EBER2 expression by about 10-fold following transfection into BJAB cells.[137] Conversely, introduction of a G at the fifth position of the AdML TATA box reduces pol II transcription of this gene by 5-fold.[137] These results suggest that in the context of these two promoters, the TATA sequence that is optimal for pol III transcription is different from that preferred by pol II. Although EBER2 is normally transcribed exclusively by pol III, changing the TATA box and two residues immediately 3' to it so as to match the AdML sequence produced a template that is transcribed by both pols II and III.[137] In this construct, pol II initiated transcription 3-4 bp downstream from the pol III initiation site.[137] Removal of the A- and B-blocks had no effect upon pol II transcription of this construct.[137] In the absence of any TATA box, EBER2 is transcribed exclusively by pol III.[137] Thus, the TATAGAG element of EBER2 may represent a specialized TATA box that serves to stimulate transcription by pol III without activating transcription by pol II.

BC1 Genes

The rat BC1 gene has A- and B-blocks, as well as upstream homologies to a TATA box (TTAAAT, -28), a proximal sequence element (PSE, -60) and two octamer sequences (-178 and -387).[16] (Fig. 2.7) These motifs are conserved in rat, mouse and Chinese hamster.[16] However, the TATA, PSE and octamer motifs are absent from BC1 of guinea pig, a more evolutionarily distant rodent.[16] Deletion or substitution of either the A- or the B-block abolishes expression in whole cell extracts prepared from rat brains.[16] These mutations are better tolerated in a HeLa nuclear extract, although still detrimental.[16] In brain extract, 53 bp of upstream sequence is sufficient for maximal transcription; expression is reduced by deletion to -33, declines further on deletion to -17, and is abolished on removal of all upstream sequences.[16] In contrast, deletion of all 5'-flanking sequences has little or no effect upon transcription in a HeLa extract.[16] Substitution of the TATA box decreases expression in the brain extract but not in the HeLa system.[16] The 5'-flanking region alone is unable to support transcription in either brain or HeLa extract.[16] When placed upstream of a $tRNA^{Leu}$ gene, the BC1 5'-flanking sequences reduce transcription in a brain extract.[16] This suggests that the positive effect of the upstream region is dependent upon some particular feature of the BC1 ICR. The BC1 B-block lacks the invariant A found in the tRNA consensus.[16] Substitution of a consensus B-block containing this A reduces transcription in brain extract unless the upstream region is removed.[16] The BC1 promoter therefore displays an unusual tissue-specific interaction between flanking and internal sequences.

7SL Genes

The 7SL RNA genes in *S. cerevisiae* and *S. pombe* have internal A- and B-blocks that conform closely to the type II consensus in both sequence and spacing.[138] However, the ICR sequences of human 7SL genes are fairly degenerate.[67] These genes resemble EBER2 in that both internal and external regions are required for significant expression. 7SL transcription in a HeLa nuclear extract is diminished 2-fold following 5' deletion to -37, 5- to 20-fold by deletion to -21, and 10- to 50-fold by deletion to -2.[10,139] Deletion to -37 has a much more severe effect in transfected HeLa cells, reducing expression by more than 20-fold.[10] An uncharacterized upstream element 5' to -66 also contributes to expression levels in vivo, but not in vitro.[10] A sequence TGACGTAA that matches the consensus binding site for ATF occurs at -51 to -44, the same position as the ATF motifs upstream of the EBER genes.[10,137] Point mutation of the 7SL ATF site mimicked the detrimental effect of 5' deletion to -37, both in vitro and in vivo.[10] It has been suggested that the sequence TAGTA at -28 to -24 may serve as a TATA box for the 7SL gene.[10,137] This possibility is supported by the fact that deletions extending to -21 or beyond have altered initiation sites, as well as reduced transcript levels.[139] Furthermore, insertions between this motif and the start site result in aberrant upstream initiation, as well as a ~10-fold decrease in efficiency.[139] The 5'-flanking sequences of the 7SL gene can be effectively replaced by those of the 7SK gene.[5]

Alu SINEs

Retroposed derivatives of 7SL, such as Alu and B1 genes, may lack the upstream promoter sequences that are important for expression of their progenitor. Accordingly, two 7SL pseudogenes and an Alu gene were found to be transcribed 50- to 100-fold less efficiently than a 7SL gene in a HeLa extract.[139] Despite the presence

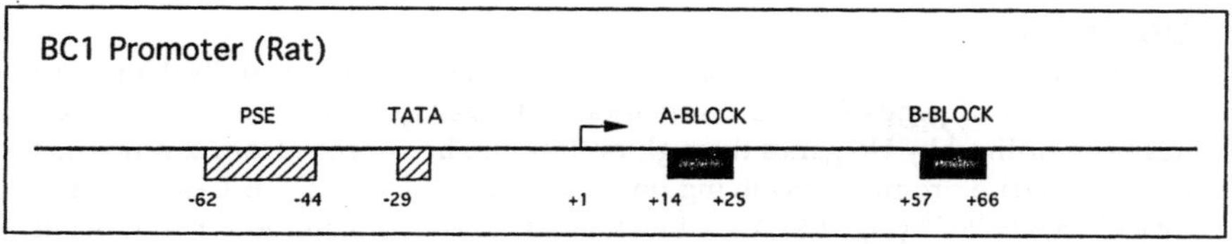

Fig. 2.7. Diagram depicting the promoter arrangement of the rat BC1 gene. The site of transcription initiation is indicated by +1. Essential promoter elements are shown as closed boxes and modulatory promoter elements are shown as cross-hatched boxes.

of nearly one million potential templates, Alu transcripts are present in only 100-1000 copies per HeLa cell, which is less than 1% of the abundance of 7SL RNA.[140,141] Domain swapping experiments demonstrated that the 5'-flanking region of the 7SL gene conferred efficient transcription upon a 7SL pseudogene or an Alu element, whereas upstream sequences of a pseudogene did not allow high level expression.[139,142] Replacement of all sequences upstream of +3 had no effect upon the transcription in vitro of an Alu gene from the α-globin complex.[143] In contrast, the A- and B-blocks are required for efficient Alu expression.[140,144,145] A downstream T-rich terminator sequence can also stimulate the level of Alu transcription.[142]

Ullu and Weiner[139] speculated that the flanking sequences of most transposed Alu genes will be incompatible with efficient transcription and that most Alu expression may result from a small subset of family members. This could help explain the considerable evidence that newly inserted Alu genes originate from a small group of "founder elements", whereas the vast majority of members are unable to undergo extensive amplification (reviewed by Deininger et al[146]). Indeed, the small subset of Alu elements that are transpositionally competent have also been shown to be preferentially transcribed in vivo.[147,148] One Alu founder element was shown to have inserted downstream of sequences that stimulate its expression by 10-fold.[149] These 5'-flanking sequences include a consensus AP1 site at -33 that doubles the level of transcription.[149] Nevertheless, sequencing of cDNAs has shown that many different Alu elements are transcribed by pol III in vivo.[140] Several studies have suggested that genomic DNA contains enough potentially active Alu genes to sustain a high level of expression and that the low transcription that occurs in vivo is due to CpG methylation and chromatin-mediated repression.[150-154] Recently transposed Alus tend to be more active templates than the ancient family members.[140,144] A representative of an older Alu subfamily was found to take longer than a younger repeat to assemble a stable transcription complex.[144] It was suggested that old Alu genes may have accumulated inactivating mutations, whereas Alus that transposed most recently are likely to have functional internal promoters; the latter may be repressed by methylation.[144] Thus, following transposition, very few Alu copies will have flanking sequences that are optimal for expression; those that do are likely to be repressed by CpG methylation and assembly into chromatin. With time, these genes will accumulate mutations that inactivate their promoters; this process will be accelerated by the rapid transition of 5 me-C to T, since Alu repeats have an unusually high CpG content.

U6 Genes

The promoters of U6 snRNA genes have evolved much more rapidly than those of most class III genes. Indeed, a remarkable diversity of promoter organization has been utilized by U6 genes through evolution. The U6 gene from *S. cerevisiae* has a tripartite promoter involving upstream, internal and downstream regulatory elements.[155,156] (Fig. 2.8) An A-block at +21 to +31 is required for significant expression both in vitro and in vivo.[156-158] A functional B-block is situated in a unique position 120 bp downstream of the coding region.[155,156] This B-block sequence is essential for transcription in vivo and in crude extracts, but is not required in a reconstituted system using purified factors.[85,155,156,158-162] Whereas intragenic promoter mutations can affect RNA stability, the external location of the U6 B-block allows functional analysis in the absence of such complications. Kaiser and Brow[162] introduced all possible single base substitutions in the U6 B-block core and analyzed the effects on expression both in yeast cells and in crude extracts. They found that the B-block sequence requirements for U6 are very similar to those of tRNA genes.[162] However, mutation of positions 6 or 7 has a more severe effect for U6 than for tRNA genes.[162] Most changes from the B-block consensus decrease U6 transcription significantly.[162] In general, the in vitro and in vivo analyses agreed well, although substitutions had a more severe effect in vitro.[162] However, mutations in positions 4 and 5 of the B-block resulted in a 15- to 50-fold reduction in the level of U6 RNA in vivo and produced a lethal phenotype.[162] These bases are conserved in all known yeast tRNA genes and have been shown to be required for efficient tRNA synthesis.[14,67]

Although for tRNA and VA genes promoter strength diminishes with interblock spacings greater than 60 bp,[79,81,132] certain deletions of extragenic U6 DNA that reduce the interblock separation from the 202 bp found in the wild-type have a negative rather than a positive effect.[156,161] However, a construct in which the B-block was placed 70 bp from the start site functioned at least 5-fold more efficiently than a comparable construct with the B-block 200 bp further downstream.[163] The B-block can also function in reverse orientation, although expression is reduced to 15% of wild-type levels.[157]

A region around -55 shows partial homology to the proximal sequence elements (PSEs) of vertebrate U6 promoters, but does not appear to contribute to expression of the yeast gene.[156,157,161] A perfect consensus TATA box (TATAAA) situated at -30 to -25 influences start site selection and stimulates transcription in vitro, although it has little effect upon expression levels in vivo.[85,156-158,161] Positioning of the start site involves an interplay between the TATA box and the A-block, although internal initiations also occur that are dictated by the unusual distance from the B-block.[155-157] The A-block is the strongest of the start site determinants in vivo, whereas the TATA box specifies a particular nucleotide within the window defined by the A-block.[161] Genetic experiments using yeast strains defective in pol II or pol III demonstrated that in the absence of the U6 downstream region, the 5'-flanking sequences direct only pol II transcription in cells.[163] Introduction of the A- and B-blocks converts the U6 gene into a pol III-specific template.[163]

In contrast to the situation in *S. cerevisiae*, the A- and B-blocks of the *S. pombe* U6 gene are both located within the transcribed region.[155] However, in this case the B-block, situated at +68 to +78, is located within a 50 bp intron.[155,164] Brow and Guthrie[155] suggested that the unprecedented B-block positions that are found in yeast U6 genes may reflect an incompatibility of B-block sequences with the function of the highly conserved U6 transcript.

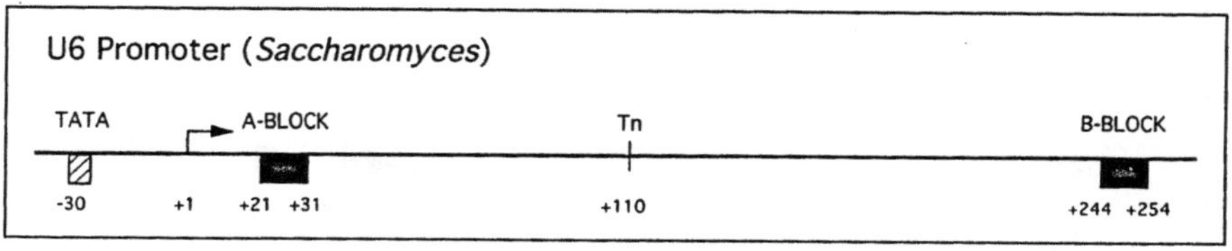

Fig. 2.8. Diagram depicting the promoter arrangement of the U6 gene of *Saccharomyces cerevisiae*. The site of transcription initiation is indicated by +1 and the site of termination is indicated by Tn. Essential promoter elements are shown as closed boxes and modulatory promoter elements are shown as cross-hatched boxes.

U6 genes in vertebrates have dispensed with B-blocks altogether. A-block homologies remain,[165,166] but these sequences may have been conserved under selection for RNA function, rather than as part of the promoter. The 5'-flanking regions of mouse and human U6 genes will continue to direct transcription both in vitro and in vivo, even after the entire coding sequence has been replaced by vector.[6-8] These upstream regions show considerable homology to the promoters of the class II snRNA genes (Fig. 2.9). The first region of homology is the proximal sequence element (PSE) located between -70 and -50.[6,7,166-168] Mutational analyses have shown this to be an important promoter element for U6 genes from human, mouse and *Xenopus*.[6,7,108,167-176] The PSEs of the human U2 and U6 promoters are identical at 13 out of 17 positions and are functionally interchangeable.[7,177]

About 150 bp upstream from the PSE, U1, U2 and U6 each contains a distal sequence element (DSE) which includes an octamer motif, and these regions have been shown to enhance U6 expression in vivo.[6,166,167,169,172,176,178] The DSEs from U2 and U6 genes are at least partially interchangeable in supporting expression.[167,169,178] In addition to octamer motifs, the human and mouse U6 DSEs contain binding sites for Staf.[122] Mutation of the Staf-binding motif causes a 50-fold decrease in the expression of human U6 following injection into *Xenopus* oocytes, but mouse U6 transcription is reduced only 2.5-fold.[122] Staf and octamer motifs can function synergistically, with a marked dependence on their relative spacing.[119,122] Instead of a Staf site, the DSE of *Xenopus* U6 contains a binding site for Sp1.[179] Optimally spaced octamer and Sp1 motifs have been shown to function cooperatively in the context of the U2 DSE.[180] Mutation of the Sp1 motif between -303 and -294 of the *Xenopus* U6 promoter reduces expression in microinjected oocytes by 4-fold, whereas mutation of the octamer site at -246 to -239 reduces expression by ~7-fold.[179] Mutation of both these motifs abolishes the activity of the DSE.[179] U6 DSEs show some positional flexibility, but do not show the extreme position- and orientation-independence generally associated with the enhancers of many class II genes.[6,167,179]

All known U6 promoters also contain a TATA motif between -30 and -25 (TTTATA in human, TTATAA in *Xenopus*), which is not present in the class II U snRNA genes.[164-166,181-183] Point mutations in these TATA sequences can severely reduce or abolish transcription by pol III.[7,8,169,170,172-174,176,184] Expression is also severely compromised by changes in the separation between the PSE and the TATA box.[173,185] Both the PSE and the TATA box are involved in start site selection, although neither element alone plays a dominant role in this regard.[172,173,185] Instead, initiation

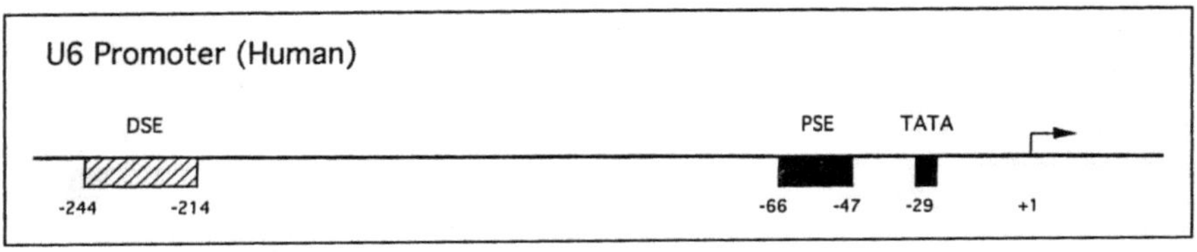

Fig. 2.9. Diagram depicting the promoter arrangement of the human U6 gene. The site of transcription initiation is indicated by +1. Essential promoter elements are shown as closed boxes and modulatory promoter elements are shown as cross-hatched boxes. Abbreviations: DSE, distal sequence element; PSE, proximal sequence element.

is positioned at a fixed distance from the compound element formed by both the PSE and the TATA box together.[185] The precise sequence around the start site can also influence the position and level of transcription.[170,185]

The conspicuous difference between U6 promoters in yeast and those in vertebrates is that the downstream B-block that is essential in the former has been replaced by an upstream PSE in the latter. TATA boxes and A-blocks are found in both situations, although the A-block is redundant in vertebrates. These observations suggest that B-blocks and PSEs may perform similar or overlapping functions. This idea is supported by the fact that the introduction of a B-block can restore basal transcription to a *Xenopus* U6 gene with a deleted PSE.[108,172] Although such a hybrid is efficient as a basal promoter, it does not respond to the presence of the DSE.[108,172] Thus, in allowing upstream activation the PSE appears to serve an additional function that cannot be performed by the B-block.[172] This ability to respond to distal regulatory elements may explain why the PSE/TATA box combination has superseded the TATA box/B-block combination in U6 genes. *Xenopus* $tRNA^{Sec}$ may represent an intermediate stage in this process of promoter evolution. In addition to a functional B-block, this gene has upstream PSE and TATA box sequences that can substitute for those of U6.[108] Synthesis of $tRNA^{Sec}$ is stimulated by sequences located between -209 and -195.[121] It is very unusual for tRNA genes to respond to elements positioned so far upstream from the start site. This ability may be conferred upon the $tRNA^{Sec}$ gene by the presence of the PSE.

In the nature of their upstream elements and their lack of internal control regions, vertebrate U6 genes resemble class II rather than classical class III genes. Nevertheless, these U6 genes seem to be transcribed by pol III, in that their transcription is inhibited by 200 μg/ml α-amanitin and by 15 μM tagetitoxin, but is not inhibited by 1 μg/ml α-amanitin.[7,165,166,169,170,173,181,186-188] However, the sensitivity of U6 transcription to α-amanitin and tagetitoxin is not precisely the same as that of tRNA synthesis, raising the possibility that different subspecies of pol III might be involved.[187,188] Low background levels of U6 transcription that appear to result from pols I and II have also been observed.[7,168,170,173] Considerable interest has focused on the question as to what features of vertebrate U6 genes allow them to be transcribed by pol III when they so closely resemble the other U snRNA genes that are all transcribed by pol II.[189] Paradoxically, it was found for both human and *Xenopus* U6 genes that the TATA region is a major determinant of polymerase specificity.[7,170,171,173] Inactivation of the U6 TATA element induces pol II transcription of human or *Xenopus* U6 genes, whereas insertion of a TATA box into the corresponding position of U2 promoters confers recognition by pol III.[7,170,173] Pol II transcription of a TATA-less U6 gene initiates at -3 and -4.[173] The choice of termination site also varies with the polymerase.[7,171] A meticulous point mutational analysis uncovered a variety of instances in which particular changes in the TATA motif have a differential effect upon pols II and III.[184] For example, mutation of the *Xenopus* U6 TATA box to TGATAA severely impairs pol II transcription but has only a moderate influence upon pol III, whereas mutation to TTGTAA has a more severe effect upon pol III than it does upon pol II.[184] TATA mutations can also have differential effects upon pol II and pol III transcription in plants.[190] These experiments show that the precise sequence requirements for a class III TATA box differ from those of a TATA box in a class II gene. However, there is considerable overlap, since TATA elements from many class II promoters can compete for pol III factors[107,191-194] and can substitute for the U6 TATA box in directing pol III transcription.[173] Therefore the context of a TATA box as well as its sequence must be important in dictating polymerase specificity for the U snRNA genes. This conclusion is supported by the fact that the U6 TATA box is compatible with pol II transcription when fused to the class II thymidine kinase promoter.[184] Furthermore, the U6 TATA box alone is insufficient to switch the polymerase specificity of the U1 gene, although it does reduce pol II transcription.[168] As well as the introduction of a TATA box, pol III transcription of U1 requires that the PSE be moved 4 bp further upstream so that its distance from the TATA box becomes the same as in the U6 promoter.[168] This result is consistent with the rigorous spacing requirement of the U6 promoter.[173,185] The PSEs of class II U snRNA promoters are typically positioned about 4 bp further downstream than are U6 PSEs.[168] This lack of positional equivalence may contribute towards the determination of polymerase specificity. Vertebrate pols II and III also have distinct preferences for sequences around the initiation site.[170,171] Therefore the final choice of polymerase is likely to reflect a complex balance between several interacting promoter elements.

A combination of two single nucleotide changes within the coding region of the wild-type human U6 gene can create an internal promoter that is nearly as active in vitro as the natural upstream promoter.[195] Indeed, a variant human U6 gene has been isolated that lacks an upstream promoter altogether.[195] This gene, named 87U6, differs from the wild-type U6 gene at ten positions within the coding region, but has no homology upstream of the start site.[195] 87U6 is transcribed as well as the wild-type gene in a HeLa extract.[195] All natural 5' or 3' flanking sequences can be replaced from 87U6 without impairing its expression.[195] Deletion and linker scanning mutagenesis mapped two internal promoter regions: one lies between +1 and +20, and the other lies between +47 and +60 and includes an A-block homology.[195] This 87U6 A-block was shown to be functional when placed in the context of a 5S gene.[195] The distinct promoter arrangement of 87U6 raises the possibility that it is regulated differently from the wild-type U6 genes.

In vitro transcription of a sea urchin U6 gene does not require any sequences downstream of the start site.[196] Maximal expression is achieved with a TATA box at -25, a PSE at -55, and an E-box (CACGTG) at -80.[196,197] The TATA box is important but not essential for expression, whereas the two upstream elements are absolutely required.[196] The U6 PSE can be replaced by PSE elements from U1 or U2 genes, even though these sequences have only 3 bp in common.[197]

In the dicotyledonous plant *Arabidopsis* the polymerase specificity of the U snRNA genes appears to be determined solely by the relative positions of the promoter elements.[198,199] A TATA box is found in all U snRNA promoters in this species, but the distance between this motif and an upstream sequence element (USE) is 32-36 bp for pol II templates and 23-26 bp for pol III templates—a difference of about one helical turn.[183,198-200] Insertion of 10 bp between these elements converts class III U snRNA promoters into class II-specific promoters, while deletion of 10 bp allows U2 transcription by pol III, provided that the distance to the start site is also shortened by 2 bp.[198,199] In monocotyledonous plants such as maize, U snRNA transcription is also dependent on TATA and USE elements.[201] However, in contrast to dicots, an additional element called MSP (for monocot-specific promoter) is also required for efficient expression in transfected maize protoplasts.[201] MSP motifs are present in one to three copies upstream of the USE in monocot snRNA gene promoters.[201] The MSPs are interchangeable between pol II and pol III templates.[201]

In several species of trypanosome, the 5' end of a functional tRNA gene is located 97 bp upstream from the U6 gene, divergently orientated.[202] Deletion of this tRNA gene or substitution of its A- or B-block abolishes expression of the linked U6 gene in vivo.[202] tRNA genes are also found 95 bp upstream of the U-snRNA B and 7SL genes, and these too are required for transcription of the linked gene.[202] Although this arrangement was found in three distantly related members of the family Trypanosomatidae, the particular tRNA type varied between species.[202] Tagetitoxin- and α-amanitin-sensitivity experiments suggest that U2 snRNA is synthesized by pol III in trypanosomes.[203] Like the other class III genes, U2 transcription requires A- and B-block elements located upstream of the initiation site.[203] In the case of U2, however, these do not appear to be part of a tRNA gene.[203] It is unclear how ICR sequences function as extragenic elements in these cases.

H1 Genes

The H1 RNA gene in *S. cerevisiae* has both internal and external promoter elements.[204] Appropriately spaced A- and B-blocks are situated within the transcribed leader that is processed out of the mature transcript.[204] Point mutation of either motif severely reduces or abolishes expression, with lethal consequences.[204] Sequences 5' to the start site stimulate in vivo expression by 10-fold.[204] A TATA box is located at -28 and PSE homologies are found upstream of this,[204] as for the yeast U6 genes. Similarly, the human gene for H1 RNA has a promoter that resembles those of vertebrate U6 genes. There is an internal A-block homology but no B-block, a TATA box at -30, a PSE at -68, and an octamer motif at -90.[205] The importance of these various elements has yet to be documented, but sequences upstream of -98 do not affect transcription in vitro.[205]

MRP Genes

The promoter arrangements of human and mouse MRP genes also closely resemble those of vertebrate U6 genes, with internal A-blocks, no B-blocks, TATA boxes at -30, PSEs at -65, and octamer motifs at -215.[12,206] Several potential Sp1 sites are situated between -220 and -400, but these are poorly conserved between species.[206] Deletion analyses showed that upstream sequences are required for transcription in HeLa cell extracts, whereas internal sequences are not, and that the region between -84 and +2 is sufficient for MRP synthesis in this system.[12] Sequences between -737 and -84 further stimulated expression following injection into frog oocytes, but had no effect in vitro.[12] The DSE of the *Xenopus* MRP promoter contains a Staf-binding site between -220 and -190; substitution of this motif reduces expression in injected oocytes to 15% of the wild-type level.[122]

Y RNA Genes

Genes encoding human and mouse Y RNAs have TATA, PSE and DSE sequences located upstream of the coding region in positions that are highly conserved with respect to other type III promoters.[207,208] Deletion of sequences upstream of -6 abolishes expression of the human Y4 gene in transient transfection assays.[207] Octamer and Staf-binding motifs are found in DSE elements of mammalian Y genes.[122,207,208] A Staf site is the only motif recognized in the DSE of the human Y4 promoter.[122] Substitution of this element decreases expression by 50-fold following injection into *Xenopus* oocytes.[122]

7SK Genes

The promoter structure of the human 7SK gene is similar to that of vertebrate U6 genes (Fig. 2.10). Sequences upstream of the 7SK start site are necessary and sufficient to direct transcription, and there is no internal promoter.[3-5,11,209,210] A TATA box between -30 and -25 is required for accurate and efficient transcription.[3,15] A PSE located between -65 and -48 is necessary for expression in transfected HeLa cells.[15] However, it is less important in vitro, since 5' deletion to -37 only reduces expression to 21% of the wild-type level.[3] Sequences between -243 and -59 enhance transcription by about 4-fold in vitro.[3,210] This effect involves a series of highly degenerate octamer elements located between -200 and -70 and can be mimicked by the introduction of a single consensus octamer motif at -90.[211,212] In contrast, expression in vivo is stimulated 10-fold in a position- and orientation-dependent manner by a region situated between -243 and -210 that is analogous to the DSEs of U snRNA genes.[11,15,212,213] Two degenerate octamer motifs within this region contribute to its activation function.[11,15] Alteration of the sequence to match the octamer consensus enhances transcription.[15,213] Another significant element in the 7SK DSE is a CACCC-box located between -223 and -219, and mutation of this site reduces expression by up to 5-fold.[11,15] Staf can bind between -219 and -193, and substitution of bases -200 to -203 reduces 7SK expression by 3-fold.[122] The relative importance of the CACCC and octamer motifs is somewhat controversial.[11,15] However, it is clear that full upstream activation involves a combination of these elements and is mediated via a cooperative interaction with the PSE.[11,15,211,212]

As with U6 genes, the TATA box of 7SK is a major determinant of polymerase specificity.[15] Following substitution of the TATA sequence, the mutant 7SK gene generates three transcripts when transfected into HeLa cells; pol III continues to initiate at +1, but at ~10% of wild-type levels, whereas pol II initiates with comparable efficiency at -3 and -4.[15] Both polymerases are stimulated by the presence of the DSE.[15]

Class II Genes

At high template concentrations the human c-myc gene can be transcribed by both pols II and III, either in vitro or in *Xenopus* oocytes.[214,215] Pol III transcription of c-myc does not require sequences upstream of -35 in the P2 promoter, but is

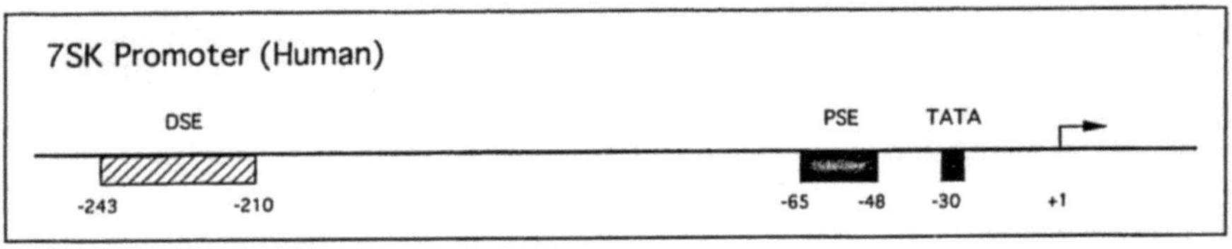

Fig. 2.10. Diagram depicting the promoter arrangement of the human 7SK gene. The site of transcription initiation is indicated by +1. Essential promoter elements are shown as closed boxes and modulatory promoter elements are shown as cross-hatched boxes. Abbreviations: DSE, distal sequence element; PSE, proximal sequence element.

absolutely dependent upon a TATA box.[215] Indeed, the TATA-dependence of pol III transcription from P2 is more stringent than that of pol II transcription from the same promoter. Transcription of c-myc by pol III is interesting from a mechanistic perspective, but is unlikely to be of physiological significance, since it is inefficient in vitro and is undetectable in isolated nuclei.[215] However, pol III transcription from the adenovirus E2E promoter has been observed in adenovirus-infected cells.[216] E2E transcription by pol III can occur in vitro with only a TATA box upstream of the start site, but binding sites for ATF and E2F further stimulate transcription by both pol II and pol III.[217]

TATA-dependent pol III transcription of the brain creatine kinase gene has also been reported.[218,219] As was the case for c-myc,[215] certain mutations in the creatine kinase TATA sequence were found to have differential effects upon transcription by pols II and III.[218,219] For example, inversion of the TATA box, so that TATAAATA became TATTTATA, reduced pol II transcription to <20% but stimulated pol III transcription to 130-150% of the wild-type level.[218] A study carried out in *Drosophila* nuclear extracts using artificial constructs containing a canonical TATA box as the sole promoter element found that pol II and pol III transcribed exclusively from opposite orientations of a common TATA box.[220] In this case, TATAAAAA directed only pol II initiation, whereas the reverse orientation, TTTTTATA, directed only pol III transcription.[220] However, this result appears to be exceptional.[221,222] Reversing the orientation of the U6 TATA box has little effect in either yeast[223] or humans.[173]

Pol III transcription of class II genes is favored by particular types of extract or the use of high template concentrations.[214,215,217] Preferential transcription by pol III can also be achieved if the template is preincubated with class III factors before addition of class II factors.[218,219] These observations suggest a competitive mechanism in which inappropriate pol III transcription of class II genes becomes possible when the availability or access of class II factors is restricted. On the other hand, pol III may actively suppress transcription by pol II. For example, pol II transcribes the *Xenopus* TFIIIA gene from +1 and -284, whereas pol III initiates at +1 and -70.[224] Pol II transcription increases if tagetitoxin is used to specifically inhibit pol III.[224] This implies that pols II and III compete at this promoter.

The cis-acting sequences that control the TFIIIA gene extend as far upstream as -1800,[225-228] and certain key elements are positioned within the 5'-untranslated region of the -284 transcript. Both the +1 and -284 sites contain consensus pol II initiator elements and a TATA box occurs at -32.[224] A sequence element at -269 to -264 stimulates transcription in oocytes but not in somatic cells[225] and may there-

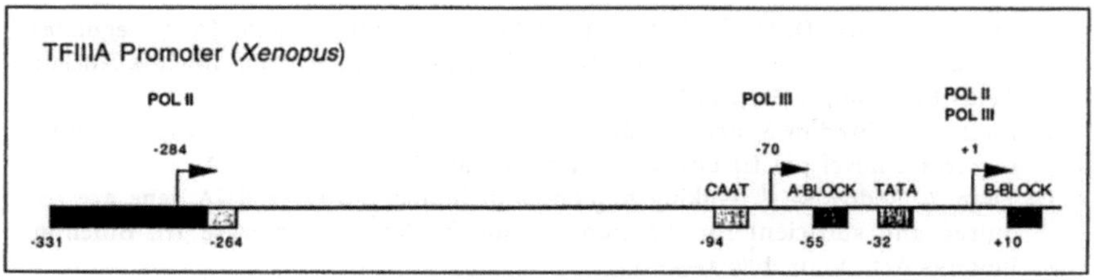

Fig. 2.11. Diagram depicting the transcription start sites and putative promoter arrangement of the *Xenopus* TFIIIA gene. Pol II can initiate transcription at -284 and +1. Pol III can initiate transcription at -70 and +1. A TATA box and CAAT box are located upstream of the +1 site. Potential A- and B-blocks are located downstream of the -70 start site. A binding site for USF is found between -269 and -264. A binding site for TFIIIA is located between -264 and -331.

fore contribute to the higher expression that is found in the former.[229] This element binds a *Xenopus* homologue of the major late transcription factor (MLTF or USF) that may be differentially modified between oocytes and somatic cells.[225,226,230] A binding site for TFIIIA itself lies between -264 and -331, raising the possibility of an autoregulatory circuit.[231] Sequences that resemble A- and B-blocks are positioned downstream of -70 in such a way that they could direct pol III initiation at this site.[224] Competition in this region between pol II and pol III factors may contribute to the reduced synthesis of TFIIIA in somatic cells (Fig. 2.11).

References

1. Ciliberto G, Raugei S, Constanzo F et al. Common and interchangeable elements in the promoters of genes transcribed by RNA polymerase III. Cell 1983; 32:725-733.
2. Shi YG, Tyler BM. All internal promoter elements of *Neurospora crassa* 5S rRNA and tRNA genes, including the A boxes, are functionally gene-specific. J Biol Chem 1991; 266:8015-8019.
3. Murphy S, Di Liegro C, Melli M. The in vitro transcription of the 7SK RNA gene by RNA polymerase III is dependent only on the presence of an upstream promoter. Cell 1987; 51:81-87.
4. Kleinert H, Benecke B-J. Transcription of human 7S K DNA in vitro and in vivo is exclusively controlled by an upstream promoter. Nucleic Acids Res 1988; 16:1319-1331.
5. Kleinert H, Gladen A, Geisler M et al. Differential regulation of transcription of human 7 S K and 7 S L RNA genes. J Biol Chem 1988; 263:11511-11515.
6. Das G, Henning D, Wright D et al. Upstream regulatory elements are necessary and sufficient for transcription of a U6 RNA gene by RNA polymerase III. EMBO J 1988; 7:503-512.
7. Lobo SM, Hernandez N. A 7 bp mutation converts a human RNA polymerase II snRNA promoter into an RNA polymerase III promoter. Genes Dev 1989; 58:55-67.
8. Kunkel GR, Pederson T. Transcription of a human U6 small nuclear RNA gene in vivo withstands deletion of intragenic sequences but not of an upstream TATATA box. Nucleic Acids Res 1989; 17:7371-7379.
9. Shaw KJ, Olsen MV. Effects of altered 5'-flanking sequences on the in vivo expression of a *Saccharomyces cerevisiae* $tRNA^{Tyr}$ gene. Mol Cell Biol 1984; 4:657-665.

10. Bredow S, Surig D, Muller J et al. Activating-transcription-factor (ATF) regulates human 7S L RNA transcription by RNA polymerase III in vivo and in vitro. Nucleic Acids Res 1990; 18:6779-6784.
11. Kleinert H, Bredow S, Benecke BJ. Expression of a human 7S K RNA gene in vivo requires a novel pol III upstream element. EMBO J 1990; 9:711-718.
12. Yuan Y, Reddy R. 5' flanking sequences of human MRP/7-2 RNA gene are required and sufficient for the transcription by RNA polymerase III. Biochim Biophys Acta 1991; 1089:33-39.
13. Sprague KU, Larson D, Morton D. 5' flanking sequence signals are required for activity of silkworm alanine tRNA genes in homologous in vitro transcription systems. Cell 1980; 22:171-178.
14. Gaeta BA, Sharp SJ, Stewart TS. Saturation mutagenesis of the *Drosophila* $tRNA^{Arg}$ gene B-box intragenic promoter element: requirements for transcription activation and stable complex formation. Nucleic Acids Res 1990; 18:1541-1548.
15. Boyd DC, Turner PC, Watkins NJ et al. Functional redundancy of promoter elements ensures efficient transcription of the human 7SK gene in vivo. J Mol Biol 1995; 253:677-690.
16. Martignetti JA, Brosius J. BC1 RNA: transcriptional analysis of a neural cell-specific RNA polymerase III transcript. Mol Cell Biol 1995; 15:1642-1650.
17. Raymond GJ, Johnson JD. The role of non-coding DNA sequences in transcription and processing of yeast tRNA. Nucleic Acids Res 1983; 11:5969-5988.
18. Wilson ET, Larson D, Young LS et al. A large region controls tRNA gene transcription. J Mol Biol 1985; 183:153-163.
19. Gabrielsen OS, Oyen TB. The requirement for the A block promoter element in tRNA gene transcription in vitro depends on the ionic environment. Nucleic Acids Res 1987; 15:5699-5713.
20. Wilson ET, Condliffe DP, Sprague KU. Transcriptional properties of BmX, a moderately repetitive silkworm gene that is an RNA polymerase III template. Mol Cell Biol 1988; 8:624-631.
21. Wolffe AP, Morse RH. The transcription complex of the *Xenopus* somatic 5 S RNA gene. A functional analysis of protein-DNA interactions outside of the internal control region. J Biol Chem 1990; 265:4592-4599.
22. Hall BD, Clarkson SG, Tocchini-Valentini G. Transcription initiation of eukaryotic transfer RNA genes. Cell 1982; 29:3-5.
23. Ciliberto G, Castagnoli L, Cortese R. Transcription by RNA polymerase III. Curr Topics Dev Biol 1983; 18:59-88.
24. Sharp SJ, Schaack J, Cooley L et al. Structure and transcription of eukaryotic tRNA genes. CRC Crit Rev Biochem 1984; 19:107-144.
25. Geiduschek EP, Tocchini-Valentini GP. Transcription by RNA polymerase III. Annu Rev Biochem 1988; 57:873-914.
26. Sollner-Webb B. Surprises in polymerase III transcription. Cell 1988; 52:153-154.
27. Murphy S, Moorefield B, Pieler T. Common mechanisms of promoter recognition by RNA polymerases II and III. Trends Genet 1989; 5:122-126.
28. Kunkel GR. RNA polymerase III transcription of genes that lack internal control regions. Biochim Biophys Acta 1991; 1088:1-9.
29. Sakonju S, Bogenhagen DF, Brown DD. A control region in the center of the 5S RNA gene directs specific initiation of transcription: I. The 5' border of the region. Cell 1980; 19:13-25.
30. Bogenhagen DF, Sakonju S, Brown DD. A control region in the center of the 5S RNA gene directs specific initiation of transcription: II. The 3' border of the region. Cell 1980; 19:27-35.

31. Brown DD, Schlissel MS. A positive transcription factor controls the differential expression of two 5S RNA genes. Cell 1985; 42:759-767.
32. Bogenhagen DF. The intragenic control region of the *Xenopus* 5S gene contains two factor A binding domains that must be aligned properly for efficient transcription initiation. J Biol Chem 1985; 260:6466-6471.
33. Pieler T, Appel B, Oei S-L et al. Point mutational analysis of the *Xenopus laevis* 5S gene promoter. EMBO J 1985; 4:1847-1853.
34. Pieler T, Oei S-L, Hamm J et al. Functional domains of the *Xenopus laevis* 5S gene promoter. EMBO J 1985; 4:3751-3756.
35. Pieler T, Hamm J, Roeder RG. The 5S gene internal control region is composed of three distinct sequence elements, organized as two functional domains with variable spacing. Cell 1987; 48:91-100.
36. Keller HJ, You QM, Romaniuk PJ et al. Additional intragenic promoter elements of the *Xenopus* 5S RNA genes upstream from the TFIIIA-binding site. Mol Cell Biol 1990; 10:5166-5176.
37. Wormington WM, Bogenhagen DF, Jordan E et al. A quantitative assay for *Xenopus* 5S RNA gene transcription in vitro. Cell 1981; 24:809-817.
38. Sands MS, Bogenhagen DF. TFIIIA binds to different domains of 5S RNA and the *Xenopus borealis* 5S RNA gene. Mol Cell Biol 1987; 7:3985-3993.
39. Fradkin LG, Yoshinaga SK, Berk AJ et al. Human transcription factor TFIIIC2 specifically interacts with a unique sequence in the *Xenopus laevis* 5S rRNA gene. Mol Cell Biol 1989; 9:4941-4950.
40. Oei SL, Pieler T. A transcription stimulatory factor binds to the upstream region of *Xenopus* 5 S RNA and tRNA genes. J Biol Chem 1990; 265:7485-7491.
41. McConkey GA, Bogenhagen DF. Transition mutations within the *Xenopus borealis* somatic 5S RNA gene can have independent effects on transcription and TFIIIA binding. Mol Cell Biol 1987; 7:486-494.
42. Xing YY, Worcel A. The C-terminal domain of transcription factor IIIA interacts differently with different 5S RNA genes. Mol Cell Biol 1989; 9:499-514.
43. Fedoroff NV, Brown DD. The nucleotide sequence of oocyte 5S DNA in *Xenopus laevis*. I. The AT-rich spacer. Cell 1978; 13:701-716.
44. Peterson RC, Doering JL, Brown DD. Characterization of two *Xenopus* somatic 5S DNA and one minor oocyte-specific DNA. Cell 1980; 20:131-141.
45. Andrews DL, Millstein L, Hamkalo BA et al. Competition between *Xenopus* satellite I sequences and pol III genes for stable transcription complex formation. Nucleic Acids Res 1984; 12:7753-7769.
46. Millstein L, Eversole-Cire P, Blanco J et al. Differential transcription of *Xenopus* oocyte and somatic-type 5S genes in a *Xenopus* oocyte extract. J Biol Chem 1987; 262:1-11.
47. Peck LJ, Millstein L, Eversole-Cire P et al. Transcriptionally inactive oocyte-type 5S RNA genes of *Xenopus laevis* are complexed with TFIIIA in vitro. Mol Cell Biol 1987; 7:3503-3510.
48. Reynolds W. Effect of sequence differences between somatic and oocyte 5S RNA genes on transcriptional efficiency in an oocyte S150 extract. Mol Cell Biol 1988; 8:5056-5058.
49. Reynolds WF, Azer K. Sequence differences upstream of the promoters are involved in the differential expression of the *Xenopus* somatic and oocyte 5S RNA genes. Nucleic Acids Res 1988; 16:3391-3403.
50. Reynolds WF. Sequences preceding the minimal promoter of the *Xenopus* somatic 5S RNA gene increase binding efficiency for transcription factors. Nucleic Acids Res 1989; 17:9381-9394.

51. Wingender E, Frank R, Blocker H et al. Complete synthesis and transcription in vitro of a gene coding for human ribosomal 5S RNA. Gene 1988; 64:77-85.
52. Sorensen PD, Frederiksen S. Characterization of human 5S rRNA genes. Nucleic Acids Res 1991; 19:4147-4151.
53. Sharp SJ, Garcia AD. Transcription of the *Drosophila melanogaster* 5S RNA gene requires an upstream promoter and four intragenic sequence elements. Mol Cell Biol 1988; 8:1266-1274.
54. Tyler BM. Transcription of *Neurospora crassa* 5S rRNA genes requires a TATA box and three internal elements. J Mol Biol 1987; 196:801-811.
55. Morton DG, Sprague KU. In vitro transcription of a silkworm 5S RNA gene requires an upstream signal. Proc Natl Acad Sci USA 1984; 81:5519-5522.
56. Selker EU, Morzycka-Wroblewska E, Stevens JN et al. An upstream signal is required for in vitro transcription of *Neurospora* 5S RNA genes. Mol Gen Genet 1986; 205:189-192.
57. Garcia AD, O' Connell AM, Sharp SJ. Formation of an active transcription complex in the *Drosophila melanogaster* 5S RNA gene is dependent on an upstream region. Mol Cell Biol 1987; 7:2046-2051.
58. Smith TPL, Young LS, Bender LB et al. Silkworm TFIIIA requires additional class III factors for committment to transcription complex assembly on a 5S RNA gene. Nucleic Acids Res 1995; 23:1244-1251.
59. Challice JM, Segall J. Transcription of the 5S rRNA gene of *Saccharomyces cerevisiae* requires a promoter element at +1 and a 14 base-pair internal control region. J. Biol Chem 1989; 264:20060-20067.
60. Taylor MJ, Segall J. Characterization of factors and DNA sequences required for accurate transcription of the *Saccharomyces cerevisiae* 5S RNA gene. J Biol Chem 1985; 260:4531-4540.
61. Lee Y, Erkine AM, Van Ryk DI et al. In vivo analyses of the internal control region in the 5S rRNA gene from *Saccharomyces cerevisiae*. Nucleic Acids Res 1995; 23:634-640.
62. Murphy MH, Baralle FE. Construction and functional analysis of a series of synthetic RNA polymerase III promoters. J Biol Chem 1984; 259:10208-10211.
63. Galli G, Hofstetter H, Birnstiel ML. Two conserved sequence blocks within eukaryotic tRNA genes are major promoter elements. Nature 1981; 294:626-631.
64. Hofstetter H, Kressmann A, Birnstiel ML. A split promoter for a eukaryotic tRNA gene. Cell 1981; 24:573-585.
65. Sharp S, DeFranco D, Dingermann T et al. Internal control regions for transcription of eukaryotic tRNA genes. Proc Natl Acad Sci USA 1981; 78:6657-6661.
66. Ciliberto G, Castagnoli L, Melton DA et al. Promoter of a eukaryotic $tRNA^{Pro}$ gene is composed of three noncontiguous regions. Proc Natl Acad Sci USA 1982; 79:1195-1199.
67. Allison DS, Goh SH, Hall BD. The promoter sequence of a yeast $tRNA^{Tyr}$ gene. Cell 1983; 34:655-664.
68. Krayev AS, Markusheva TV, Kramerov DA et al. Ubiquitous transposon-like repeats B1 and B2 of the mouse genome: B2 sequencing. Nucleic Acids Res 1982; 10:7461-7475.
69. Folk WR, Hofsteter H, Birnstiel ML. Some bacterial tRNA genes are transcribed by eukaryotic RNA polymerase III. Nucleic Acids Res 1982; 10:7153-7162.
70. Gruissem W, Prescott D, Greenberg BM et al. Transcription of *E. coli* and Euglena chloroplast tRNA gene clusters and processing of polycistronic transcripts in a HeLa cell-free system. Cell 1982; 30:81-92.
71. Newman AJ, Ogden RC, Abelson J. tRNA gene transcription in yeast: effects of specified base substitutions in the intragenic promoter. Cell 1983; 35:117-125.

72. Traboni C, Ciliberto G, Cortese R. Mutations in box B of the promoter of a eukaryotic $tRNA^{Pro}$ gene affect rate of transcription, processing, and stability of the transcripts. Cell 1984; 36:179-187.
73. Reyes VM, Newman A, Abelson J. Mutational analysis of the coordinate expression of the yeast $tRNA^{Arg}$-$tRNA^{Asp}$ gene tandem. Mol Cell Biol 1986; 6:2436-2442.
74. Nichols M, Bell J, Klekamp MS et al. Multiple mutations of the first gene of a dimeric tRNA gene abolish in vitro tRNA gene transcription. J Biol Chem 1989; 264:17084-17090.
75. Di Segni G, McConaughy BL, Shapiro RA et al. *TAP1*, a yeast gene that activates the expression of a tRNA gene with a defective internal promoter. Mol Cell Biol 1993; 13:3424-3433.
76. Folk WR, Hofstetter H. A detailed mutational analysis of the eukaryotic $tRNA^{Met}$ gene promoter. Cell 1983; 33:585-593.
77. Ciampi MS, Melton DA, Cortese R. Site-directed mutagenesis of a tRNA gene: base alterations in the coding region affect transcription. Proc Natl Acad Sci USA 1982; 79:1388-1392.
78. Willis I, Hottinger H, Pearson D et al. Mutations affecting excision of the intron from a eukaryotic dimeric tRNA precursor. EMBO J 1984; 3:1573-1580.
79. Dingermann T, Sharp S, Schaack J et al. Stable transcription complex formation of eukaryotic tRNA genes is dependent on a limited separation of the two intragenic control regions. J Biol Chem 1983; 258:10395-10402.
80. Baker RE, Camier S, Sentenac A et al. Gene size differentially affects the binding of yeast transcription factor τ to two intragenic regions. Proc Natl Acad Sci USA 1987; 84:8768-8772.
81. Fabrizio P, Coppo A, Fruscoloni P et al. Comparative mutational analysis of wildtype and stretched $tRNA^{Leu3}$ gene promoters. Proc Natl Acad Sci USA 1987; 84:8763-8767.
82. Zasloff M, Santos T, Romeo P et al. Transcription and precursor processing of normal and mutant human $tRNA_i^{Met}$ genes in a homologous cell-free system. J Biol Chem 1982; 257:7857-7863.
83. Hipskind RA, Clarkson SG. 5'-flanking sequences that inhibit in vitro transcription of a *Xenopus laevis* tRNA gene. Cell 1983; 34:881-890.
84. Joazeiro CAP, Kassavetis GA, Geiduschek EP. Alternative outcomes in assembly of promoter complexes: the roles of TBP and a flexible linker in placing TFIIIB on tRNA genes. Genes Dev 1996; 10:725-739.
85. Chalker DL, Sandmeyer SB. Sites of RNA polymerase III transcription initiation and *Ty3* integration at the U6 gene are positioned by the TATA box. Proc Natl Acad Sci USA 1993; 90:4927-4931.
86. Fruscoloni P, Zamboni M, Panetta G et al. Mutational analysis of the transcription start site of the yeast $tRNA^{Leu3}$ gene. Nucleic Acids Res 1995; 23:2914-2918.
87. Johnson JD, Raymond GJ. Three regions of a yeast $tRNA_3^{Leu}$ gene promote RNA polymerase III transcription. J Biol Chem 1984; 259:5990-5994.
88. Raymond KC, Raymond GJ, Johnson JD. In vivo modulation of yeast tRNA gene expression by 5'-flanking sequences. EMBO J 1985; 4:2649-2656.
89. Zecherle GN, Whelen S, Hall BD. Purines are required at the 5' ends of newly initiated RNAs for optimal RNA polymerase III gene expression. Mol Cell Biol 1996; 16:5801-5810.
90. Koski RA, Allison DS, Worthington M et al. An in vitro RNA polymerase III system from *S. cerevisiae*: effects of deletions and point mutations upon SUP4 gene transcription. Nucleic Acids Res 1982; 10:8127-8143.
91. Kassavetis GA, Blanco JA, Johnson TE et al. Formation of open and elongating transcription complexes by RNA polymerase III. J Mol Biol 1992; 226:47-58.

92. Gouilloud E, Clarkson SG. A dispersed tyrosine tRNA gene from *Xenopus laevis* with high transcriptional activity in vitro. J Biol Chem 1986; 261:486-494.
93. Stutz F, Gouilloud E, Clarkson SG. Oocyte and somatic tyrosine tRNA genes in *Xenopus laevis*. Genes Dev 1989; 3:1190-1198.
94. Larson D, Bradford-Wilcox J, Young LS et al. A short 5' flanking region containing conserved sequences is required for silkworm alanine tRNA gene activity. Proc Natl Acad Sci USA 1983; 80:3416-3420.
95. Palida F, Hale C, Sprague KU. Transcription of a silkworm $tRNA^{AlaC}$ gene is directed by two AT-rich upstream sequence elements. Nucleic Acids Res 1993; 21:5875-5881.
96. Young LS, Takahashi N, Sprague KU. Upstream sequences confer distinctive transcriptional properties on genes encoding silkgland-specific $tRNA^{Ala}$. Proc Natl Acad Sci USA 1986; 83:374-378.
97. Sharma S, Gopinathan KP. Transcriptional silencing of a $tRNA^{Gly1}$ copy from within a multigene family is modulated by distal *cis* elements. Proc Natl Acad Sci USA 1996; 271:28146-28153.
98. Marschalek R, Dingermann T. Identification of a protein factor binding to the 5'-flanking region of a tRNA gene and being involved in modulation of tRNA gene transcription in vivo in *Saccharomyces cerevisiae*. Nucleic Acids Res 1988; 16:6737-6752.
99. DeFranco D, Sharp S, Soll D. Identification of regulatory sequences contained in the 5'-flanking region of *Drosophila* lysine $tRNA_2$ genes. J Biol Chem 1981; 256:12424-12429.
100. Dingermann T, Burke DJ, Sharp S et al. The 5' flanking sequences of Drosophila $tRNA^{Arg}$ genes control their in vitro transcription in a *Drosophila* cell extract. J Biol Chem 1982; 257:14738-14744.
101. Schaack J, Sharp S, Dingermann T et al. The extent of a eukaryotic tRNA gene: 5'- and 3'-flanking sequence dependence for transcription and stable complex formation. J Biol Chem 1984; 259:1461-1467.
102. Stewart TS, Soll D, Sharp S. Point mutations in the 5' ICR and anticodon region of a *Drosophila* $tRNA^{Arg}$ gene decrease in vitro transcription. Nucleic Acids Res 1985; 13:435-447.
103. StLouis D, Spiegelman GB. Steady-state kinetic analysis of transcription of cloned $tRNA^{Ser}$ genes from *Drosophila melanogaster*. Eur J Biochem 1985; 148:305-313.
104. Lofquist A, Sharp S. The 5'-flanking sequences of *Drosophila melanogaster* $tRNA^{Asn5}$ genes differentially arrest RNA polymerase III. J Biol Chem 1986; 261:14600-14606.
105. Sajjadi F, Spiegelman GB. The modulatory element TNNCT affects transcription of a *Drosophila* $tRNA^{Val4}$ gene without affecting transcription complex stability. Nucleic Acids Res 1989; 17:755-766.
106. Taneja R, Gopalkrishnan R, Gopinathan KP. Regulation of glycine tRNA gene expression in the posterior silk glands of the silkworm *Bombyx mori*. Proc Natl Acad Sci USA 1992; 89:1070-1074.
107. Sharma S, Gopinathan KP. Role of TATATAA element in the regulation of $tRNA^{Gly1}$ gene expression in *Bombyx mori* is position dependent. J Mol Biol 1996; 262:396-406.
108. Carbon P, Krol A. Transcription of the *Xenopus laevis* selenocysteine $tRNA^{(Ser)Sec}$ gene: a system that combines an internal B box and upstream elements also found in U6 snRNA genes. EMBO J 1991; 10:599-606.
109. McBryant SJ, Kassavetis GA, Gottesfeld JM. Repression of vertebrate RNA polymerase III transcription by DNA binding proteins located upstream from the transcription start site. J Mol Biol 1995; 250:315-326.

110. Arnold GJ, Schmutzler C, Thomann U et al. The human $tRNA^{Val}$ gene family: organization, nucleotide sequences and homologous transcription of three single-copy genes. Gene 1986; 44:287-297.
111. Morry MJ, Harding JD. Modulation of transcriptional activity and stable complex formation by 5'-flanking regions of mouse $tRNA^{His}$ genes. Mol Cell Biol 1986; 6:105-115.
112. Arnold GJ, Gross HJ. Unrelated leader sequences can efficiently promote human tRNA gene transcription. Gene 1987; 51:237-246.
113. Arnold GJ, Schmutzler C, Gross HJ. Functional dissection of 5' and 3' extragenic control regions of human $tRNA^{Val}$ genes reveals two different regulatory effects. DNA 1988; 7:87-97.
114. Capone JP. Modulation of the phenotypic expression of a human serine tRNA gene by 5'-flanking sequences. DNA 1988; 7:459-468.
115. Gonos ES, Goddard JP. The role of the 5'-flanking sequence of a human $tRNA^{Glu}$ gene in modulation of its transcriptional activity in vitro. Biochem J 1990; 272:797-803.
116. Huibregtse JM, Engelke DR. Genomic footprinting of a yeast tRNA gene reveals stable complexes over the 5'-flanking region. Mol Cell Biol 1989; 9:3244-3252.
117. Rooney RJ, Harding JD. Transcriptional activity and factor binding are stimulated by separate and distinct sequences in the 5' flanking region of a mouse $tRNA^{Asp}$ gene. Nucleic Acids Res 1988; 16:2509-2521.
118. Lee BJ, Kang SG, Hatfield D. Transcription of *Xenopus* selenocysteine $tRNA^{Ser}$ (Formerly designated opal suppressor phosphoserine tRNA) gene is directed by multiple 5'-extragenic regulatory elements. J Biol Chem 1989; 264:9696-9702.
119. Myslinski E, Schuster C, Krol A et al. Promoter strength and structure dictate module composition in RNA polymerase III transcriptional activator elements. J Mol Biol 1993; 234:311-318.
120. Myslinski E, Schuster C, Huet J et al. Point mutations 5' to the tRNA selenocysteine TATA box alter RNA polymerase III transcription by affecting the binding of TBP. Nucleic Acids Res 1993; 21:5852-5858.
121. Myslinski E, Krol A, Carbon P. Optimal $tRNA^{(Ser)Sec}$ gene activity requires an upstream SPH motif. Nucleic Acids Res 1992; 20:203-209.
122. Schaub M, Myslinski E, Schuster C et al. Staf, a promiscuous activator for enhanced transcription by RNA polymerases II and III. EMBO J 1997; 16:173-181.
123. Allison DS, Hall BD. Effects of alterations in the 3' flanking sequence on in vivo and in vitro expression of the yeast SUP4-o $tRNA^{Tyr}$ gene. EMBO J 1985; 4:2657-2664.
124. Wang Z, Roeder RG. TFIIIC1 acts through a downstream region to stabilize TFIIIC2 binding to RNA polymerase III promoters. Mol Cell Biol 1996; 16:6841-6850.
125. Gutierrez-Hartmann A, Lieberburg I, Gardner D et al. Transcription of two classes of rat growth hormone gene-associated repetitive DNA: differences in activity and effects of tandem repeat structure. Nucleic Acids Res 1984; 12:7153-7173.
126. White RJ, Stott D, Rigby PWJ. Regulation of RNA polymerase III transcription in response to F9 embryonal carcinoma stem cell differentiation. Cell 1989; 59:1081-1092.
127. Fowlkes DM, Shenk T. Transcriptional control regions of the adenovirus VA1 RNA gene. Cell 1980; 22:405-413.
128. Guilfoyle R, Weinmann R. Control region for adenovirus VARNA transcription. Proc Natl Acad Sci USA 1981; 78:3378-3382.
129. Wu G-J, Railey JF, Cannon RE. Defining the functional domains in the control region of the adenovirus type 2 specific VARNA1 gene. J Mol Biol 1987; 194:423-442.

130. Rohan RM, Ketner G. A comprehensive collection of point mutations in the internal promoter of the adenoviral VA_I gene. J Biol Chem 1987; 262:8500-8507.
131. Railey JF, Wu G-J. Organization of multiple regulatory elements in the control region of the adenovirus type 2-specific VARNA1 gene: fine mapping with linker-scanning mutants. Mol Cell Biol 1988; 8:1147-1159.
132. Cannon RE, Wu G-J, Railey JF. Functions of and interactions between the A and B blocks in adenovirus type 2-specific *VARNA1* gene. Proc Natl Acad Sci USA 1986; 83:1285-1289.
133. Thimmappaya B, Jones N, Shenk T. A mutation which alters initiation of transcription by RNA polymerase III on the Ad5 chromosome. Cell 1979; 18:947-954.
134. Vilalta A, Kickhoefer VA, Rome LH et al. The rat vault RNA gene contains a unique RNA polymerase III promoter composed of both external and internal elements that function synergistically. J Biol Chem 1994; 269:29752-29759.
135. Rosa MD, Gottlieb E, Lerner M et al. Striking similarities are exhibited by two small Epstein-Barr virus-encoded ribonucleic acids and the adenovirus-associated ribonucleic acids VAI and VAII. Mol Cell Biol 1981; 1:785-796.
136. Howe JG, Shu M-D. Epstein-Barr virus small RNA (EBER) genes: unique transcription units that combine RNA polymerase II and III promoter elements. Cell 1989; 57:825-834.
137. Howe JG, Shu M-D. Upstream basal promoter element important for exclusive RNA polymerase III transcription of the EBER 2 gene. Mol Cell Biol 1993; 13:2655-2665.
138. Willis IM. RNA polymerase III. Genes, factors and transcriptional specificity. Eur J Biochem 1993; 212:1-11.
139. Ullu E, Weiner AM. Upstream sequences modulate the internal promoter of the human 7SL RNA gene. Nature 1985; 318:371-374.
140. Liu W-M, Maraia RJ, Rubin CM et al. Alu transcripts: cytoplasmic localization and regulation by DNA methylation. Nucleic Acids Res 1994; 22:1087-1095.
141. Liu W-M, Chu W-M, Choudary PV et al. Cell stress and translational inhibitors transiently increase the abundance of mammalian SINE transcripts. Nucleic Acids Res 1995; 23:1758-1765.
142. Chu WM, Liu WM, Schmid CW. RNA polymerase III promoter and terminator elements affect *Alu* RNA expression. Nucleic Acids Res 1995; 23:1750-1757.
143. Perez-Stable C, Ayres TM, Shen C-KJ. Distinctive sequence organization and functional programming of an *Alu* repeat promoter. Proc Natl Acad Sci USA 1984; 81:5291-5295.
144. Liu W-M, Schmid CW. Proposed roles for DNA methylation in *Alu* transcriptional repression and mutational inactivation. Nucleic Acids Res 1993; 21:1351-1359.
145. Thorey IS, Cĕcna G, Reynolds W et al. *Alu* sequence involvement in transcriptional insulation of the keratin 18 gene in transgenic mice. Mol Cell Biol 1993; 13:6742-6751.
146. Deininger PL, Batzer MA, Hutchison CA et al. Master genes in mammalian repetitive DNA amplification. Trends Genet 1992; 8:307-311.
147. Matera G, Hellman U, Hintz MF et al. Recently transposed *Alu* repeats result from multiple source genes. Nucleic Acids Res 1990; 18:6019-6023.
148. Matera G, Hellman U, Schmid CW. A transpositionally and transcriptionally competent *Alu* subfamily. Mol Cell Biol 1990; 10:5424-5432.
149. Chesnokov I, Schmid CW. Flanking sequences of an Alu source stimulate transcription in vitro by interacting with sequence-specific transcription factors. J Mol Evol 1996; 42:30-36.
150. Schmid CW. Human Alu subfamilies and their methylation revealed by blot hybridization. Nucleic Acids Res 1991; 19:5613-5617.

151. Englander EW, Wolffe AP, Howard BH. Nucleosome interactions with a human Alu element. J Biol Chem 1993; 268:19565-19573.
152. Kochanek S, Renz D, Doerfler W. DNA methylation in the *Alu* sequences of diploid and haploid primary human cells. EMBO J 1993; 12:1141-1151.
153. Englander EW, Howard BH. Nucleosome positioning by human Alu elements in chromatin. J Biol Chem 1995; 270:10091-10096.
154. Russanova VR, Driscoll CT, Howard BH. Adenovirus type 2 preferentially stimulates polymerase III transcription of *Alu* elements by relieving repression: a potential role for chromatin. Mol Cell Biol 1995; 15:4282-4290.
155. Brow DA, Guthrie C. Transcription of a yeast U6 snRNA gene requires a polymerase III promoter element in a novel position. Genes Dev 1990; 4:1345-1356.
156. Eschenlauer JB, Kaiser MW, Gerlach VL et al. Architecture of a yeast U6 RNA gene promoter. Mol Cell Biol 1993; 13:3015-3026.
157. Burnol A-F, Margottin F, Schultz P et al. Basal promoter and enhancer element of yeast U6 snRNA gene. J Mol Biol 1993; 233:644-658.
158. Marsolier M-C, Tanaka S, Livingstone-Zatchej M et al. Reciprocal interferences between nucleosomal organization and transcriptional activity of the yeast *SNR6* gene. Genes Dev 1995; 9:410-422.
159. Moenne A, Camier S, Anderson G et al. The U6 gene of *Saccharomyces cerevisiae* is transcribed by RNA polymerase C (III) in vivo and in vitro. EMBO J 1990; 9:271-277.
160. Burnol A-F, Margottin F, Huet J et al. TFIIIC relieves repression of U6 snRNA transcription by chromatin. Nature 1993; 362:475-477.
161. Gerlach VL, Whitehall SK, Geiduschek EP et al. TFIIIB placement on a yeast U6 RNA gene in vivo is directed primarily by TFIIIC rather than by sequence-specific DNA contacts. Mol Cell Biol 1995; 15:1455-1466.
162. Kaiser MW, Brow DA. Lethal mutations in a yeast U6 RNA gene B block promoter element identify essential contacts with transcription factor-IIIC. J Biol Chem 1995; 270:11398-11405.
163. Roberts S, Colbert T, Hahn S. TFIIIC determines RNA polymerase III specificity at the TATA-containing yeast U6 promoter. Genes Dev 1995; 9:832-842.
164. Tani T, Ohshima Y. The gene for the U6 small nuclear RNA in fission yeast has an intron. Nature 1989; 337:87-90.
165. Kunkel GR, Maser RL, Calvet JP et al. U6 small nuclear RNA is transcribed by RNA polymerase III. Proc Natl Acad Sci USA 1986; 83:8575-8579.
166. Krol A, Carbon P, Ebel J-P et al. *Xenopus tropicalis* U6 snRNA genes transcribed by pol III contain the upstream promoter elements used by pol II dependent U snRNA genes. Nucleic Acids Res 1987; 15:2463-2478.
167. Kunkel GR, Pederson T. Upstream elements required for efficient transcription of a human U6 RNA gene resemble those of U1 and U2 genes even though a different polymerase is used. Genes Dev 1988; 2:196-204.
168. Lescure A, Carbon P, Krol A. The different positioning of the proximal sequence element in the *Xenopus* RNA polymerase II and III snRNA promoters is a key determinant which confers RNA polymerase III specificity. Nucleic Acids Res 1991; 19:435-441.
169. Carbon P, Murgo S, Ebel J-P et al. A common octamer motif binding protein is involved in the transcription of U6 snRNA by RNA polymerase III and U2 snRNA by RNA polymerase II. Cell 1987; 51:71-79.
170. Mattaj IW, Dathan NA, Parry HD et al. Changing the RNA polymerase specificity of U snRNA gene promoters. Cell 1988; 55:435-442.
171. Lobo SM, Ifill S, Hernandez N. Cis-acting elements required for RNA polymerase II and III transcription in the human U2 and U6 snRNA promoters. Nucleic Acids Res 1990; 18:2891-2899.

172. Parry HD, Mattaj IW. Positive and negative functional interactions between promoter elements from different classes of RNA polymerase III-transcribed genes. EMBO J 1990; 9:1097-1104.
173. Lobo SM, Lister J, Sullivan ML et al. The cloned RNA polymerase II transcription factor IID selects RNA polymerase III to transcribe the human U6 gene in vitro. Genes Dev 1991; 5:1477-1489.
174. Simmen KA, Bernues J, Parry HD et al. TFIID is required for in vitro transcription of the human U6 gene by RNA polymerase III. EMBO J 1991; 10:1853-1862.
175. Simmen KA, Waldschmidt R, Bernues J et al. Proximal sequence element factor binding and species specificity in vertebrate U6 snRNA promoters. J Mol Biol 1992; 223:873-884.
176. Danzeiser DA, Urso O, Kunkel GR. Functional characterization of elements in a human U6 small nuclear RNA gene distal control region. Mol Cell Biol 1993; 13:4670-4678.
177. Parry HD, Tebb G, Mattaj IW. The *Xenopus* U2 gene PSE is a single, compact element required for transcription initiation and 3' end formation. Nucleic Acids Res 1989; 17:3633-3644.
178. Bark C, Weller P, Zabielski J et al. A distant enhancer element is required for polymerase III transcription of a U6 RNA gene. Nature 1987; 328:356-359.
179. Lescure A, Tebb G, Mattaj IW et al. A factor with Sp1 DNA-binding specificity stimulates *Xenopus* U6 snRNA in vivo transcription by RNA polymerase III. J Mol Biol 1992; 228:387-394.
180. Janson L, Pettersson U. Cooperative interactions between transcription factors Sp1 and OTF-1. Proc Natl Acad Sci USA 1990; 87:4732-4736.
181. Das G, Henning D, Reddy R. Structure, organization, and transcription of *Drosophila* U6 small nuclear RNA genes. J Biol Chem 1987; 262:1187-1193.
182. Brow DA, Guthrie C. Spliceosomal RNA U6 is remarkably conserved from yeast to mammals. Nature 1988; 334:213-218.
183. Waibel F, Filipowicz W. U6 snRNA genes of Arabidopsis are transcribed by RNA polymerase III but contain the same two upstream promoter elements as RNA polymerase II-transcribed U-snRNA genes. Nucleic Acids Res 1990; 18:3451-3458.
184. Simmen KA, Mattaj IW. Complex requirements for RNA polymerase III transcription of the *Xenopus* U6 promoter. Nucleic Acids Res 1990; 18:5649-5657.
185. Goomer RS, Kunkel GR. The transcriptional start site for a human U6 small nuclear RNA gene is dictated by a compound promoter element consisting of the PSE and the TATA box. Nucleic Acids Res 1992; 20:4903-4912.
186. Reddy R, Henning D, Das G et al. The capped U6 small nuclear RNA is transcribed by RNA polymerase III. J Biol Chem 1987; 262:75-81.
187. Steinberg TH, Mathews DE, Durbin RD et al. Tagetitoxin: a new inhibitor of eukaryotic transcription by RNA polymerase III. J Biol Chem 1990; 265:499-505.
188. Waldschmidt R, Wanandi I, Seifart KH. Identification of transcription factors required for the expression of mammalian U6 genes in vitro. EMBO J 1991; 10:2595-2603.
189. Dahlberg JE, Lund E. How does III x II make U6? Science 1991; 254:1462-1463.
190. Heard DJ, Kiss T, Filipowicz W. Both *Arabidopsis* TATA binding protein (TBP) isoforms are functionally identical in RNA polymerase II and III transcription in plant cells: evidence for gene-specific changes in DNA binding specificity of TBP. EMBO J 1993; 12:3519-3528.
191. White RJ, Jackson SP, Rigby PWJ. A role for the TATA-box-binding protein component of the transcription factor IID complex as a general RNA polymerase III transcription factor. Proc Natl Acad Sci USA 1992; 89:1949-1953.

192. White RJ, Rigby PWJ, Jackson SP. The TATA-binding protein is a general transcription factor for RNA polymerase III. J Cell Science 1992; 16 (Supp): 1-7.
193. Radebaugh CA, Matthews JL, Geiss GK et al. TATA box-binding protein (TBP) is a constituent of the polymerase I-specific transcription initiation factor TIF-IB (SL1) bound to the rRNA promoter and shows differential sensitivity to TBP-directed reagents in polymerase I, II, and III transcription factors. Mol Cell Biol 1994; 14:597-605.
194. McBryant SJ, Meier E, Leresche A et al. TATA-box DNA binding activity and subunit composition of RNA polymerase III transcription factor IIIB from *Xenopus laevis*. Mol Cell Biol 1996; 16:4639-4647.
195. Tichelaar JW, Knerer B, Vrabel A et al. Transcription of a variant human U6 small nuclear RNA gene is controlled by a novel, internal RNA polymerase III promoter. Mol Cell Biol 1994; 14:5450-5457.
196. Li J-M, Parsons RA, Marzluff WF. Transcription of the sea urchin U6 gene in vitro requires a TATA-like box, a proximal sequence element, and sea urchin USF, which binds an essential E box. Mol Cell Biol 1994; 14:2191-2200.
197. Li J-M, Haberman RP, Marzluff WF. Common factors direct transcription through the proximal sequence elements (PSEs) of the embryonic sea urchin U1, U2, and U6 genes despite minimal sequence similarity among the PSEs. Mol Cell Biol 1996; 16:1275-1281.
198. Waibel F, Filipowicz W. RNA-polymerase specificity of transcription of Arabidopsis U snRNA genes determined by promoter element spacing. Nature 1990; 346:199-202.
199. Kiss T, Marshallsay C, Filipowicz W. Alteration of the RNA polymerase specificity of U3 snRNA genes during evolution and in vitro. Cell 1991; 65:517-526.
200. Vankan P, Filipowicz W. A U-snRNA gene-specific upstream element and a -30 "TATA box" are required for transcription of the U2 snRNA gene of *Arabidopsis thaliana*. EMBO J 1989; 8:3875-3882.
201. Connelly S, Marshallsay C, Leader D et al. Small nuclear RNA genes transcribed by either RNA polymerase II or RNA polymerase III in monocot plants share three promoter elements and use a strategy to regulate gene expression different from that used by their dicot plant counterparts. Mol Cell Biol 1994; 14:5910-5919.
202. Nakaar V, Dare AO, Hong D et al. Upstream tRNA genes are essential for expression of small nuclear and cytoplasmic RNA genes in trypanosomes. Mol Cell Biol 1994; 14:6736-6742.
203. Fantoni A, Dare AO, Tschudi C. RNA polymerase III-mediated transcription of the trypanosome U2 small nuclear RNA gene is controlled by both intragenic and extragenic regulatory elements. Mol Cell Biol 1994; 14:2021-2028.
204. Lee JY, Evans CF, Engelke DR. Expression of RNase P RNA in *Saccharomyces cerevisiae* is controlled by an unusual RNA polymerase III promoter. Proc Natl Acad Sci USA 1991; 88:6986-6990.
205. Baer M, Nilsen TW, Costigan C et al. Structure and transcription of a human gene for H1 RNA, the RNA component of human RNase P. Nucleic Acids Res 1990; 18:97-103.
206. Topper JN, Clayton DA. Characterization of human MRP/Th RNA and its nuclear gene: full length MRP/Th RNA is an active endoribonuclease when assembled as an RNP. Nucleic Acids Res 1990; 18:793-799.
207. Maraia RJ, Sasaki-Tozawa N, Driscoll CT et al. The human Y4 small cytoplasmic RNA gene is controlled by upstream elements and resides on chromosome 7 with all other hY scRNA genes. Nucleic Acids Res 1994; 22:3045-3052.

208. Farris AD, Gross JK, Hanas JS et al. Genes for murine Y1 and Y3 Ro RNAs have class 3 RNA polymerase III promoter structures and are unlinked on mouse chromosome 6. Gene 1996; 174:35-42.
209. Murphy S, Tripodi M, Melli M. A sequence upstream from the coding region is required for the transcription of the 7SK RNA genes. Nucleic Acids Res 1986; 14:9243-9260.
210. Kruger W, Benecke B-J. Structural and functional analysis of a human 7S K RNA gene. J Mol Biol 1987; 195:31-41.
211. Murphy S, Pierani A, Scheidereit C et al. Purified octamer binding transcription factors stimulate RNA polymerase III-mediated transcription of the 7SK RNA gene. Cell 1989; 59:1071-1080.
212. Murphy S, Yoon JB, Gerster T et al. Oct-1 and Oct-2 potentiate functional interactions of a transcription factor with the proximal sequence element of small nuclear RNA genes. Mol Cell Biol 1992; 12:3247-3261.
213. Kleinert H, Assert R, Benecke BJ. A single base pair deletion from the inactive octamer-like motif of the 7S K distal sequence element brings full functionality in vivo. J Biol Chem 1991; 266:23872-23877.
214. Chung J, Sussman DJ, Zeller R et al. The *c-myc* gene encodes superimposed RNA polymerase II and III promoters. Cell 1987; 51:1001-1008.
215. Bentley DL, Brown WL, Groudine M. Accurate, TATA box-dependent polymerase III transcription from promoters of the *c-myc* gene in injected *Xenopus* oocytes. Genes Dev 1989; 3:1179-1189.
216. Huang W, Pruzan R, Flint SJ. In vivo transcription from the adenovirus E2 early promoter by RNA polymerase III. Proc Natl Acad Sci USA 1994; 91:1265-1269.
217. Pruzan R, Chatterjee PK, Flint SJ. Specific transcription from the adenovirus E2E promoter by RNA polymerase III requires a subpopulation of TFIID. Nucleic Acids Res 1992; 20:5705-5712.
218. Mitchell MT, Hobson GM, Benfield PA. TATA box-mediated polymerase III transcription in vitro. J Biol Chem 1992; 267:1995-2005.
219. Mitchell MT, Benfield PA. TATA box-mediated in vitro transcription by RNA polymerase III. J Biol Chem 1993; 268:1141-1150.
220. Wang Y, Stumph WE. RNA polymerase II/III transcription specificity determined by TATA box orientation. Proc Natl Acad Sci USA 1995; 92:8606-8610.
221. Huang W, Wong JM, Bateman E. TATA elements direct bi-directional transcription by RNA polymerases II and III. Nucleic Acids Res 1996; 24:1158-1163.
222. Wang Y, Jensen RC, Stumph WE. Role of TATA box sequence and orientation in determining RNA polymerase II/III transcription specificity. Nucleic Acids Res 1996; 24:3100-3106.
223. Whitehall SK, Kassavetis GA, Geiduschek EP. The symmetry of the yeast U6 RNA gene's TATA box and the orientation of the TATA-binding protein in yeast TFIIIB. Genes Dev 1995; 9:2974-2985.
224. Martinez E, Lagna G, Roeder RG. Overlapping transcription by RNA polymerases II and III of the *Xenopus* TFIIIA gene in somatic cells. J Biol Chem 1994; 269:25692-25698.
225. Hall RK, Taylor WL. Transcription factor IIIA gene expression in *Xenopus* oocytes utilizes a transcription factor similar to the major late transcription factor. Mol Cell Biol 1989; 9:5003-5011.
226. Scotto KW, Kaulen H, Roeder RG. Positive and negative regulation of the gene for transcription factor IIIA in *Xenopus laevis* oocytes. Genes Dev 1989; 3:651-662.
227. Pfaff SL, Hall RK, Hart GC et al. Regulation of the *Xenopus laevis* transcription factor IIIA gene during oogenesis and early embryogenesis: negative elements repress the O-TFIIIA promoter in embryonic cells. Dev Biol 1991; 145:241-254.

228. Pfaff SL, Taylor WL. Characterization of a *Xenopus* oocyte factor that binds to a developmentally regulated cis-element in the TFIIIA gene. Dev Biol 1992; 151:306-316.
229. Kim SH, Darby MK, Joho KE et al. The characterization of the TFIIIA synthesized in somatic cells of *Xenopus laevis*. Genes Dev 1990; 4:1602-1610.
230. Kaulen H, Pognonec P, Gregor PD et al. The *Xenopus* B1 factor is closely related to the mammalian activator USF and is implicated in the developmental regulation of TFIIIA gene expression. Mol Cell Biol 1991; 11:412-424.
231. Hanas JS, Smith JF. Identification of a TFIIIA binding site on the 5' flanking region of the TFIIIA gene. Nucleic Acids Res 1990; 18:2923-2928.

CHAPTER 3

RNA Polymerase III

Pol III is the largest of the nuclear RNA polymerases, with an aggregate molecular weight of 600-700 kD (reviewed by Thuriaux and Sentenac[1-3]). This is, perhaps, surprising since pol II transcribes a much larger and more diverse set of templates, and so might have been expected to show the greatest complexity. However, several transcription factors have been described that bind tightly and specifically to pol II.[4-6] These include TFIIF, which is involved in recruiting pol II to initiation complexes, reducing its binding to nonspecific sites, and suppressing pausing during elongation.[7,8] It has been found that TFIIF is not required for VA_I transcription.[9] Its function in the class III system, and those of the other pol II-associated factors, may possibly be performed by the additional subunits that copurify with pol III.

Biochemical Characterization

The three classes of nuclear RNA polymerases that occur in eukaryotes were originally separated by chromatography on DEAE Sephadex.[10] They were classified according to their distinct chromatographic properties, salt requirements, template preferences, and especially by their differential sensitivity to the toxin α-amanitin, a cyclic octapeptide produced by the poisonous *Amanita* mushrooms.[10-12] α-amanitin inhibits transcript elongation by interfering with the translocation process, but does not prevent formation of the first phosphodiester bond.[13] In mammals, pol II is the most sensitive to α-amanitin (50% inhibition at 25 ng/ml), whereas pol III displays intermediate sensitivity (50% inhibition at 20 μg/ml) and pol I is completely resistant.[14] In *Saccharomyces*, pol III is highly resistant, whereas pol I shows intermediate sensitivity (50% inhibition at 300-600 μg/ml) and pol II is again extremely susceptible (50% inhibition at 1 μg/ml α-amanitin).[15,16] These differential sensitivities to α-amanitin are the standard tool for diagnosing which polymerase is responsible for transcribing any given template. A bacterial phytotoxin called tagetitoxin is also available; this is produced by *Pseudomonas syringae* and inhibits pol III in preference to pols I and II.[17]

The physical properties of the nuclear RNA polymerases are well conserved throughout eukaryotes, a fact which has greatly facilitated their purification from a diverse range of organisms. Pol III has been purified from a wide variety of organisms, including human,[18-20] mouse,[21,22] frog,[23,24] silkworm,[25] fruitfly,[26] wheat,[27,28] and yeast.[16,29,30] Overall, there is considerable similarity in the composition of pol III from different organisms. (Table 3.1) Some of the apparent differences between species may result from the dissociation of subunits during purification. However, human pol III is stable to washing with 2M urea or 0.3% sarkosyl and the enzyme

RNA Polymerase III Transcription, Second Edition, by Robert J. White.

Table 3.1. Polypeptide Composition of RNA Polymerase III from Various Eukaryotes (sizes in kD)

Yeast	Wheat	Fruitfly	Silkworm	Frog	Mouse	Human
160	150	154	155	155	155	155
128	130	135	136	138	138	135
82	94	62	67	94	89	82
53	55	58	62	68	70	62
40	38	38	49	41	53	53
37	30	32	39	39	49	40
34	28	31	36	33	41	39
31	25	27.2	31	29	33/32	38
27	24.5	26.5	28	24	29	36
25	20.5	21.5	18	23	19	32
23	20	17.5		19		29
19	19.5					20
14.5	17.8					18
11	17					15
10						12
10						10

core can withstand 4 M urea.[20] Insects appear to lack an 80-90 kD pol III subunit that is found in all other cases examined.[25,26] Pol III from mouse can be resolved chromatographically into two distinct forms.[14] This heterogeneity results from the alternate use of C32 or C33 subunits.[22] The murine III_A form is localized predominantly in the nucleoplasm whereas the III_B form is mainly cytoplasmic, however, the functional significance of this is unclear.[14]

Like pols I and II, pol III is composed of two large subunits, of around 160 and 130 kD, and a complex series of smaller components with sizes ranging from 10 to 90 kD. The largest polypeptide of pol III is related immunologically to the largest polypeptides of pols I and II, and the second largest polypeptides of pols I, II and III are also immunologically related to each other.[31,32] Although pol III has many unique subunits, others are also found in pol I or in all three of the nuclear RNA polymerases. (Table 3.2) Indeed, 11 of the 12 subunits of *Saccharomyces* pol II are related or identical to a subunit in pols I and/or III.[33] A partner may be found for the 12th pol II subunit once genes have been cloned for all the components of pols I and III. The convention is to name a subunit with a letter and a number, such that the letter describes which polymerases contain that subunit (A for pol I, B for pol II, and C for pol III) and the number denotes its apparent size (in kD). For example, the 40 kD subunit that forms part of pols I and III, but not of pol II, is referred to as AC40.

Table 3.2. Subunits of *S. cerevisiae* Pols I, II and III

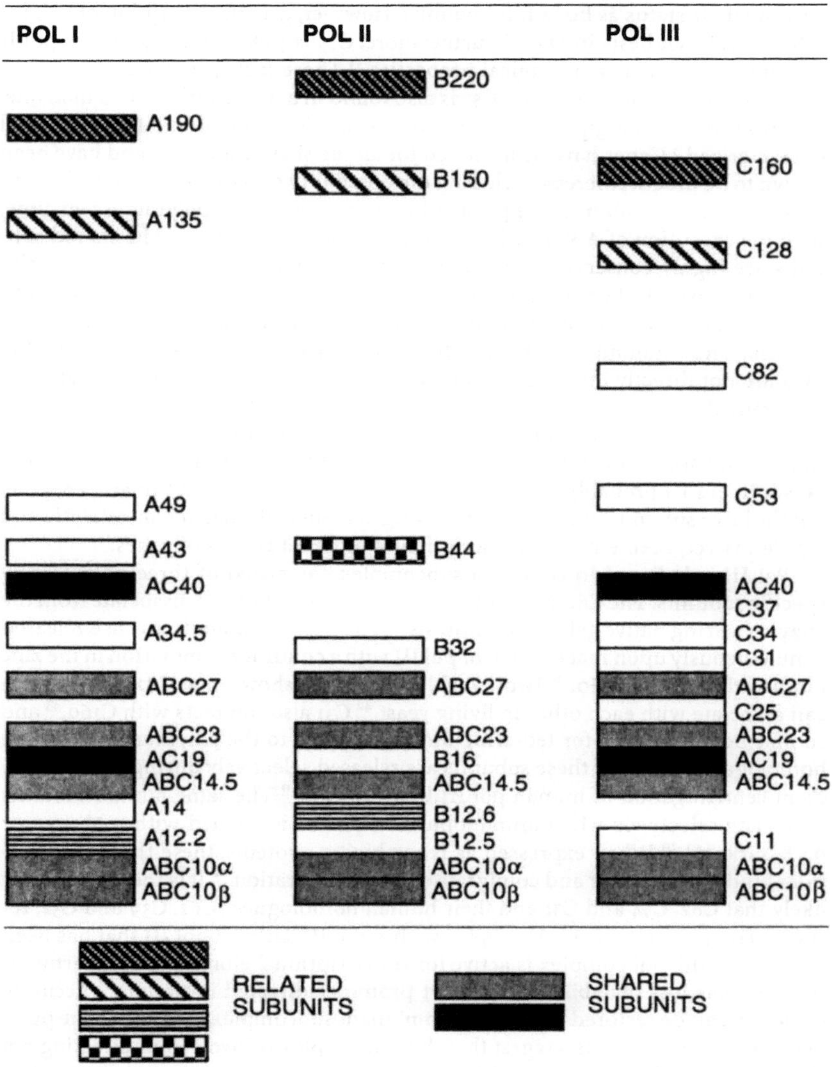

The structural core of pol III in *S. cerevisiae* consists of C160, C128, AC40 and AC19. The two large subunits show considerable homology to the β and β' subunits of prokaryotic RNA polymerases,[34-37] while the small core subunits, which are also found in pol I, display some homology with a short N-proximal segment of the prokaryotic α subunit.[38] Five subunits that are found in all the nuclear RNA polymerases (ABC 27, 23, 14.5, 10α and 10β) and up to seven pol III-specific subunits associate with the evolutionarily conserved core of the yeast pol III enzyme. C37 and C25 are found in submolar ratios and may be absent from some preparations that nevertheless support accurate transcription,[39,40] which raises a question regarding their status as bona fide subunits. However, genetic disruption of C25 inhibits tRNA synthesis in vivo.[33] Furthermore, C25 is 50% homologous to a pol II subunit with similar biochemical properties.[33] These facts provide a strong case for considering it part of pol III. C53 is also found in a molar ratio of less than one, but this reflects its high susceptibility to proteolytic degradation.[3] C11 is not well characterized.[3] Genes have been cloned for all the shared subunits and have been shown to be unique, thereby excluding the possibility of polymerase-specific variants.[3] The shared subunits are present in one copy each per polymerase enzyme, with the exception of ABC27, which has a stoichiometry of two.[3] The shared subunits are highly conserved through evolution, with 40-75% amino acid identity between yeast and human homologues.[41] Furthermore, four of the shared subunits from yeast can be functionally replaced by their human counterparts.[41] The fact that these subunits are missing from prokaryotic polymerases suggests that they are not directly involved in catalytic function, but their roles have yet to be determined.

The large subunits together with the shared subunits are likely to constitute the functional core of the polymerase, with roles in the basic steps of transcription. Pols I, II and III probably evolved from a common ancestral multimeric enzyme, but the large subunits must then have diverged to meet the particular catalytic and regulatory requirements for transcription of different classes of genes.

Pol III is believed to contain a subcomplex composed of three polymerase-specific subunits. The C82, C34 and C31 subunits of yeast pol III dissociate from the enzyme during native gel electrophoresis.[16] These three subunits are also released simultaneously upon inactivation of pol III with a conditional mutation in the zinc binding domain of C160.[40] Two-hybrid analysis has shown that C82, C34 and C31 can associate with each other in living yeast.[42] C31 also interacts with C160,[43] and so may be responsible for tethering the subcomplex to the polymerase core. The homologues of each of these subunits are released selectively during sucrose gradient centrifugation of human pol III in 0.5 M KCl.[20] The same human subunits dissociate selectively when immunopurified pol III is washed with 4 M urea or 0.5% sarkosyl.[20] When expressed as recombinant proteins, these three polypeptides bind to each other and comigrate during gel filtration.[20] It is therefore highly likely that C82, C34 and C31 and their human homologues (C62, C39 and C32, respectively) associate in a subcomplex within pol III. Human pol III that has been depleted of this subcomplex is active for transcriptional elongation and termination, but has lost the ability to support promoter-directed initiation.[20] Accurate initiation can be restored when the recombinant subcomplex is added to the purified core.[20] These results suggest that the subcomplex is involved in directing pol III to the preinitiation complex. This contention is supported strongly by the fact that yeast C34 and human C39 bind directly to the initiation factor TFIIIB.[20,42,44]

UV crosslinking with *Drosophila* pol III identified the two largest subunits as interacting with DNA, whereas nascent RNA bound predominantly to the largest subunit.[45] In humans, the two largest subunits are labeled by a photoaffinity analog incorporated into nascent transcripts, with the largest subunit the more heavily labeled.[46] Ten subunits of yeast pol III can be photocrosslinked to the SUP4 $tRNA^{Tyr}$ gene following TFIIIB- and TFIIIC-dependent recruitment.[47-49] C160, C128, C34

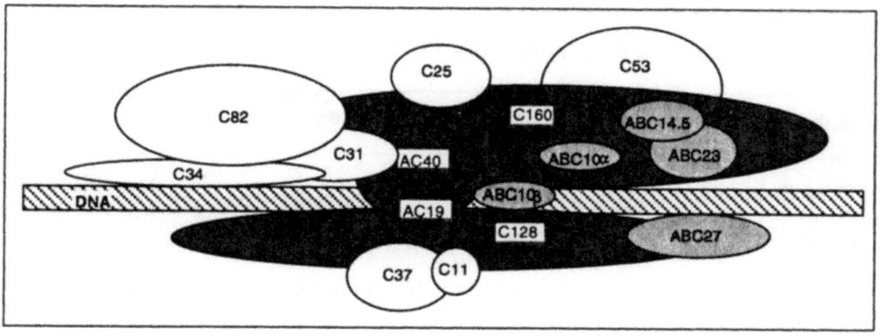

Fig. 3.1. Postulated subunit arrangement of a transcribing yeast pol III molecule.

and ABC27 are all relatively extended along the surface of the DNA, whereas other subunits, such as C82 and C53, are quite localized in their accessibility to crosslinking.[47] C34 projects furthest upstream from the yeast polymerase and is therefore in a position to interact with TFIIIB.[47,49] In contrast, ABC27 appears to be situated at the leading edge of the enzyme prior to initiation.[47] (Fig. 3.1) During elongation, multiple subunits are accessible to specific crosslinking from a photoactive nucleotide situated in the middle of the transcription bubble, with the C160, C128 and C34 subunits labeled most strongly.[47] C128 can be labeled by a nucleoside triphosphate analog placed in the initiator nucleotide substrate site.[50]

Genetic Characterization

Definitive evidence that a copurifying polypeptide represents an integral component of the polymerase rather than a contaminant can be difficult to obtain.[1,22,24,51] Ideally, this requires disassembly and reconstitution, genetic analysis, or the demonstration that an antibody directed against such a polypeptide can affect polymerase function. Most progress in this regard has been made with the yeast enzyme, initially by the use of subunit-specific antibodies,[1,31,32,39,52] and more recently with the cloning of genes for the majority of the 16 copurifying subunits (reviewed in refs. 2, 3 and 53). (Table 3.3) Several of the genes were cloned by screening expression libraries with antibodies raised against purified subunits,[54] while partial sequencing of purified polypeptides provided the sequence information used to screen for other subunit genes.[20,36,55,56] All of the pol III subunit genes that have been cloned from yeast have been found to be both unique and essential for viability.[3,53] This contrasts with the situation for pol II, in which certain subunits are dispensable for growth.[55,57]

Mutations in C160, C82, C53, C34, C31 and C25 have all been shown to inhibit tRNA synthesis.[33,56,58-62] However, in each case expression of 5S RNA was much less inhibited. The reason why tRNA synthesis is more sensitive than 5S gene transcription to partial inactivation of pol III is not known. One possibility is that 5S genes are better able to recruit limiting amounts of polymerase, perhaps due to the different composition of their transcription complexes.

Table 3.3. Subunits of *S. cerevisiae* Pol III

Subunit	Gene	Features
Core Subunits		
C160	RPC160/RPO31	β′ homology; Zn binding
C128	RET1	β homology; Zn binding; NTP binding
AC40	RPC40	α homology
AC19	RPC19	α homology; phosphorylated
Common Subunits		
ABC27	RPB5	basic
ABC23	RPB6	acidic; phosphorylated
ABC14.5	RPB8	acidic
ABC10α	RPC10	Zn binding
ABC10β	RPB10	Zn binding
Unique Subunits		
C82	RPC82	leucine zipper
C53	RPC53	hydrophilic
C37	not published	
C34	RPC34	hydrophilic
C31	RPC31	acidic tail
C25	YKL1	RPB7 homology
C11	not published	

C160

The gene for the largest subunit of pol II was first cloned in *Drosophila* by P element transposon tagging.[63] It was found to cross-hybridize at low stringency to the yeast gene for C160, the largest pol III subunit, thereby providing a route to its cloning.[64] This gene, which is called RPO31 or RPC160, was also cloned independently by immunological screening.[54] Mutations in RPC160 have been shown to affect synthesis specifically of tRNA, H1 RNA and U6 snRNA.[58,60,65]

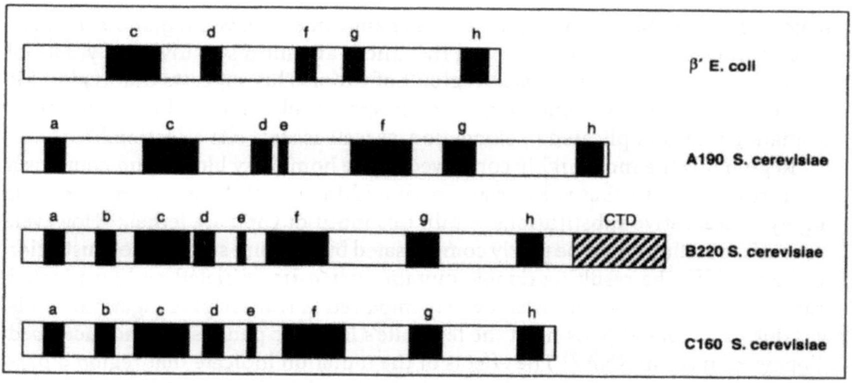

Fig. 3.2. Regions of homology between the largest subunits of *E. coli* RNA polymerase and pols I, II and III from *S. cerevisiae*. Homology blocks are shown in black. The C-terminal domain (CTD) of B220 is also indicated.

Immunological studies originally disclosed the presence of conserved epitopes in the large subunits of pols I, II and III.[52] Cloning of the genes revealed that C160 contains eight regions of significant homology to B220, the largest subunit of yeast pol II[3,34,36] (Fig. 3.2). The relative positions of the homologous regions in C160 and B220 are almost identical, and the similarity of the predicted gene products is 35% overall and 43-71% within the major homology blocks.[34] Most of these sequence blocks are also homologous to A190, the largest subunit of yeast pol I, and to β', the largest subunit of *Escherichia coli* RNA polymerase, again in the same relative positions.[34,36] These subunits are therefore likely to perform equivalent functions. Indeed, the C-terminal half of domain f of C160 can be exchanged for that of B220 without affecting pol III function.[66] Blocks c, d and f are conserved in all RNA polymerases that have been analyzed, and are therefore likely to contribute to the "active site", which may be regarded as the catalytic region and the binding sites for DNA and RNA. Each of the six major homology blocks is longer when C160 and B220 are compared with each other than when they are compared with the β' sequence.[34] Two of the major homology blocks also display limited similarity to the DNA-binding domain of prokaryotic DNA polymerases.[34] It is likely that the conserved regions of the two largest subunits cooperate in forming a cleft which contains the active site, each making functionally distinct contributions. For yeast pol I, there is genetic evidence for a direct interaction between region a and the second largest subunit.[67]

C160 and B220 are less related to A190 than they are to each other.[36] For example, homology block b is found in C160, B220 and in archaebacterial polymerases, but is missing from A190.[3] This has led to the suggestion that two nuclear RNA polymerases may have existed at one stage of evolution, one being the precursor of pol I and the other being the precursor of pols II and III.[36] Paradoxically, however, pol III shares more common subunits with pol I than it does with pol II.[1,53]

Conditional mutations in the largest subunit of RNA polymerases frequently map in the conserved regions.[68-70] Mutations affecting nucleotide binding by yeast pol II[71] and transcription termination by *E. coli* RNA polymerase[72,73] map to the major homology blocks. A single residue change in conserved region f confers α-amanitin resistance upon mouse pol II.[74] The α-amanitin sensitivity of yeast pol III can be altered by mutations in region f of C160.[3] This suggests that region f is involved in transcript elongation, which is specifically blocked by α-amanitin.[13] Domain g is also implicated in elongation, as well as start site selection.[66]

Region d is the most highly conserved of the homology blocks and contains a seven residue motif that is invariant in all multimeric RNA polymerases.[75] Even highly conservative substitutions within this motif of C160 are lethal.[75] However, one such mutation could be partly compensated by a second substitution just prior to the motif.[75] The resulting double mutant, *rpc160-112*, initiated and terminated transcription normally, but was severely impaired in transcript elongation.[75] This was due to increased pausing at the template's intrinsic pause sites and increased slippage of nascent RNA.[75] The effects of the mutation indicate that region d participates in the active site near the growing point of the transcript.[75]

Domain f contains a high density of clustered charged residues. Double alanine mutations introduced into this region do not affect polymerase assembly but are nevertheless generally lethal.[66] However, mutant C160-270, which bears a D829A R830A double substitution in domain f, was found to be unaffected in formation of the first phosphodiester bond.[66] This mutant enzyme showed increased pausing, although the rate of RNA elongation between pause sites was unimpaired.[66] The mutation was found to decrease the transition from abortive to processive transcription and increase the cleavage of nascent transcripts.[66] As a consequence, tRNA synthesis was reduced by 2- to 10-fold.[66] It was proposed that the polymerase adopts two alternative conformational states, one during active elongation and the other during pausing and abortive transcription.[66] The C160-270 mutation may affect the transition between these states.[66]

The conserved N-terminal domain a of the largest subunit of all eukaryotic and archaebacterial RNA polymerases is rich in cysteine and histidine residues and shares a minimal consensus sequence of $CX_2CX_{6\text{-}12}CXGHXGX_{24\text{-}37}CX_2C$.[40] In vitro mutagenesis studies of yeast C160 defined this invariant motif as being strictly essential for pol III function.[40] All mutations introduced into the conserved cysteines and histidines within this motif were lethal.[40] Two mutations in residues adjacent to invariant cysteines produced conditional temperature-sensitive phenotypes.[40] One of the thermosensitive mutations generated an unstable polymerase in which subunits C82, C34 and C31 were readily dissociated from the enzyme core.[40] These observations suggest that domain a of C160 plays an important role in maintaining the structural integrity of pol III. This is consistent with the idea that eukaryotic RNA polymerases contain a minimal structure similar to the $\alpha_2\beta\beta'$ bacterial core enzyme together with the shared subunits. The enzyme-specific subunits would then be assembled subsequently and, as such, might also be the most liable to dissociate. The conserved cysteines and histidines are likely to bind zinc, since A190, B220 and C160 have all been shown to bind $^{65}Zn(II)$ in vitro, as have bacterially-expressed domain a peptides from B220 and C160.[40,76] Zinc ions may play a structural role in maintaining the spatial organization of the enzyme. Indeed, it has been proposed that the largest and second largest subunits interact through their zinc-binding motifs, perhaps by co-chelation of shared zinc ions, since a suppressor of a conditional mutation in domain a of A190 maps to the zinc-binding domain of the second-largest subunit A135.[67] As well as the two largest subunits, several of the small subunits of yeast pols I, II and III can bind zinc in vitro.[76] A catalytic role for zinc has not been excluded.

C128

A point mutation in C128, the second largest subunit of yeast pol III, reduces the efficiency of transcription termination both in vitro and in vivo.[77] The gene for C128, named RET1, was cloned by complementation of an ochre suppression phenotype in cells carrying this mutation.[37] RET1 is an essential gene encoding a 1149 amino acid polypeptide with extensive homology to the second-largest subunits of other RNA polymerases.[37] For instance, C128 is identical to B150 at 440 positions.[37] Seventeen regions of high sequence homology are shared by the second-largest subunits of RNA polymerases from yeast, *Drosophila* and archaebacteria, while thirteen of these are also found in polymerases from eubacteria and chloroplasts.[37] Zinc finger motifs of the form $CX_2CGX_{7\text{-}24}CX_2C$ are conserved in the eukaryotic and archaebacterial genes.[37] The conserved regions include the nucleotide binding sites of the *E. coli* β subunit.[37] Many mutations introduced in the conserved regions of C128 can affect transcription termination, suggesting that these sites may be involved in the response of pol III to termination signals.[78] Comparable mutations can alter termination by the *E. coli* RNA polymerase.[78] Termination can also be affected by substitutions in the nonconserved regions of this subunit.[78] Substitutions in C128 that affect polymerization have also been obtained.[78]

AC40

RPC40, the gene for the yeast AC40 subunit, was cloned by screening an expression library with antibodies directed against isolated AC40.[54] RPC40 has the potential to code for a 335 amino acid polypeptide with an isoelectric point (pI) of 5.6.[79] A temperature-sensitive RPC40 mutant was shown by immunoprecipitation to be defective in assembly of pols I and III in vivo, whereas pol II was synthesized normally.[79] AC40 is likely to function at an early stage in pol I and III assembly, since no stable subcomplexes were observed following its inactivation at the nonpermissive temperature.[79] The B44 subunit appears to be the pol II counterpart of AC40, since these subunits display sequence homology and temperature-sensitive mutations in B44 can block pol II assembly.[80] A short region of homology exists between AC40, B44 and the α subunit of *E. coli* RNA polymerase.[38] Mutagenesis confirmed that this α motif is essential for the function of AC40.[81] Equivalent substitutions in B44 and the *E. coli* α subunit resulted in similar phenotypes, which suggests that the α motif performs related functions in each case.[81]

A cDNA encoding the murine homologue of AC40 has been isolated through the purification of mouse pol I.[82] Peptide sequences were obtained from the 40 kD subunit and these were used to design primers for screening a cDNA library.[82] The mouse gene is present in a single copy and encodes a 355 amino acid polypeptide which is 44% identical to AC40 from *Saccharomyces*.[82]

AC19

The other core subunit found in pols I and III of yeast is AC19. Antibodies against this polypeptide inhibit both nonspecific transcription and tRNA synthesis in vitro.[31,39] RPC19, the gene for AC19, was cloned by screening a genomic library with oligonucleotides corresponding to the sequences of peptides generated by tryptic digestion of the purified subunit.[38] RPC19 is unique, essential, and encodes a 143 amino acid protein with a pI of 4.5, in which the N-terminal region is significantly acidic.[38] AC19 is phosphorylated in vivo, as are ABC23 and possibly also C53.[83] Consensus sites for casein kinase II and protein kinase C are found in the AC19 sequence.[38]

The short sequence motif shared between AC40, B44 and the *E. coli* α subunit is also found in AC19.[38] Whereas AC40 and AC19 are present in single copies in pols I and III, the B44 and α subunits both occur in pairs in their respective polymerases.[38] It is therefore possible that AC40 and AC19 together share the functions performed by two B44 or two α subunits. In support of this, a conditional defect in AC40 can be corrected by overexpressing AC19, and vice versa.[3] However, mutagenesis of the putative α motif in AC19 has no effect on growth, unlike the case for AC40.[81] Furthermore, AC19 displays no significant homology to AC40, B44 or bacterial α outside its putative α motif. Indeed, AC19 shows greater similarity to the L subunit of RNA polymerase from *Sulfolobus acidocaldarius*.[81] AC19 may therefore have evolved from an archaeal gene, unlike AC40 and B44 which are likely to have had prokaryotic origins.[81] Nevertheless, AC40 and AC19 interact strongly with each other in vivo, as shown by two-hybrid analysis and by extragenic suppression of mutations.[81]

The gene encoding the murine homologue of AC19 (mRPA16) was cloned through a yeast two-hybrid screen using as bait the mouse equivalent of AC40 (mRPA40).[84] Its product has a mass of 16 kD and is 45% identical and 83% similar to AC19 from *S. cerevisiae*.[84] The α motif is especially well conserved.[84] Substitutions in the α motif of mRPA40 disrupt its interaction with mRPA16.[84]

ABC27, ABC23 and ABC14.5

The yeast genes for the ABC27, ABC23 and ABC14.5 subunits that are found in each of the nuclear RNA polymerases were cloned by screening genomic libraries with oligonucleotides corresponding to peptide sequences obtained from purified subunits.[55] Each of these genes is present in a single copy and has no closely related homologues detectable by hybridization.[55] The coding sequence of ABC23 is located only 233 bp away from that of TFIIIA and the two are divergently transcribed, making an overlap of regulatory elements seem likely.[85,86] This gene was also independently cloned as a suppressor of a temperature-sensitive mutation in the largest subunit of pol II.[87] Although the genes for the common subunits are all required for viability, none of them contains recognizable sequence motifs for binding nucleic acids or nucleotides and so they may not be directly involved in the catalytic steps of transcription.[55] The absence of homologous subunits in bacterial RNA polymerases also argues against a direct catalytic role. Possible alternative functions could include structural roles or coordinating the activity of pols I, II and III. The predicted sequences show ABC27 to be very basic (pI 10.15), whereas ABC23 and ABC14.5 are both quite acidic (pI values of 5.15 and 4.28, respectively).[55]

The N-terminal 42 residues of ABC23 are not required for growth and are not conserved in the archaebacterial counterparts.[88] Genetic analyses suggest that ABC23 interacts functionally with B220, since nonlethal mutations in these two subunits give synthetic lethality when combined.[87] Immunoelectron microscopy indicates that ABC23 is located close to the groove in polymerase that is thought to represent the DNA-binding site.[89] Consistent with this, antibody access to ABC23 is impeded by DNA binding.[39] However, gel retardation and surface plasmon resonance experiments demonstrated that the presence of ABC23 does not affect the ability of pol I to bind to DNA.[89] Mutations in ABC23 were found to prevent the assembly of pols I and II and to destabilize their largest subunits.[88] A biochemical analysis provided evidence that ABC23 is required for the structural and functional integrity of pol I.[89] An inactive form of pol I that lacked ABC23, A43 and A14 was unable to synthesize the first phosphodiester bond, but this activity could be restored using recombinant ABC23.[89] Electron microscopy revealed that the association of ABC23 induces a major conformational change in pol I.[89] Although pol III was not tested in these studies, it is likely that ABC23 performs a similar role in all three nuclear RNA pols and interacts with the largest subunit in each case.

The small shared subunits are highly conserved during evolution. Indeed, antibodies against yeast ABC23 cross-reacted with plant, insect, crustacean and mammalian RNA polymerases.[52] ABC23 from *S. cerevisiae* shows only a few conservative substitutions when compared with its homologues from *S. pombe*, *D. melanogaster*, and *H. sapiens*, apart from the N-terminal region, which is variable in size, although consistently acidic.[41] An *S. cerevisiae* strain that is defective for ABC23 can be fully complemented by the equivalent subunits from *S. pombe*, *D. melanogaster* or *H. sapiens*.[41] Like its yeast counterparts, human ABC23 can also complement a mutation in B220, indicating that the conservation extends to extragenic suppression properties.[41] The human homologue of ABC14.5 is 33% identical to the *S. cerevisiae* subunit and also complements its deficiency.[41] In contrast, the human homologue of ABC27 cannot substitute for its yeast equivalent, despite their displaying 44% identity and 80% similarity.[41] Archaebacteria have homologues of ABC27, ABC23 and ABC10β, but do not contain homologues of ABC14.5 or ABC10α.[89]

ABC10α and ABC10β

The *S. cerevisiae* genes for the small common subunits ABC10α and ABC10β were cloned on the basis of microsequence data.[90,91] These subunits each contain 70 amino acid residues and comigrate on SDS-polyacrylamide gels, but are unrelated and can be separated by reverse-phase chromatography.[81,91] Like the two largest subunits, ABC10α and β can both bind zinc.[91] Conditional mutations in either AC40 or AC19 can be rescued by overexpressing ABC10β, which suggests that these three subunits interact in vivo.[81] Genes for these subunits were cloned from other species by PCR screening of cDNA libraries.[41]

The ABC10α proteins from *S. pombe* and from humans have 63 and 58 amino acid residues, respectively.[41] They are both closely related to ABC10α from *S. cerevisiae* in their C-terminal 25 residues, whereas the remainder of these subunits is poorly conserved in both size and sequence.[41] Nevertheless, each can complement an *S. cerevisiae* strain without ABC10α.[41] In each of these species, the subunit is highly basic (pI 9.3-10.2) and contains a canonical $CX_2CX_{13}CX_2C$ zinc-binding motif.[41] This is consistent with the ability of ABC10α to bind zinc in vitro.[81]

ABC10β is the RNA polymerase subunit that is most highly conserved between yeast and man.[41] The human version has 67 amino acids and is 70% identical to the *S. cerevisiae* subunit.[41] It is also 45% identical to the N subunit of RNA polymerase from the archaebacterium *Sulfolobus acidocaldarius* and 22% identical to the vaccinia virus enzyme.[41] A defect in yeast ABC10β can be complemented by the human version but not by the archael N subunit.[41] ABC10β binds zinc in vitro but lacks a canonical zinc-binding motif.[81] However, six residues that are conserved between the yeast, archael and viral subunits form a CX_2CGX_nCCR motif that may constitute an atypical zinc finger.[41]

C82

Antibodies against C82, the third largest subunit of yeast pol III, inhibit transcription of tRNA genes.[39] RPC82, which encodes C82, was cloned by screening libraries with oligonucleotides designed on the basis of peptide sequence data.[56] The 655 amino acid gene product has no strong sequence similarity to other known proteins.[56] However, a heptad repeat of four leucine residues lies close to the C-terminus in a region predicted to form an α-helix.[56] Such a structure is characteristic of the leucine zipper dimerization domain.[92,93] This region of C82 may therefore be involved in interactions with other subunits. Of eight different subunit genes tested on multicopy plasmids, only RPC31, which encodes C31, was able to suppress a temperature-sensitive mutation in RPC82.[56] This suggests that a specific and direct interaction occurs between the C82 and C31 subunits. Heptad repeats of leucines have not been observed in C31, but are found in C34 and ABC23.[56] As mentioned above, C82, C34 and C31 can associate with each other in two-hybrid analyses[42] and are released simultaneously upon inactivation of pol III with a conditional mutation in the zinc binding domain of C160.[40] They are also dissociated from the enzyme during native gel electrophoresis.[16] It therefore seems highly likely that C82, C34 and C31 associate in a subcomplex within pol III. An analogous complex is found in human pol III.[20]

A cDNA encoding the human homologue of C82 was cloned from a HeLa cell library using probes designed on the basis of peptide sequence data.[20] The human subunit, C62, has a pI of 8.35 and contains 533 amino acid residues.[20] Phylogenetic conservation is poor, with only 20% identity and 42% similarity between man and yeast.[20] Leucine repeats were not found in human C62.[20]

C53

Part of RPC53, the gene for yeast C53, was obtained by screening an expression library with subunit-specific antibodies.[54] The remainder of the gene was isolated from a genomic library by screening with this partial clone.[61] The 424 amino acid predicted gene product has a pI of 9.2 and is highly hydrophilic throughout its length, except for a 50 residue C-terminal tail.[61] This suggests that C53 lies on the surface of the polymerase—an idea consistent with its relative ease of dissociation[40] and extreme susceptibility to proteolysis.[1,39] The N-terminal 125 amino acids are highly basic, and are followed by alternating clusters of acidic and basic residues.[61] All the essential functions of C53 can be performed by the C-terminal 165 amino acids, although the N-terminal 192 residues increase the stability of this subunit and its affinity for the rest of the polymerase.[94] The temperature-sensitive phenotype of cells containing only the C-terminal domain of C53 can be suppressed by a multicopy plasmid carrying RPC160.[94] This implies a direct interaction between C160 and the C-terminal 165 residues of C53.

The human homologue of C53 is encoded by the BN51 gene.[61,95] The BN51 product has a mass of 44 kD with a high proportion of charged residues (18% basic and 19% acidic).[95] Overall, there is 23% identity between *Saccharomyces* C53 and human BN51.[95] The C-terminal 136 amino acids of C53 and BN51 share 25% identity and 78% similarity, and this region is essential for activity in both cases.[61,95] When introduced into hamster cells carrying temperature-sensitive BN51, yeast C53 can weakly complement the deficiency.[95] BHK cells harboring a temperature-sensitive BN51 mutation downregulate the synthesis of tRNA and 5S rRNA when shifted to the nonpermissive temperature.[95,96] This loss of pol III activity is accompanied by increased transcription of a tubulin gene by pol II and large rRNA genes by pol I.[95] When tested for VA_I transcription, extracts prepared from BN51 mutant cells were more susceptible to heat inactivation than extracts from control cells.[95] Antibodies against BN51 immunoprecipitate pol III but not pols I or II.[96]

The human BN51 gene was isolated originally by its ability to complement the temperature-sensitive G1 phase arrest of a BHK cell mutant.[97] As well as causing a rapid inhibition in tRNA synthesis, temperature-sensitive mutations in *Saccharomyces* C53 also result in preferential G1 arrest.[61,94] The most likely explanation for this striking phenotype is that the loss of pol III activity causes a decrease in protein synthesis. This may result in a failure to accumulate unstable proteins such as cyclins that are required for passage through G1. Low doses of cycloheximide have been shown to inhibit entry into S phase, as has isoleucine deprivation.[98,99] The cell cycle arrest associated with the BN51 mutation is accompanied by a decrease in protein synthesis of only 13%.[97] When induced by cycloheximide, a similar reduction in protein accumulation is sufficient to cause a 5-fold drop in S phase entry.[98]

C34

Part of RPC34, the gene for yeast C34, was cloned by screening an expression library with anti-C34 antibodies.[62] The fragment isolated in this way was then used to obtain the full sequence from a genomic library.[62] RPC34 is a single copy gene that is essential for viability but is not sufficiently conserved to cross-hybridize with fission yeast DNA.[62] Its predicted 317 amino acid gene product displays no significant homology to other sequenced proteins.[62] It is hydrophilic, with 54% polar residues and a calculated pI of 5.5.[62] This suggests that C34 is positioned on the surface of the pol III molecule, an idea supported by the fact that anti-C34 antibodies strongly inhibit tRNA synthesis in vitro.[39] Such antibodies have much less effect upon nonspecific transcription of polyd(A-T), implying that C34 may interact with class III transcription factors.[39] Indeed, C34 has been shown by two-hybrid analysis to interact with the yBRF subunit of TFIIIB in vivo.[42] Direct binding has been demonstrated in vitro using recombinant C34 and yBRF.[44] The C-terminal 120 amino acid residues of C34 are required both for viability[62] and for binding to yBRF.[44] Photocrosslinking reveals that C34 projects upstream from the bulk of pol III in a preinitiation complex.[47,49] It is therefore ideally placed to interact with TFIIIB. C34 is likely to direct pol III to promoter-bound TFIIIB via the C34/C82/C31 subcomplex.[42]

Peptide sequences from purified HeLa cell pol III were used to design probes which allowed the isolation of a cDNA encoding the human homologue of C34.[20] This subunit, C39, has a pI of 6.35 and contains 317 amino acid residues.[20] It is 27% identical and 50% similar to its *S. cerevisiae* homologue.[20] A GST fusion protein containing residues 84-317 binds directly to human BRF and TBP.[20]

C31

Part of RPC31, the gene for yeast C31, was originally isolated by immunological screening of an expression library.[54] This fragment was then used as a probe to screen a genomic library and thereby obtain the full sequence.[59] RPC31 is unique and essential for viability, but may not be highly conserved across species, since no homologue was detected in *S. pombe* by cross-hybridization at low stringency.[59] The 251 amino acid gene product has a pI of 4.5 with a highly acidic C-terminus in which 30 of the last 51 residues are aspartates or glutamates.[59] This acidic tail of C31 is reminiscent of the highly phosphorylated C-terminus of the largest subunit of pol II. Deletion of 22 residues from this C-terminal tail is sufficient to produce a lethal phenotype.[59] Deletion of 16 residues generates a conditional mutant that is impaired in tRNA synthesis in vivo and fails to grow at 37°C.[43] The mutation does not affect the integrity of the polymerase, since its subunit composition is unchanged.[43] The mutant enzyme transcribes nonspecific templates with wild-type efficiency, but is compromised in its ability to carry out accurate transcription of tRNA genes.[43] The mutation inhibits initiation, without affecting elongation, termination or recycling of the polymerase.[43] It may therefore impair the interaction with transcription factors or the formation of an open complex. The conditional growth defect that results from this C-terminal truncation can be suppressed specifically by overexpressing C160.[43] This suggests that C160 and C31 may interact with each other. Increased dosage of C128, C82, C53, AC40, C34, AC19 or ABC10β did not suppress the mutation in C31.[43]

A cDNA encoding the human homologue of C31 was isolated on the basis of peptide sequence data.[20] The human subunit, C32, consists of 233 amino acid residues and has a pI of 4.54.[20] It resembles its yeast equivalent in having an acidic C-terminal tail.[20] However, overall there is only 25% identity and 51% similarity in this polypeptide between man and *S. cerevisiae*.[20]

C25

The gene encoding C25 was cloned by serendipity.[33] It is essential for cell growth and viability, and its inactivation causes a specific reduction in tRNA synthesis in vivo.[33] The predicted gene product contains 212 amino acid residues and displays 50% homology and 25% identity to the RPB7 subunit of pol II.[33] Both C25 and RPB7 are readily dissociable from their respective polymerases and are not required for certain transcription assays in vitro.[33] Despite this strong resemblance, overexpression of RPB7 cannot compensate for the loss of C25 in supporting growth.[33]

Promoters of Subunit Genes

The promoter regions of the subunit genes frequently display homologous sequence motifs which may provide a means for coordinate regulation. Binding sites for the ABF1 transcription factor are found in the promoters of several subunit genes,[33,37,38,56] and ABF1 has been shown to regulate RPC160 and RPC40.[100] An element referred to as the PAC box is often located close to ABF1 sites in the promoters of pol I and III subunit genes, but not in pol II subunit genes.[33,38,56,61] A third motif that occurs frequently upstream of pol I and III subunit genes is the RPG box.[37,38,61,79] This is also associated with the genes for most ribosomal proteins and for elongation factor EF1α, where it is required for efficient expression in vivo.[101-104] The RPG box is recognized by the transcription factor TUF (GRF1/RAP) and has been shown to coordinate the synthesis of the ribosomal proteins.[102,104] Since pols I and III contribute the rRNA components of the ribosome, these promoter sequences may coordinate production of both the RNA and the protein constituents of the translation machinery.

References

1. Sentenac A. Eukaryotic RNA polymerases. CRC Crit Rev Biochem 1985; 18:31-90.
2. Sentenac A, Riva M, Thuriaux P et al. Yeast RNA polymerase subunits and genes. In: McKnight SL, Yamamoto KR, eds. Transcriptional Regulation. Cold Spring Harbor, NY: Cold Spring Harbor Laboratory 1992:27-53.
3. Thuriaux P, Sentenac A. Yeast nuclear RNA polymerases. In: The Molecular and Cellular Biology of the Yeast *Saccharomyces*: Gene Expression. Cold Spring Harbor, NY: Cold Spring Harbor Laboratory 1992; II:1-48.
4. Greenblatt J. RNA polymerase-associated transcription factors. Trends Biochem Sci 1991; 16:408-411.
5. Orphanides G, Lagrange T, Reinberg D. The general transcription factors of RNA polymerase II. Genes Dev 1996; 10:2657-2683.
6. Roeder RG. The role of general initiation factors in transcription by RNA polymerase II. Trends Biochem Sci 1996; 21:327-334.
7. Flores O, Lu H, Killeen M et al. The small subunit of transcription factor IIF recruits RNA polymerase II into the preinitiation complex. Proc Natl Acad Sci USA 1991; 88:9999-10003.
8. Killeen MT, Greenblatt JF. The general transcription factor RAP30 binds to RNA polymerase II and prevents it from binding nonspecifically to DNA. Mol Cell Biol 1992; 12:30-37.
9. Burton ZF, Killeen M, Sopta M et al. RAP30/74: a general initiation factor that binds to RNA polymerase II. Mol Cell Biol 1988; 8:1602-1613.
10. Roeder RG, Rutter WJ. Multiple forms of DNA-dependent RNA polymerase in eukaryotic organisms. Nature 1969; 224:234-237.
11. Kedinger C, Gniazdowski M, Mandel JL et al. α-Amanitin: a specific inhibitor of one of two DNA-dependent RNA polymerase activities from calf thymus. Biochem Biophys Res Commun 1970; 38:165-171.
12. Chambon P. Eukaryotic nuclear RNA polymerases. Annu Rev Biochem 1975; 44:613-635.
13. de Mercoyrol L, Job C, Job D. Studies on the inhibition by α amanitin of single-step addition reactions and productive RNA synthesis catalyzed by wheat-germ RNA polymerase II. Biochem J 1989; 258:165.
14. Schwartz LB, Sklar VEF, Jaehning SJ et al. Isolation and partial characterization of the multiple forms of deoxyribonucleic acid-dependent ribonucleic acid polymerase in mouse myeloma MOPC 315. J Biol Chem 1974; 249:5889-5897.
15. Schultz LD, Hall BD. Transcription in yeast: α-amanitin sensitivity and other properties which distinguish between RNA polymerases I and III. Proc Natl Acad Sci USA 1976; 73:1029-1033.
16. Valenzuela P, Hager GL, Weinberg F et al. Molecular structure of yeast RNA polymerase III: demonstration of the tripartite transcription system in lower eukaryotes. Proc Natl Acad Sci USA 1976; 73:1024-1028.
17. Steinberg TH, Mathews DE, Durbin RD et al. Tagetitoxin: a new inhibitor of eukaryotic transcription by RNA polymerase III. J Biol Chem 1990; 265:499-505.
18. Jaehning JA, Woods PS, Roeder RG. Purification, properties, and subunit structure of DNA-dependent RNA polymerase III from uninfected and adenovirus-2-infected KB cells. J Biol Chem 1977; 252:8762-8771.
19. Wang Z, Roeder RG. TFIIIC1 acts through a downstream region to stabilize TFIIIC2 binding to RNA polymerase III promoters. Mol Cell Biol 1996; 16:6841-6850.
20. Wang Z, Roeder RG. Three human RNA polymerase III-specific subunits form a subcomplex with a selective function in specific transcription initiation. Genes Dev 1997; 11:1315-1326.

21. Sklar VEF, Schwartz LB, Roeder RG. Distinct molecular structure of nuclear class I, II and III DNA-dependent RNA polymerases. Proc Natl Acad Sci USA 1975; 72:348-352.
22. Sklar VEF, Roeder RG. Purification and subunit structure of DNA-dependent RNA polymerase III from the mouse plasmacytoma, MOPC 315. J Biol Chem 1976; 251:1064-1073.
23. Engelke DR, Shastry BS, Roeder RG. Multiple forms of DNA-dependent RNA polymerases in *Xenopus laevis*. Rapid purification and structural and immunological properties. J Biol Chem 1983; 258:1921-1931.
24. Roeder RG. Multiple forms of DNA-dependent RNA polymerases in *Xenopus laevis*. Properties, purification, and subunit structure of class III RNA polymerases. J Biol Chem 1983; 258:1932-1941.
25. Sklar VEF, Jaehning JA, Gage LP et al. Purification and subunit structure of DNA-dependent RNA polymerase III from the posterior silk gland of *Bombyx mori*. J Biol Chem 1976; 251:2794-2800.
26. Gundelfinger E, Saumweber H, Dallendorfer A et al. RNA polymerase III from *Drosophila hydei* pupae. Purification and partial characterization. Eur J Biochem 1980; 111:395-401.
27. Teissere M, Penon P, Azou Y et al. RNA polymerase III from wheat embryos. Purification by affinity and hydrophobic chromatography, characterization and molecular properties. FEBS Lett 1977; 82:77-81.
28. Jendrisak J. Purification and subunit structure of DNA-dependent RNA polymerase III from wheat germ. Plant Physiol 1981; 67:438-444.
29. Hager GL, Holland MJ, Rutter WJ. Isolation of RNA polymerases I, II and III from *Saccharomyces cerevisiae*. Biochemistry 1977; 16:1-8.
30. Wandzilak TM, Benson RW. *Saccharomyces cerevisiae* DNA-dependent RNA polymerase III: a zinc metalloenzyme. Biochemistry 1978; 17:426-431.
31. Buhler J-M, Huet J, Davies KE et al. Immunological studies of yeast nuclear RNA polymerase at the subunit level. J Biol Chem 1980; 255:9949-9954.
32. Breant B, Huet J, Sentenac A et al. Analysis of yeast RNA polymerases with subunit-specific antibodies. J Biol Chem 1983; 258:11968-11973.
33. Sadhale PP, Woychik NA. C25, an essential RNA polymerase III subunit related to the RNA polymerase II subunit RPB7. Mol Cell Biol 1994; 14:6164-6170.
34. Allison LA, Moyle M, Shales M et al. Extensive homology among the largest subunits of eukaryotic and prokaryotic RNA polymerases. Cell 1985; 42:599-610.
35. Sweetser D, Nonet M, Young RA. Prokaryotic and eukaryotic RNA polymerases have homologous core subunits. Proc Natl Acad Sci USA 1987; 84:1192-1196.
36. Memet S, Saurin W, Sentenac A. RNA polymerases B and C are more closely related to each other than to RNA polymerase A. J Biol Chem 1988; 263: 10048-10051.
37. James P, Whelen S, Hall BD. The RET1 gene of yeast encodes the second-largest subunit of RNA polymerase III. Structural analysis of the wild-type and ret1-1 mutant alleles. J Biol Chem 1991; 266:5616-5624.
38. Dequard CM, Riva M, Carles C et al. RPC19, the gene for a subunit common to yeast RNA polymerases A (I) and C (III). J Biol Chem 1991; 266:15300-15307.
39. Huet J, Riva M, Sentenac A et al. Yeast RNA polymerase C and its subunits. Specific antibodies as structural and functional probes. J Biol Chem 1985; 260: 15304-15310.
40. Werner M, Hermann LDS, Treich I et al. Effect of mutations in a zinc-binding domain of yeast RNA polymerase C (III) on enzyme function and subunit association. Mol Cell Biol 1992; 12:1087-1095.

41. Shpakovski GV, Acker J, Wintzerith M et al. Four subunits that are shared by the three classes of RNA polymerase are functionally interchangeable between *Homo sapiens* and *Saccharomyces cerevisiae*. Mol Cell Biol 1995; 15:4702-4710.
42. Werner M, Chaussivert N, Willis IM et al. Interaction between a complex of RNA polymerase III subunits and the 70-kDa component of transcription factor IIIB. J Biol Chem 1993; 268:20721-20724.
43. Thuillier V, Stettler S, Sentenac A et al. A mutation in the C31 subunit of *Saccharomyces cerevisiae* RNA polymerase III affects transcription initiation. EMBO J 1995; 14:351-359.
44. Khoo B, Brophy B, Jackson SP. Conserved functional domains of the RNA polymerase III general transcription factor BRF. Genes Dev 1994; 8:2879-2890.
45. Gundelfinger ED. Interaction of nucleic acids with the DNA-dependent RNA polymerases of *Drosophila*. FEBS Lett 1983; 157:133-138.
46. Bartholomew B, Meares CF, Dahmus ME. Photoaffinity labeling of RNA polymerase III transcription complexes by nascent RNA. J Biol Chem 1990; 265:3731-3737.
47. Bartholomew B, Durkovich D, Kassavetis GA et al. Orientation and topography of RNA polymerase III in transcription complexes. Mol Cell Biol 1993; 13:942-952.
48. Bartholomew B, Braun BR, Kassavetis GA et al. Probing close DNA contacts of RNA polymerase III transcription complexes with the photoactive nucleoside 4-thiodeoxythymidine. J Biol Chem 1994; 269:18090-18095.
49. Persinger J, Bartholomew B. Mapping the contacts of yeast TFIIIB and RNA polymerase III at various distances from the major groove of DNA by DNA photoaffinity labeling. J Biol Chem 1996; 271:33039-33046.
50. Riva M, Schaffner AR, Sentenac A et al. Active site labeling of the RNA polymerases A, B, and C from yeast. J Biol Chem 1987; 262:14377-14380.
51. Geiduschek EP, Tocchini-Valentini GP. Transcription by RNA polymerase III. Annu Rev Biochem 1988; 57:873-914.
52. Huet J, Sentenac A, Fromageot P. Spot-immunodetection of conserved determinants in eukaryotic RNA polymerases. J Biol Chem 1982; 257:2613-2618.
53. Gabrielsen OS, Sentenac A. RNA polymerase III (C) and its transcription factors. Trends Biochem Sci 1991; 16:412-416.
54. Riva M, Memet S, Micouin J-Y et al. Isolation of structural genes for yeast RNA polymerases by immunological screening. Proc Natl Acad Sci USA 1986; 83:1554-1558.
55. Woychik NA, Liao SM, Kolodziej PA et al. Subunits shared by eukaryotic nuclear RNA polymerases. Genes Dev 1990; 4:313-323.
56. Chiannilkulchai N, Stadler R, Riva M et al. *RPC28* encodes the highly conserved, third-largest subunit of RNA polymerase C (III) from *Saccharomyces cerevisiae*. Mol Cell Biol 1992; 12:4433-4440.
57. Young RA. RNA polymerase II. Annu Rev Biochem 1991; 60:689-715.
58. Gudenus R, Mariotte S, Moenne A et al. Conditional mutants of *RPC160*, the gene encoding the largest subunit of RNA polymerase C in *Saccharomyces cerevisiae*. Genetics 1988; 119:517-526.
59. Mosrin C, Riva M, Beltrame M et al. The RPC31 gene of *Saccharomyces cerevisiae* encodes a subunit of RNA polymerase C (III) with an acidic tail. Mol Cell Biol 1990; 10:4737-4743.
60. Lee JY, Evans CF, Engelke DR. Expression of RNase P RNA in *Saccharomyces cerevisiae* is controlled by an unusual RNA polymerase III promoter. Proc Natl Acad Sci USA 1991; 88:6986-6990.

61. Mann C, Micouin J-Y, Chiannilkulchai N et al. *RPC53* encodes a subunit of *Saccharomyces cerevisiae* RNA polymerase C (III) whose inactivation leads to a predominantly G_1 arrest. Mol Cell Biol 1992; 12:4314-4326.
62. Stettler S, Mariotte S, Riva M et al. *RPC34* encodes an essential and specific subunit of yeast RNA polymerase C (III). J Biol Chem 1992; 267:21390-21395.
63. Searles LL, Jokerst RS, Bingham PM et al. Molecular cloning of sequences from a *Drosophila* RNA polymerase II locus by P element transposon tagging. Cell 1982; 31:585-592.
64. Ingles CJ, Himmelfarb HJ, Shales M et al. Identification, molecular cloning, and mutagenesis of *Saccharomyces cerevisiae* RNA polymerase genes. Proc Natl Acad Sci USA 1984; 81:2157-2161.
65. Moenne A, Camier S, Anderson G et al. The U6 gene of *Saccharomyces cerevisiae* is transcribed by RNA polymerase C (III) in vivo and in vitro. EMBO J 1990; 9:271-277.
66. Thuillier V, Brun I, Sentenac A et al. Mutations in the α-amanitin conserved domain of the largest subunit of yeast RNA polymerase III affect pausing, RNA cleavage and transcriptional transitions. EMBO J 1996; 15:618-629.
67. Yano R, Nomura M. Suppressor analysis of temperature-sensitive mutations of the largest subunit of RNA polymerases I in *Saccharomyces cerevisiae*: a suppressor gene encodes the second-largest subunit of RNA polymerase I. Mol Cell Biol 1991; 11:754-764.
68. Himmelfarb HJ, Simpson EM, Friesen JD. Isolation and characterization of temperature-sensitive RNA polymerase II mutants of *Saccharomyces cerevisiae*. Mol Cell Biol 1987; 7:2155-2164.
69. Wittekind M, Dodd J, Yu L et al. Isolation and characterization of temperature-sensitive mutations in *RPA190*, the gene encoding the largest subunit of RNA polymerase I from *Saccharomyces cerevisiae*. Mol Cell Biol 1988; 8:3997-4008.
70. Scafe C, Martin C, Nonet M et al. Conditional mutations occur predominantly in highly conserved residues of RNA polymerase II subunits. Mol Cell Biol 1990; 10:1270-1275.
71. Riva M, Carles C, Sentenac A et al. Mapping the active site of yeast RNA polymerase B (II). J Biol Chem 1990; 265:16498-16503.
72. Jin DJ, Gross CA. Mapping and sequencing of mutations in the *Escherichia coli rpoB* gene that leads to rifampicin resistance. J Mol Biol 1988; 202:45-58.
73. Landick R, Stewart J, Lee DN. Amino acid changes in conserved regions of the β-subunit of *Escherichia coli* RNA polymerase alter transcription pausing and termination. Genes Dev 1990; 4:1623-1636.
74. Bartolomei MS, Corden JL. Localization of an alpha-amanitin resistance mutation in the gene encoding the largest subunit of mouse RNA polymerase II. Mol Cell Biol 1987; 7:586-594.
75. Dieci G, Hermann-Le Denmat S, Lukhtanov E et al. A universally conserved region of the largest subunit participates in the active site of RNA polymerase III. EMBO J 1995; 14:3766-3776.
76. Treich I, Riva M, Sentenac A. Zinc-binding subunits of yeast RNA polymerases. J Biol Chem 1991; 266:21971-21976.
77. James P, Hall BD. *ret1-1*, a yeast mutant affecting transcription termination by RNA polymerase III. Genetics 1990; 125:293-303.
78. Shaaban SA, Krupp BM, Hall BD. Termination-altering mutations in the second-largest subunit of yeast RNA polymerase III. Mol Cell Biol 1995; 15:1467-1478.
79. Mann C, Buhler J-M, Treich I et al. *RPC40*, a unique gene for a subunit shared between yeast RNA polymerases A and C. Cell 1987; 48:627-637.

80. Kolodziej P, Young RA. RNA polymerase II subunit RPB3 is an essential component of the mRNA transcription apparatus. Mol Cell Biol 1989; 9:5387-5394.
81. Lalo D, Carles C, Sentenac A et al. Interactions between three common subunits of yeast RNA polymerases I and III. Proc Natl Acad Sci USA 1993; 90:5524-5528.
82. Song CZ, Hanada K, Yano K et al. High conservation of subunit composition of RNA polymerase I(A) between yeast and mouse and the molecular cloning of mouse RNA polymerase I 40-kDa subunit RPA40. J Biol Chem 1994; 269: 26976-26981.
83. Bell GI, Valenzuela P, Rutter WJ. Phosphorylation of yeast DNA-dependent RNA polymerases in vivo and in vitro. J Biol Chem 1977; 252:3082-3091.
84. Yao Y, Yamamoto K, Nishi Y et al. Mouse RNA polymerase I 16-kDa subunit able to associate with 40-kDa subunit is a homolog of yeast AC19 subunit of RNA polymerases I and III. J Biol Chem 1996; 271:32881-32885.
85. Archambault J, Milne CA, Schappert KT et al. The deduced sequence of the transcription factor TFIIIA from *Saccharomyces cerevisiae* reveals extensive divergence from *Xenopus* TFIIIA. J Biol Chem 1992; 267:3282-3288.
86. Woychik NA, Young RA. Genes encoding transcription factor IIIA and the RNA polymerase common subunit RPB6 are divergently transcribed in *Saccharomyces cerevisiae.* Proc Natl Acad Sci USA 1992; 89:3999-4003.
87. Archambault J, Schappert KT, Friesen JD. A suppressor of an RNA polymerase II mutation of *Saccharomyces cerevisiae* encodes a subunit common to RNA polymerases I, II, and III. Mol Cell Biol 1990; 10:6123-6131.
88. Nouraini S, Archambault J, Friesen JD. Rpo26p, a subunit common to yeast RNA polymerases, is essential for the assembly of RNA polymerases I and II and for the stability of the largest subunits of these enzymes. Mol Cell Biol 1996; 16:5985-5996.
89. Lanzendorfer M, Smid A, Klinger C et al. A shared subunit belongs to the eukaryotic core RNA polymerase. Genes Dev 1997; 11:1037-1047.
90. Woychik NA, Young RA. RNA polymerase II subunit RPB10 is essential for yeast cell viability. J Biol Chem 1990; 265:17816-17819.
91. Carles C, Treich I, Bouet F et al. Two additional common subunits, ABC10 alpha and ABC10 beta, are shared by yeast RNA polymerases. J Biol Chem 1991; 266:24092-24096.
92. Landschulz WH, Johnson PF, McKnight SL. The leucine zipper: a hypothetical structure common to a new class of DNA binding proteins. Science 1988; 240:1759-1764.
93. Abel T, Maniatis T. Action of leucine zippers. Nature 1989; 341:24-25.
94. Chiannilkulchai N, Moenne A, Sentenac A et al. Biochemical and genetic dissection of the *Saccharomyces cerevisiae* RNA polymerase C53 subunit through the analysis of a mitochondrially mis-sorted mutant construct. J Biol Chem 1992; 267:23099-23107.
95. Ittmann M, Ali J, Greco A et al. The gene complementing a temperature-sensitive cell cycle mutant of BHK cells is the human homologue of the yeast *RPC53* gene, which encodes a subunit of RNA polymerase C (III). Cell Growth and Differentiation 1993; 4:503-511.
96. Jackson AJ, Ittmann M, Pugh BF. The BN51 protein is a polymerase (pol)-specific subunit of RNA pol III which reveals a link between pol III transcription and pre-rRNA processing. Mol Cell Biol 1995; 15:94-101.
97. Ittmann M, Greco A, Basilico C. Isolation of the human gene that complements a temperature-sensitive cell cycle mutation in BHK cells. Mol Cell Biol 1987; 7:3386-3393.

98. Brooks RF. Continuous protein synthesis is required to maintain the probability of entry into S phase. Cell 1977; 12:311-317.
99. Rossow PW, Riddle VGH, Pardee AB. Synthesis of labile, serum-dependent protein in early G1 controls animal cell growth. Proc Natl Acad Sci USA 1979; 76:4446-4450.
100. Della Seta F, Treich I, Buhler JM et al. ABF1 binding sites in yeast RNA polymerase genes. J Biol Chem 1990; 265:15168-15175.
101. Leer JR, Van Raamsdonk-Duin M, Mager WH et al. Conserved sequences upstream of yeast ribosomal protein genes. Curr Genet 1985; 9:273-277.
102. Huet J, Cottrelle P, Cool M et al. A general upstream binding factor for genes of the yeast translational apparatus. EMBO J 1985; 4:3539-3547.
103. Woudt LP, Smit AB, Mager WH et al. Conserved sequence elements upstream of the gene encoding yeast ribosomal protein L25 are involved in transcription activation. EMBO J 1986; 5:1037-1040.
104. Vignais M-L, Huet J, Buhler J-M et al. Contacts between the factor TUF and RPG sequences. J Biol Chem 1990; 265:14669-14674.

CHAPTER 4

Transcription Factors Utilized by RNA Polymerase III

Purified pol III initiates transcription randomly.[1-4] Accurate and specific initiation requires the assistance of transcription factors in order to recruit the polymerase to the appropriate start sites of the appropriate sets of genes.[5] Segall et al[6] found that at least two chromatographically separable HeLa cell factors are required to reconstitute specific transcription of tRNA and VA genes. These general factors were designated TFIIIB and TFIIIC, because they elute from phosphocellulose in the PC-B (0.1-0.35 M KCl) and PC-C (0.35-0.6 M KCl) fractions, respectively.[6] (Fig. 4.1) The expression of many other class III templates, such as B1, B2 and vRNA genes, is also supported by a combination of these two fractions, but not by either fraction alone, suggesting that there are similar basal factor requirements for many class III genes.[7-9] In addition to TFIIIB and TFIIIC, transcription of 5S genes requires a gene-specific factor called TFIIIA, that is found in the phosphocellulose flow-through (PC-A) fraction.[6,10] Corresponding fractions have been defined in extracts of cells from mouse,[11] frog,[12] silkworm,[13] fruitfly[14,15] and yeast.[16,17] Phosphocellulose fractions from human and *Xenopus* may be interchanged with one another, implying a conservation of function.[12] However, the corresponding human and *Drosophila* fractions are not compatible.[15,18]

The crude step fractions employed in the early studies have since been found to contain many factors besides TFIIIA, TFIIIB and TFIIIC that can also play a role in pol III transcription. This chapter will describe the factors that are involved in basal expression of class III genes. Factors which regulate the basal transcription machinery will be described in chapter 8. The division between basal and regulatory factors is often vague, and several examples could be justifiably included in either chapter; in such cases, the decision has been made on pragmatic grounds. It is also convenient to categorize factors into gene-specific and general groups. The general factors will be described first.

General Factors

TBP

The most general factor of all is the TATA-binding protein (TBP), which is utilized by pols I, II and III and may therefore be involved in the expression of all nuclear genes.[19-26] This has been shown to be the case in both yeast and man, and is likely to apply to many, if not all, eukaryotes. The role of TBP was originally

RNA Polymerase III Transcription, Second Edition, by Robert J. White.

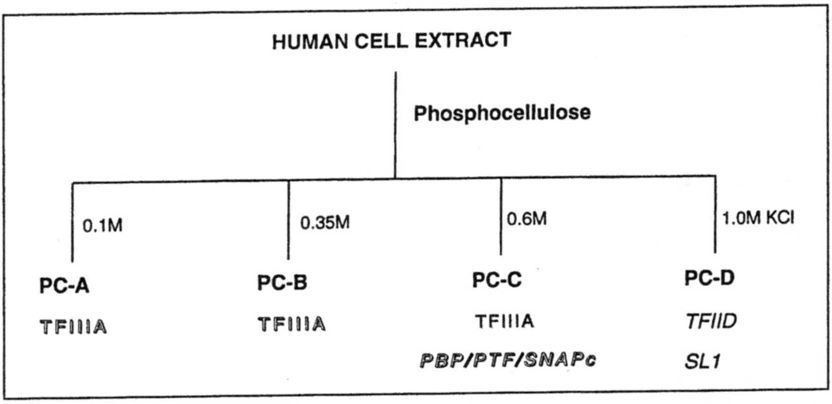

Fig. 4.1. Fractionation properties of various human transcription factors when chromatographed on phosphocellulose. Factors utilized by pol III are shadowed. Those used by pol I or pol II are italicized. KCl concentrations at which the factors elute from phosphocellulose are indicated.

assumed to be restricted to class II genes with TATA boxes in their promoters, and only subsequently was its full importance appreciated (reviewed by White and Jackson[22]).

A distinct TATA-binding factor was originally postulated for pol III transcription, on the grounds that the U6 TATA box is a major determinant of pol III specificity and has similar but distinct sequence requirements from the TATA boxes associated with class II genes.[27-29] However, Margottin et al[30] found that the TATA-binding factor required by a yeast U6 gene copurifies with TBP and can support pol II transcription from the adenovirus major late (AdML) promoter. They also found that cloned TBP can replace this factor in supporting U6 transcription by pol III.[30] This pioneering discovery provided the first example of a basal factor that is utilized by more than one RNA polymerase. Cloned TBP was soon shown to support transcription of a human U6 gene in a variety of reconstitution assays.[31,32]

Perhaps more unexpected was the subsequent discovery that TBP is also required for transcription of class III genes without TATA boxes. This was initially suggested by the observation that TATA box sequences from class II promoters can compete specifically for a human factor that is required by TATA-less class III genes.[33,34] This was shown to be the case for 5S, tRNA, VA, Alu, B1 and B2 genes, as well as U6 and EBER2.[33,34] The competition could be relieved by the addition of pure recombinant TBP.[33,34] Subsequent studies showed that transcription of tRNA and 5S genes in extracts of *Acanthamoeba*,[35] *Bombyx*[36] or *Xenopus*[37] cells is also competed specifically by TATA sequences. Further evidence came from the observation that recombinant TBP restores transcription of TATA-less pol III templates in systems that have been depleted of TBP by fractionation or heat-treatment.[33,34,38] Furthermore, immunodepletion with anti-TBP antibodies inhibits pol III transcription.[39-46] These biochemical experiments are supported by genetic analyses. Expression of U6 and tRNA genes was found to be rapidly inhibited at the nonpermissive temperature in yeast strains with temperature-sensitive TBP mutations.[47] This is not an indirect consequence of inhibiting pol II transcription, since a mutation in the largest subunit of pol II does not produce the same rapid response in class III gene activity.[47] Extracts made from yeast expressing mutant TBP are unable to transcribe tRNA and 5S genes.[48,49] This deficiency can be overcome by the addition of recombinant TBP, thereby proving that the effect of the TBP mutations upon pol III transcription is a direct one.[48,49] In addition, the detrimental effect of TATA box mutations upon U6 expression in yeast extracts or transfected *Arabidopsis* protoplasts can be relieved by compensatory changes in the DNA-binding domain of TBP.[50-52] Overexpression of TBP in *Drosophila* S-2 cells stimulates transcription of U6 and tRNA genes.[53] This is not an indirect effect of increased pol II activity, since a TBP mutant that supports pol II but not pol III transcription has no effect on U6 or tRNA genes when overexpressed in S-2 cells.[53] These combined data establish beyond doubt that TBP is required for transcription of both TATA-containing and TATA-less class III genes in a diversity of organisms.

TBP was originally purified[54-56] and then cloned[57-62] from yeast. It was found to be identical to a previously identified suppressor of Ty insertion mutations.[63] TBP genes have now also been cloned from many other species, including humans,[64-66] mice,[67] frogs,[68] flies,[69,70] plants,[71,72] plasmodia,[73] protozoa[74] and archaebacteria.[75] In most cases there is a single gene for TBP, although *Arabidopsis* and maize have two separate TBP genes.[71,72] No functional difference has been detected between the two TBP genes of plants.[50] In addition to its TBP gene, *Drosophila* has a TBP-related gene of restricted expression and unknown function.[76] All TBP genes encode a small (27-38 kD) polypeptide which has an N-terminal region that is variable in both size and sequence, but which also has a C-terminal domain that is very highly conserved (Fig. 4.2). Indeed, TBP is the most conserved eukaryotic transcription factor. The C-terminal 180 residues of human TBP are 100% identical in mice[67] and frogs[68] and at least 80% conserved in yeast.[62] Large blocks of sequence from the N-terminus of TBP are also well conserved among vertebrates.[77]

Conflicting results have been obtained concerning the significance of the less conserved N-terminal domain of TBP. In yeast, mutations in TBP that affect basal transcription by pols I, II or III all map to the C-terminal domain.[47-49,78-80] Furthermore, the conserved C-terminal region alone is sufficient for yeast cell viability.[79,81-84] This domain has also been shown to support in vitro transcription by pol II in yeast, *Drosophila* and human systems.[66,69,77,78,85-87] Similarly, in the presence of HeLa cell factors the C-terminal 180 residues of human TBP were found to be sufficient for pol I synthesis of rRNA[88] and pol III transcription of tRNA, VA, B2[43] and U6 genes.[86] In contrast, Mittal and Hernandez[77] found that deletion of the N-terminal 96 residues of human TBP abolishes transcription of the human U6 gene. Trivedi et al[53] observed that overexpression in *Drosophila* cells of the C-terminal domain of *Drosophila* TBP activated two TATA-containing pol II templates but not U6 or tRNA genes. One study described a monoclonal antibody against an N-terminal epitope of human TBP that inhibits transcription in HeLa extracts of TATA-containing but not TATA-less templates, regardless of the polymerase.[45] Thus, U6 transcription was repressed by the antibody whereas tRNA synthesis was not.[45] A peptide comprising the epitope recognized by this antibody was also inhibitory to TATA-directed transcription.[45] Another study using HeLa extracts found that antibodies against the N-terminus of TBP can inhibit expression of TATA-less VA, 5S and 7SL genes, as well as U6.[39] In an *Acanthamoeba* system, antiserum that recognizes the TBP N-terminus can inhibit transcription by all three RNA polymerases, including 5S and tRNA expression.[35] It has been suggested that the N-terminal

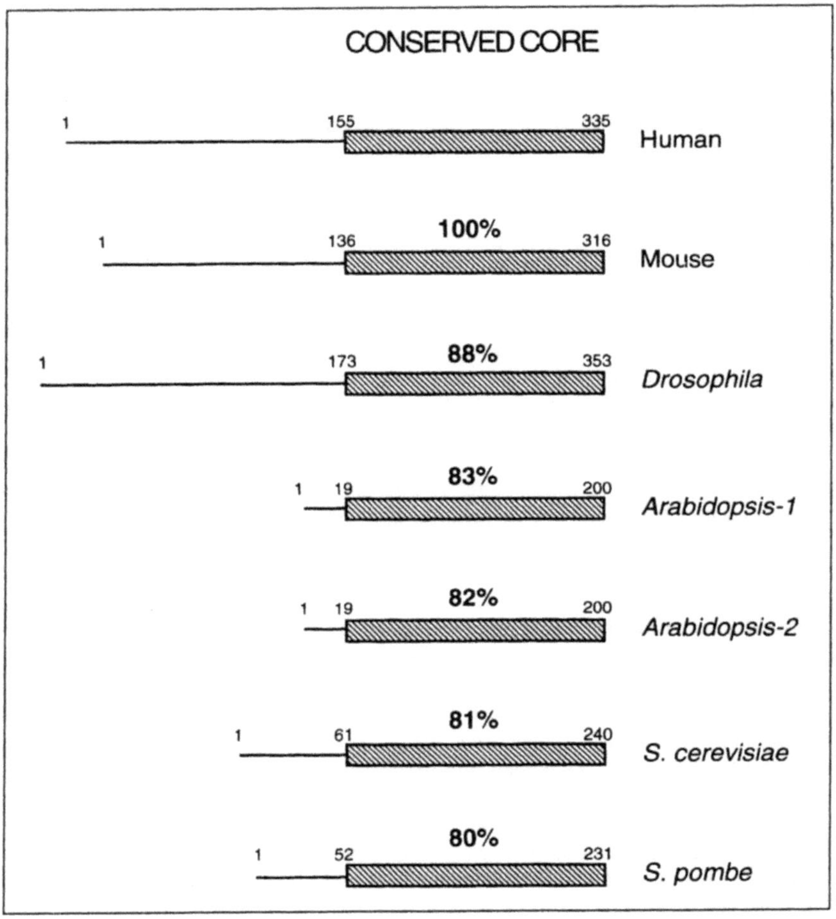

Fig. 4.2. Evolutionary conservation of the C-terminal domain of TBP. The primary structure of TBP molecules from a variety of species is schematized, with the conserved 180 C-terminal amino acid residues shaded. The degree of amino acid identity of the conserved domains relative to the human sequence is shown in bold. The positions of the N- and C-terminal domains within the primary structures are indicated. The N-terminal domains differ considerably in both size and sequence.

region of TBP can modulate the activity of the C-terminal domain.[45,77,89] Indeed, binding to the AdML or U6 TATA boxes is more efficient when the N-terminal 55 residues are removed from TBP.[77,89] In the case of U6, recruitment of full-length TBP is facilitated by cooperative interactions with the factor occupying the PSE; this effect requires sequences within the N-terminal 96 residues.[77]

The C-terminal domain of TBP contains two direct repeats of 66-67 residues that are nearly 40% identical and flank a short basic region (Fig. 4.3). Point mutagenesis has demonstrated that the basic region is involved in transcription by pols I, II and III.[80,90] However, individual substitutions within this region can have

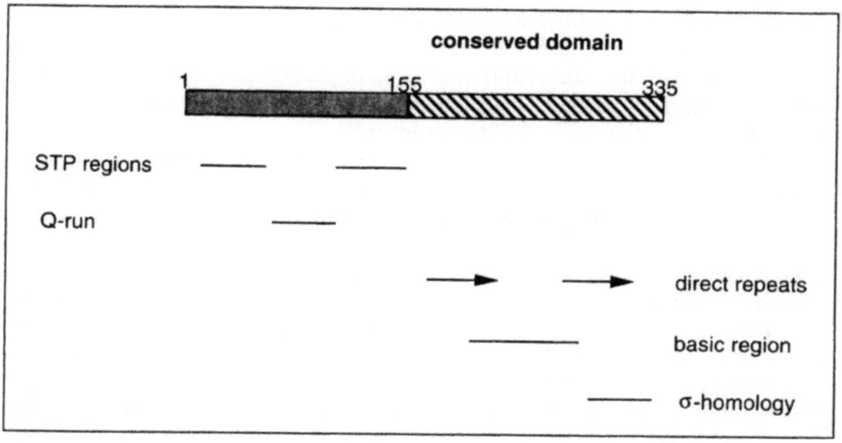

Fig. 4.3. Sequence features of human TBP. The N-terminal 155 residues that are not required for transcription are drawn shaded. This region includes a run of 34 consecutive glutamines (Q-run). The Q-run is flanked by two regions that are rich in serine, threonine and proline residues (STP regions). The essential C-terminal 180 amino acids are drawn striped. This domain includes an imperfect direct repeat of 77-78 amino acid residues. A region that is rich in basic residues separates and partially overlaps the direct repeats. A region with homology to prokaryotic sigma factors lies near the C-terminus.

differential effects upon the three polymerases, suggesting that the sites of interaction may be overlapping but distinct.[80,90] Mutational analyses localized the TATA box-binding function to the direct repeats of TBP.[50,83,91-93] TATA binding occurs through contacts in the minor groove[94,95] and induces DNA bending[89,96] TATA binding is slow, temperature-dependent and is discouraged by the nonconserved N-terminus of the TBP polypeptide.[77,89,97,98]

The structure of *Arabidopsis* TBP was determined by X-ray crystallography.[99] The C-terminal domain folds into a highly symmetrical α/β structure resembling a saddle that can sit astride DNA.[99] It is divided into two 88-89 amino acid regions which are only 31% homologous in sequence but are topologically identical, with the α-carbon backbones following the same path in space.[99] The concave underside of the saddle is a curved, 10-stranded, antiparallel β-sheet that is wide enough to accommodate the DNA helix.[99] All of the point mutations that have been shown to affect DNA-binding by TBP map to this concave region of the saddle.[99] In contrast, the convex surface of the saddle is lined with α-helices, and point mutations that specifically affect the interaction of TBP with other proteins map to this region.[99] X-ray crystallography has also revealed the structures of *Arabidopsis* or *Saccharomyces* TBP when bound to TATA DNA sequences.[100,101] These cocrystals demonstrated a highly localized and dramatic distortion of the eight base pairs of the TATA element.[100,101] The incoming and outgoing duplexes are sharply angled (by 100°) and displaced (by 1.8 nm) with respect to each other.[100,101] TBP induces 90° kinks at each end of the TATA box and drastic unwinding in between.[100,101] The TATA sequence is bent severely towards the major groove, allowing a widened shallow minor groove to contact most of the concave surface of TBP.[100,101] The TA dinucleotide is easily deformable and relatively unstable. Furthermore, few functional groups are accessible from the minor groove. Thus, the ability of a sequence to suffer the necessary distortion may be as important as base-specific interactions for recognition by TBP. This might explain how TBP is able to bind sequences that bear little relation to the TATA consensus.[102]

TBP fractionates extremely heterogeneously when an extract is chromatographed and can be detected in multiple fractions, suggesting that it may exist in a variety of different complexes, each with distinct chromatographic behavior.[42,44,86,103-107] Indeed, it has been suggested that all of the TBP in a HeLa extract exists in complexes with other proteins.[103] The polypeptides that interact with TBP in such complexes are referred to as TBP-associated factors (TAFs) (reviewed in refs. 19,21-24,26,108,109). Several different TBP-TAF complexes have been described. One of these is TFIID, which contains approximately 10 TAFs and supports transcription by pol II.[19,21,23,26,108,109] Another TBP-containing complex is SL1 (also called TIF-IB), which contains 3 TAFs and is required by pol I.[110-112] Both TFIID and SL1 elute from phosphocellulose in the high salt PC-D fraction.[110,113] (Fig. 4.1) However, pol III transcription requires a TBP-containing complex that elutes in PC-B.[39,41-44,105,114] Thus, VA_I transcription in an extract that has been immunodepleted using anti-TBP antibodies can be reconstituted by the addition of PC-B, but not by PC-C or PC-D.[42,43] Furthermore, a PC-B fraction that has been immunodepleted of its TBP-containing complexes is unable to support transcription of class III genes, whereas PC-C that has been treated in this way is not compromised.[39,41,42,44] Pol III transcription of the adenovirus E2E gene also requires a form of TBP that is present in PC-B but not in PC-D.[40] There are at least two distinct TBP-containing complexes in PC-B.[43,44] One of these is called B-TFIID and may have a role in basal pol II transcription.[44,103,115] The second is TFIIIB, as shown by its ability to reconstitute transcription of a variety of class III genes in the presence of PC-C.[39,42-44] TBP cofractionates with TFIIIB through a variety of chromatographic procedures[30,37,39,41,43,44,114,116,117] and TFIIIB can be affinity-purified by virtue of its association with TBP.[37,42,43,105,106,117] Therefore, there is at least one TBP-containing complex for each of the nuclear RNA polymerases and transcription by a given polymerase requires the appropriate complex.[39,41-43,86,105,110]

The TAFs present in each complex appear to be different in each case.[42,105,110] The TAFs may influence the DNA-binding specificity of TBP. This could explain why pol II and pol III have different preferences for particular TATA sequences.[50,51,118-120] It has even been suggested that the TAFs present in TFIIIB may prevent TBP from associating with a class II promoter.[42,105] This is probably not the case, since oligonucleotides containing TATA sequences from class II promoters compete for pol III transcription.[33-37] Indeed, one study concluded that TFIIIB is permissive for pol II.[44] However, the pol I TAFs do prevent TBP from binding canonical TATA boxes.[35,121] Such observations have lead to a model in which TBP achieves its polymerase specificity by associating with distinct sets of TAFs in order to assemble polymerase-specific TBP-containing complexes.[19-24] (Fig. 4.4) The TBP-containing complex that is required for pol III transcription is TFIIIB.

TFIIIB

TFIIIB is the central class III initiation factor since it alone is sufficient to recruit the polymerase and specify the site at which transcription begins.[122] Klekamp and Weil[123] purified TFIIIB from yeast by ion-exchange chromatography. A 60 kD

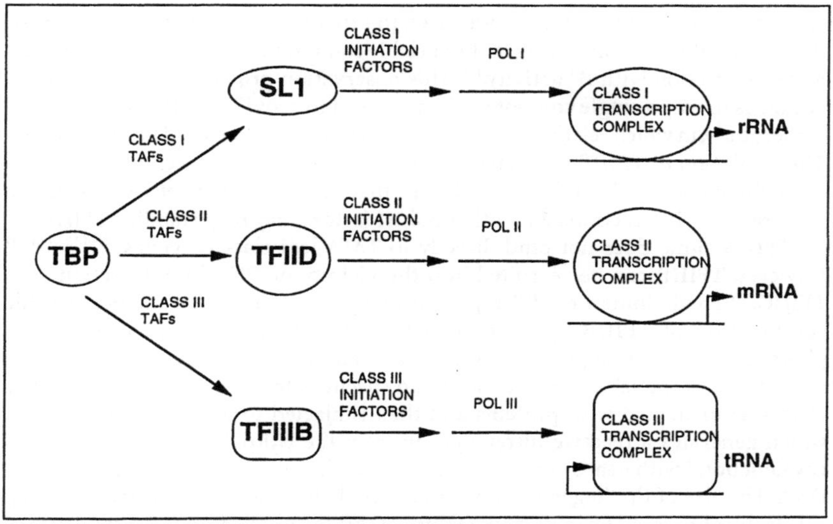

Fig. 4.4. Schematic model of TBP assembly into class-specific transcription complexes. TBP first associates with alternative sets of TAFs to generate distinct complexes (SL1, TFIID or TFIIIB). These, in turn, associate with the appropriate combinations of initiation factors at promoters to assemble class-specific complexes which serve to recruit the appropriate RNA polymerase to the appropriate sets of genes.

polypeptide constituted 30% of the total protein of the most highly purified fraction.[123] This preparation was capable of reconstituting tRNA transcription in the presence of partially purified TFIIIC and pol III.[123] An antibody raised against the gel-fractionated 60 kD polypeptide inhibited tRNA transcription when preincubated with a complete extract, whereas preimmune serum did not.[123] The inhibitory effect of the antibody was blocked by preincubation with a sample of the 60 kD protein that had been purified by two-dimensional gel electrophoresis.[123] This antibody also inhibited tRNA transcription in a human system.[123] The 60 kD yeast polypeptide is very glycine-rich and highly asymmetric, with dimensions of about 2.2 x 40 nm.[124] It was not possible to reconstitute transcription by renaturing this polypeptide after gel-purification to homogeneity.[123] Furthermore, 100-150 molecules of 60 kD polypeptide were required per template molecule.[123] Although this may merely reflect inactivation during purification, it is more likely that additional components are required for TFIIIB activity and that the bulk of these were removed during fractionation.

When Kassavetis et al[125] purified yeast TFIIIB using a protocol modified from that of Klekamp and Weil,[123] a 60 kD species was not a major constituent of their most highly purified fraction, which had undergone five chromatographic steps and yet still showed ~30 bands in an SDS-polyacrylamide gel. Nevertheless, this TFIIIB preparation had up to 60-fold greater specific activity than that of Klekamp and Weil.[125] Further chromatography on a cation exchange resin (Mono S) split this fraction into two components, termed B' and B".[126] Both of these fractions are necessary for tRNA transcription in the presence of purified TFIIIC and pol III.

However, when a crude fraction was used to provide the TFIIIC and polymerase, then B" was dispensable, owing to the variable contamination of the crude complementing fraction with B" activity.[126] This observation provided a possible explanation as to why a single component had sufficed to provide TFIIIB activity in the earlier study, in which relatively crude complementing fractions had been employed. The resolution of TFIIIB into separate components during ion exchange chromatography under mild conditions indicates that the complex is relatively labile in solution.[126,127] Interactions with other pol III factors may help stabilize TFIIIB.

TFIIIB alone does not bind directly to TATA-less class III genes.[18,38,123,128-132] However, TFIIIB can be recruited into the vicinity of DNA by interaction with TFIIIC.[125] Bartholomew et al[133] exploited this property to probe the polypeptide composition of TFIIIB by photocrosslinking. This involved incorporating a photoactive nucleotide [N-p-azidobenzoyl)-3-aminoallyl]-deoxyuridine (N_3RdU) and adjacent radiolabeled nucleotides into specific sites within a promoter.[133] The DNA was incubated with proteins and then subjected to ultraviolet irradiation, which generates a reactive nitrene group from the N_3RdU. The nitrene forms a covalent bond with nearby polypeptides, thereby "tagging" them with radiolabeled DNA. The size of the "tagged" polypeptide can then be determined by denaturing gel electrophoresis. The important feature of this technique is that the structure of N_3RdU puts the reactive nitrene on a 0.9 to 1.0 nm tether and this allows the space outside the DNA helix to be probed for nearby proteins.[133] Other photoactive reagents have since been used to probe distinct spatial volumes around a promoter.[134,135]

This approach allowed the detection of two polypeptides (of 70 and 90 kD) within the TFIIIB fraction that are recruited to the tRNA promoter by interaction with TFIIIC.[136] These polypeptides represented only minor components of the TFIIIB preparation.[126] The specificity of crosslinking was demonstrated by its dependence upon both a functional promoter and the presence of TFIIIC.[136] The 70 and 90 kD polypeptides display distinct V8 protease digestion patterns, indicating that one is not a simple degradation or modification product of the other.[136] The 70 kD polypeptide was found in the B' Mono S fraction, whereas the 90 kD polypeptide was found in the B" fraction.[126] Bartholomew et al suggested that the previously reported 60 kD polypeptide may be (substantially) identical to their 70 kD component or a degradation product of their 90 kD component.[136] However, antibodies against the 60 kD polypeptide[123] did not recognize the 70 kD B' subunit.[137]

The gene encoding the 70 kD subunit of yeast TFIIIB was cloned as a suppressor which, at high copy number, can rescue certain temperature-sensitive mutations in TBP.[137,138] It was also isolated independently as a suppressor of a mutation in the A-block of a tRNA gene; this demonstrates that an impaired interaction between TFIIIC and the A-block can be compensated by an elevated concentration of TFIIIB.[139] The gene has been variously named BRF1,[138] TDS4,[137] and PCF4.[139] I shall refer to the gene as BRF1 and its protein product as yBRF. Mutations in BRF1 showed it to be essential for growth.[137,138] Disruption of BRF1 resulted in a rapid decline in the in vivo expression of tRNA, but not of class I and II genes.[137,139] Extracts made from BRF1 mutants were defective in transcription by pol III, but were wild-type for transcription by pols I and II.[138] Transcription of tRNA and 5S genes in BRF-deficient extracts could be reconstituted by the addition of recombinant yBRF.[138] Recombinant yBRF also stimulates class III transcription in wild-type extracts, indicating that it is normally a limiting factor.[138,139] Furthermore, yBRF has also been shown to be limiting for tRNA transcription in vivo.[139]

BRF1 encodes a protein of 596 amino acids, with a predicted molecular mass of 67 kD and a pI of ~6.9.[137-139] A potential zinc finger with the sequence Cys-X_2-Cys-X_{17}-Cys-X_2-Cys occurs at the N-terminus (residues 4-28).[137,139] The C-terminal half of yBRF has no obvious similarity to other known proteins.[137-139] It is, however, highly charged, with 21% acidic residues (glutamate or aspartate), 17% basic residues (lysine or arginine), and another 14% serine and threonine.[137,139] The N-terminal half of the protein is homologous to the general pol II initiation factor TFIIB.[137-139]

TFIIB is a ubiquitous factor required for pol II transcription. It has been shown to interact with promoter-bound TBP and also with pol II, and is therefore thought to serve as a bridge between TBP and the pol II enzyme during preinitiation complex assembly.[140,141] Human TFIIB is a single polypeptide of 33 kD.[142] TFIIB genes have been cloned from many species, including human,[142,143] *Xenopus*,[144] *Drosophila*,[145,146] and *S. cerevisiae*.[147] The TFIIB proteins vary in size between 316 and 345 amino acids,[142-147] and are therefore much smaller than BRF. The human, *Xenopus* and *Drosophila* proteins are at least 79% identical, whereas yeast TFIIB (SUA7) is only 35% homologous to the human factor. This divergence between yeast TFIIB and that of higher organisms may reflect differences in the mechanisms employed in initiation site selection.[145,147] TFIIB resembles TBP in containing clusters of basic residues and a large imperfect repeat. These similarities may reflect a common origin for TFIIB and TBP. NMR[148] and X-ray crystallography[149] have revealed that each repeat of TFIIB consists of five α-helices packed together into a compact globular domain. The structure of these domains bears strong similarity to that of cyclin A.[148,149] The N-terminus of TFIIB has been shown by NMR to adopt a zinc-ribbon structure.[150]

The N-terminal 320 amino acids of yBRF are 23% identical to either human or yeast TFIIB, with an overall similarity of 44% when allowing for conservative changes.[137-139] There is only one large gap in the alignment, perhaps indicating a nonconserved domain.[137] However, yBRF cannot substitute for TFIIB in supporting pol II transcription in a fractionated yeast system.[138] Three regions are conserved between yBRF and the various TFIIB proteins.[137-139] One consists of 19 residues near the N-termini, which includes the putative zinc ribbon, and the other two contain imperfect direct repeats of 76 amino acids.[137-139] (Fig. 4.5) The repeat regions of yBRF are 25% and 28% identical to the corresponding regions of human TFIIB.[139] These repeats constitute a TR domain, which is also found in the cyclin box of cyclins and the pocket B region of RB and its relatives.[151] Clusters of basic residues that overlap the direct repeats of TFIIB have been predicted to form amphipathic helices.[143,145] The corresponding regions of yBRF contain fewer basic amino acids and are not predicted by helical wheel analysis to fold into similar structures.[139] TFIIB contains a region with homology to a highly conserved domain of prokaryotic sigma factors that is thought to contact the RNA polymerase.[142-145,147] However, the conserved residues of this sigma homology are largely absent from the corresponding yBRF sequence. The direct repeat region of TFIIB has been shown to contact TBP and RNA polymerase directly,[152-155] and this is also the case for yBRF.[156] The zinc ribbon region of TFIIB has been shown to bind TFIIF[154]

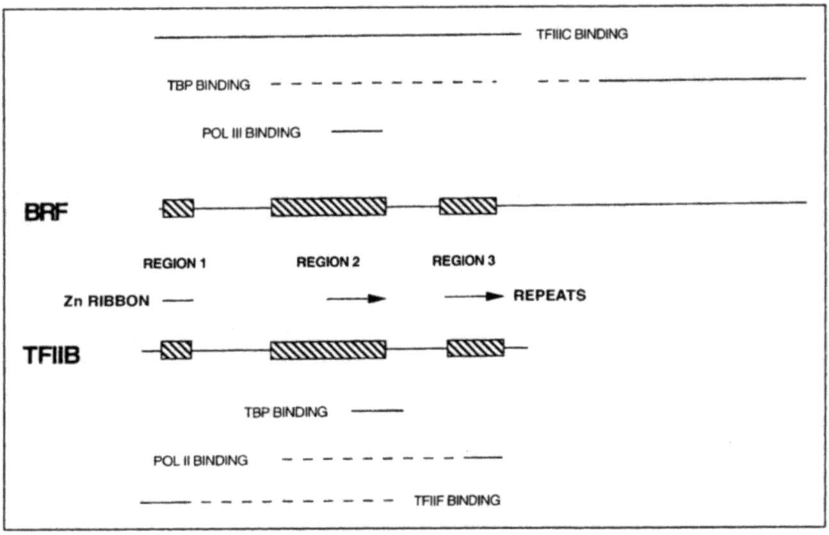

Fig. 4.5. Sequence features of yeast BRF and human TFIIB. The three regions of significant homology are drawn striped. The locations of the zinc ribbon and imperfect direct repeats shared by both polypeptides are indicated. Also indicated are the regions of TFIIB that have been shown to be required for interactions with TBP, pol II and TFIIF, and the regions of yBRF that have been shown to be required for interactions with TBP, pol III and TFIIIC.

and has been implicated as a target for glutamine-rich activation domains.[157] Since TFIIF does not appear to be involved in pol III transcription,[158] this region may have been conserved in yBRF as a target for transcriptional activators.

Mutation of the N-terminal 25 amino acids of yBRF, including half of the putative zinc ribbon, does not affect viability.[138] However, deletion of residues 1 to 40, which completely removes the first region of homology with TFIIB, is lethal.[138] Colbert and Hahn[138] found that a strain in which the C-terminal 50 amino acids of yBRF had been deleted was still viable, although loss of the last 100 residues was lethal. In contrast, Buratowski and Zhou[137] reported that deletion of the last 35 amino acids completely abolishes function in vivo.

In order to look for conserved sequences, Khoo et al[156] cloned yBRF from two yeast species that are evolutionarily distant from *S. cerevisiae*. Low stringency hybridization allowed the isolation of yBRF from *Kluveromyces lactis*, which diverged from *S. cerevisiae* ~10^8 years ago.[156] A PCR approach using degenerate primers was required to clone yBRF from *Candida albicans*, an even more phylogenetically remote yeast.[156] Despite the evolutionary distance, *C. albicans* yBRF could complement a temperature-sensitive mutation in *S. cerevisiae* yBRF.[156] The BRF genes from *K. lactis* and *C. albicans* encode polypeptides of 556 and 553 amino acid residues, respectively.[156] The *K. lactis* and *C. albicans* yBRFs are 62% and 47% identical to that of *S. cerevisiae*, and each contains the putative zinc ribbon followed by the imperfect direct repeats. A three way comparison of the regions of homology is shown in Figure 4.6. The sequence similarities are reflected in very similar hydr-

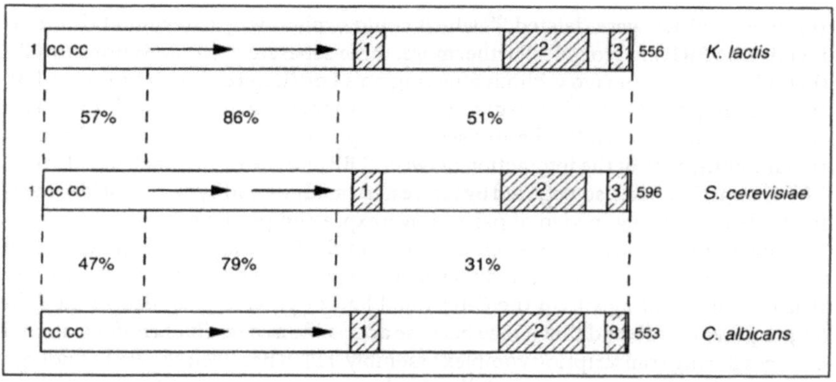

Fig. 4.6. Schematic representation of the sequence conservation of yBRF from the yeast species *K. lactis*, *S. cerevisiae* and *C. albicans*. Percentages refer to amino acid identities in the regions indicated. CC CC indicates the putative zinc ribbon. The arrows indicate direct repeats. C-terminal homology blocks 1, 2 and 3 are indicated.

opathy profiles and secondary structure predictions.[156] The most highly conserved region encompasses the two direct repeats, which are 73% identical in the three yBRFs.[156] The C-terminal region that is not seen in TFIIB is also well maintained, although several gaps must be introduced to optimize the alignment.[156] The most striking feature of this region is its high charge density; for example, in *K. lactis* yBRF, 20% of residues in this region are acidic and 19% are basic.[156] Three areas of strong conservation occur within this region; the longest is the second, which covers 77 residues and is 80% identical between *S. cerevisiae* and *K. lactis* and 43% identical between *S. cerevisiae* and *C. albicans*.[156]

Western immunoblotting analysis detects both yBRF and TBP in the B' fraction of TFIIIB, but not in purified TFIIIC or pol III preparations.[116,137,138] Furthermore, antibodies against either recombinant TBP or recombinant yBRF will specifically supershift a TFIIIB/tDNA complex in a gel retardation assay, indicating that both polypeptides are present in this functional complex.[116,137] A TFIIIB/tDNA complex containing human TBP migrates more slowly than one containing yeast TBP, consistent with the greater size of the human TBP polypeptide; this provides additional evidence for the presence of TBP in the TFIIIB/tDNA complex.[38] Recombinant TBP has been shown to bind directly to recombinant yBRF.[127,156,159] Furthermore, the allele specific suppression of TBP mutations by BRF1 suggests that TBP and yBRF interact directly in vivo. Kassavetis et al[116] showed that recombinant TBP and recombinant yBRF together are necessary and sufficient to reconstitute all of the known properties of the B' fraction in transcription, gel retardation, footprinting and photocrosslinking assays. These observations provide strong evidence that TBP and yBRF interact functionally to form part of TFIIIB.

Khoo et al[156] used GST pulldown assays to demonstrate that two separate regions of yBRF can bind to TBP. One of these maps to the direct repeat region; this was expected since the direct repeats of TFIIB also bind TBP.[152,154,155] Quite unexpected, however, was the discovery that TBP binds with even higher affinity to the C-terminal region of yBRF.[156] This interaction was prevented if the C-terminal 165 residues of yBRF were deleted,[156] which could explain why C-terminal deletions inactivate yBRF in vivo.[137,138] Furthermore, three separate point mutations in TBP that inhibit pol III in vivo without affecting pols I or II, were found to prevent TBP from binding the C-terminal region of yBRF, whereas association with the direct repeats was unimpaired.[156] These observations provide strong support for the functional significance of the interaction between TBP and the C-terminal half of yBRF. This interaction was also detected by far western analysis and by two-hybrid screening to detect binding in vivo of pairs of overexpressed proteins.[160]

Intact yBRF binds TBP less efficiently than the C-terminal region alone, and with comparable affinity to the direct repeat region.[156] This raises the possibility that the TBP-binding site in the C-terminal half of yBRF is occluded in the full-length polypeptide, and that it may become accessible as a result of conformational changes during transcription complex assembly. TFIIB has been shown to undergo conformational rearrangements following interaction with other factors.[161,162]

B" activity was found to comigrate with the 90 kD polypeptide in an SDS-polyacrylamide gel.[116] A crude B" fraction was boiled in SDS, size-fractionated under denaturing conditions, the gel was cut into slices, and the proteins were eluted and then renatured.[116] All of the transcriptional and DNA-binding properties of B" were recovered from the gel slice corresponding to molecules of 85-95 kD.[116] This suggested that the 90 kD polypeptide identified by photocrosslinking is the sole active component of B", although other polypeptides of a similar size could not be formally excluded.

Definitive evidence that the 90 kD subunit is responsible for the B" activity came with the cloning of its gene.[163-165] Two groups independently obtained peptide sequences from highly purified B" and used these to design primers which allowed PCR amplification of genomic DNA encoding the 90 kD polypeptide.[163,164] The gene was also isolated as a multicopy suppressor of temperature-sensitive mutations in the two largest subunits of TFIIIC, τ_{138} and τ_{131}.[165] It was named TFC5 and has been shown to be essential for viability.[163-165] It encodes a 594 amino acid protein with a calculated mass of ~68 kD that migrates anomalously as a ~90 kD band on SDS/PAGE.[163-165] Between residues 415 and 472, the gene product displays homology to the DNA-binding domain of Myb.[166] The shared motif is also found in N-CoR, SWI3 and ADA2, and has been referred to as the SANT domain.[166] Outside this region, there is no significant similarity to other known proteins apart from a glutamate-rich stretch between residues 329 and 357.[163,164] Recombinant B" was able to replace purified yeast B" in DNA-binding and transcription assays with either tRNA or U6 genes.[163-165]

B" can retain its activities when extensively truncated.[163,167] A core domain retaining only 176 amino acids (residues 224-400) is sufficient to support U6 transcription, whereas a larger region is required for tRNA synthesis.[167] Two distinct domains (residues 270-305 and 390-460) that are both necessary for tRNA expression can function on an either/or basis in the case of U6.[167] Protein footprinting experiments suggest that B" is folded such that these two domains are in close proximity when TFIIIB is assembled onto DNA.[167] The remarkable resistance of B" to deletion mutagenesis may be accounted for if the multiple protein/protein and protein/DNA interactions that are made by this polypeptide in the transcription complex allow it to compensate for the loss of any individual contact.

Recombinant B" binds weakly to TBP alone and 25-fold more efficiently to a complex of TBP plus yBRF; no interaction was seen between B" and yBRF in the absence of TBP.[164] Prior binding of TFIIB to a TBP/TATA complex prevents the association of B".[164] This suggests that B" contacts surfaces of TBP and/or DNA that overlap regions which interact with TFIIB. An alternative explanation is that TFIIB induces conformational changes in TBP/DNA that prevent the binding of B"; this seems improbable since crystallography has revealed no such rearrangements.[149] TFIIB contacts the C-terminal stirrup of TBP, helix H1' and the C-terminus of TBP, as well as DNA both upstream and downstream of the TATA box.[149] B" is likely to interact with one or more of these sites.

TFIIIB that has been reconstituted from recombinant TBP, yBRF and B" is capable of supporting both TFIIIC-dependent and TATA-dependent DNA binding and transcription.[163-165] This would suggest that these three polypeptides represent the sole essential components of yeast TFIIIB. However, the factor reconstituted from recombinant TBP, yBRF and B" was found to be less active than native TFIIIB.[163,165] For example, Kassavetis et al found that completely recombinant TFIIIB is somewhat less active in recruiting pol III than is TFIIIB containing natural B".[163] It could be that the recombinant polypeptides are not modified or folded correctly when expressed in bacteria. Alternatively, some component(s) may be missing from the reconstituted system. The missing component(s) could be inessential but stimulatory, or essential and present as a contaminant in the complementing fractions used in these assays. A potential candidate for a stimulatory subunit is a 17 kD polypeptide that interacts specifically with the SUP4 tRNA promoter in association with TFIIIB, but is not stably bound and is not required for transcription.[168] Alternatively, the putative missing component may be a factor that is distinct from TFIIIB. Ruth et al[165] found that transcription reconstituted using TBP, yBRF and B" could be raised to high levels by the addition of a fraction containing TFIIIE. TFIIIE has been shown to be present in both native B' and B" fractions.[169,170] It is therefore a prime candidate for the component that is deficient when fully recombinant TFIIIB is used. Even if TBP, yBRF and B" are sufficient to support efficient basal transcription in the presence of TFIIIC and TFIIIE, then the possibility remains that native TFIIIB associates with additional regulatory subunits that are not involved in basal transcription. In the pol II system, the TFIID TAFs perform regulatory functions that are dispensable for the transcription of many class II genes.[109,171]

TAFs may influence the DNA-binding specificity of TBP. The clearest example of this is in the pol I system, where SL1 TAFs prevent TBP from binding canonical TATA boxes.[35,121] More subtle effects could explain why pol II and pol III have different preferences for particular TATA sequences.[50,51,118-120,172,173] It has even been suggested that the TAFs present in TFIIIB may prevent TBP from associating with a class II promoter.[42,105] Several studies contradict this, since oligonucleotides containing TATA sequences from class II promoters compete for pol III transcription.[33-37] Roberts et al[174] showed that a TATA/TBP complex interacts equally well with yBRF and TFIIB, regardless of whether the TATA sequence came from a pol II or a pol III template. yBRF and B" can block the binding of TFIIB or TFIIA to TBP, and vice versa.[164] This implies that the binding sites for yBRF and B" on TBP overlap with those for TFIIB and TFIIA.

A putative gene encoding a homologue of yBRF has been identified in *Caenorhabditis elegans*.[175] The predicted product, CeBRF, is 26% identical and 35% similar to yBRF from *S. cerevisiae*.[175] However, it is much longer than the yBRFs, having 759 amino acid residues.[175] The strongest homology between these two proteins is within the N-terminal 270 residues, which are 38% identical and 48% similar.[175] As in yeast, CeBRF contains a putative zinc ribbon near its N-terminus.[175] In CeBRF, the zinc-binding motif has a spacer length of 15 residues (Cys-X_2-Cys-X_{15}-Cys-X_2-Cys); this is the same as TFIIB, but is shorter than the spacer found in each of the available yBRF sequences (Cys-X_2-Cys-X_{17}-Cys-X_2-Cys).[175] Like the other BRF proteins, CeBRF contains an imperfect direct repeat that constitutes a TR domain.[175] Of the three areas of strong homology that are shared by the yBRFs in their C-terminal regions,[156] only the second and third are conserved in *C. elegans*.[175] These are clearly likely to have functional significance, and may be involved in binding TBP. The C-terminal extension of CeBRF that is absent from the yBRFs contains an imperfect direct repeat of 46 and 48 residues that is rich in proline and lysine.[175]

Gottesfeld and colleagues have purified TFIIIB from extracts of *Xenopus* eggs.[37] Following ion-exchange chromatography, TFIIIB activity was purified further by native gel electrophoresis in association with a TATA-containing oligonucleotide.[37] The TATA-bound complex was found to contain predominant polypeptides of 75 and 92 kD in equal stoichiometry.[37] *Xenopus* TFIIIB was also fractionated by electrophoresis in SDS gels, and polypeptides were renatured from excised gel slices; species of around 75 and 92 kD were able to reconstitute tRNA transcription in the presence of PC-C and recombinant TBP.[37] The same renatured species were found to stabilize the interaction of TBP with a TATA box, as determined in band shift assays.[37] These data suggest that TFIIIB from *X. laevis* is minimally composed of TBP plus two TAFs, as is the case in *S. cerevisiae*.

The composition of mammalian TFIIIB is far less well characterized (reviewed by Rigby[24]). Seifart's group first purified TFIIIB from HeLa cells using essentially the same procedure as had been used in yeast, involving a series of ion-exchange columns.[130] A 60 kD polypeptide was found to constitute 90% of the protein in the most highly purified fraction.[130] This preparation could reconstitute transcription of tRNA, VA, and 5S genes in the presence of PC-C and PC-A fractions.[130] Furthermore, TFIIIB activity was found to sediment at around 60 kD following glycerol gradient centrifugation of either crude or highly purified human fractions.[130] However, it is now clear that human TFIIIB, like its counterpart in yeast, forms a complex with TBP in solution. This has been shown by the ability of anti-TBP antibodies to immunoprecipitate TFIIIB activity under nondenaturing conditions.[39,42-44,105,117,176] Furthermore, human TFIIIB will specifically bind to the conserved core domain of TBP when immobilized on an inert matrix.[43,117] Additional evidence for a stable association between these factors in higher organisms comes from the cofractionation of TFIIIB with a subpopulation of TBP molecules during a variety of chromatographic procedures.[39,41-44,114,177] It therefore seems likely that the 60 kD polypeptide originally identified is, at best, only one component of a larger TFIIIB complex.

Several groups have reported that human TFIIIB will split into two components during gradient chromatography on Mono Q.[39,44,105] Lobo et al[39] found that these eluted at 380 mM and 480 mM KCl and named them 0.38M-TFIIIB and 0.48M-TFIIIB, respectively. TBP was detected in the former, but not in the latter fraction.[39] The 0.48M-TFIIIB fraction was sufficient to reconstitute VA_I transcription in combination with PC-C.[39] However, if PC-C was first immunodepleted of its endogenous TBP, then the 0.38M-TFIIIB fraction also had to be included.[39] This indicates that 0.38M-TFIIIB was also present in the PC-C fractions used in this

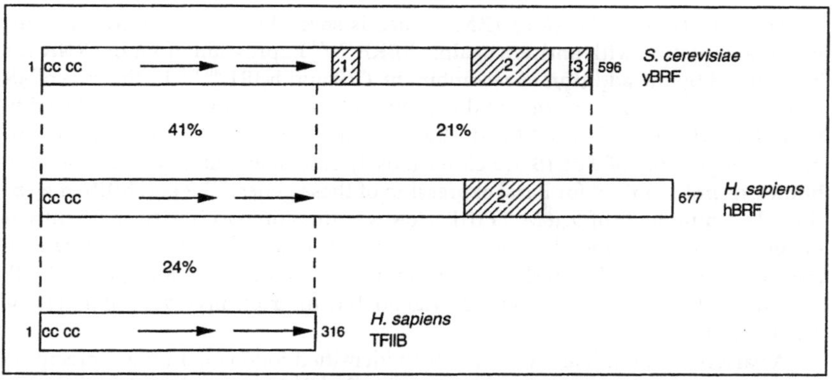

Fig. 4.7. Schematic representation of the sequence conservation between yBRF from *S. cerevisiae*, hBRF and TFIIB from *H. sapiens*. Percentages refer to amino acid identities in the regions indicated. CC CC indicates the putative zinc ribbon. The arrows indicate direct repeats. C-terminal homology blocks 1, 2 and 3 are indicated.

study. The active component of the 0.38M-TFIIIB fraction could be specifically immunodepleted using anti-TBP antibodies.[39] Furthermore, the 0.38M-TFIIIB fraction was sufficient to reconstitute transcription of VA_I and 5S genes in a TBP-immunodepleted extract, whereas recombinant TBP was not.[39] These results suggest that one or more TAF polypeptides exist in a complex with TBP in the 0.38M-TFIIIB complex and that these are required for VA_I and 5S transcription.

Using anti-TBP antibodies, a TAF of 88-90 kD can be immunoprecipitated specifically from fractions containing TFIIIB or the 0.38M-TFIIIB subcomplex.[46,178] Peptide sequencing allowed the isolation of cDNAs encoding this TAF.[46,178] It was found to contain 677 amino acid residues and have a calculated pI of 5.1.[46] The N-terminal 280 residues are 24% identical to human TFIIB and 41% identical to yBRF from *S. cerevisiae*.[46] (Fig. 4.7) As a consequence, this TAF was named hBRF.[46] It has also been referred to as TFIIIB90,[178] but this is likely to cause confusion with the 90 kD yeast B" TAF, to which it is unrelated. Like CeBRF and the three available yBRF sequences, hBRF contains a zinc ribbon motif and two direct repeats.[46,178] C-terminal to residue 280, hBRF displays little homology to the yBRFs. A notable exception is the region between hBRF residues 448 and 502, which is 29% identical to conserved region II of yBRF from *S. cerevisiae*.[46] This may constitute a TBP-binding domain. When compared to CeBRF, hBRF is 36% identical and 55% similar overall; its N-terminal 275 residues are 55% identical and 64% similar to those of CeBRF.[175] Residues 436-651 of hBRF also display 20% identity to the HMG boxes of chicken HMG-2[46,178] The corresponding region of CeBRF shows no homology to HMG-2.[175] A potential leucine zipper extends between residues 149 and 170 of hBRF and is also seen in CeBRF.[175,178] However, the coiled coil probability within this region is low, and both contain at least one substitution within a critical leucine.[175] Like its yeast homologues, hBRF contains two binding sites for TBP—a weak one in its N-terminal half and a strong one in its C-terminal half.[178] Deletion of either half abolished transcriptional activity.[178]

Transcription of VA, 5S or tRNA genes is severely compromised following immunodepletion with antisera against hBRF.[46,178] Expression of VA or tRNA can be restored by the addition of recombinant TBP and hBRF.[46,178] In the case of 5S, reconstitution of transcription in the immunodepleted extract required the addition of TFIIIA as well as TBP and hBRF.[178] These observations suggest that any other components of TFIIIB are either loosely associated, displaced by the antibodies, or not required for basal expression of these genes. TBP and hBRF may be the sole components of 0.38M-TFIIIB. However, this complex has an apparent size of 300 kD, as determined by gel-filtration on Superose 12.[39] It may therefore contain additional TAFs that perform nonessential regulatory roles. Many of the TAFs present in TFIID are dispensable for basal pol II transcription (reviewed in refs. 19, 21, 23, 26, 108).

A variety of candidate TAFs have been identified following immunoprecipitation of TFIIIB with anti-TBP antibodies.[39,42,46,105,114] However, hBRF is so far the only mammalian TFIIIB TAF which has had its functional significance confirmed unequivocally. Some of the candidate TAFs appear substoichiometric to TBP and hBRF, which may indicate that they are loosely associated. Less tightly bound TAFs may interact with TBP only indirectly, perhaps via other TAFs. It is unclear which, if any, are responsible for the 0.48M-TFIIIB activity. It has been suggested that the 172 kD TAF reported by Taggart et al[42] corresponds to the abundant 170 kD TAF that forms part of B-TFIID and is unrelated to TFIIIB.[44,178]

The TFIIIB requirement of type III promoters is different from that of types I and II. Several studies have found that TFIIIB fractions which support VA_I expression are inactive for U6 or 7SK.[39,46,114,176-178] For example, Lobo et al[39] observed that 0.38M-TFIIIB is required to reconstitute VA_I transcription in TBP-immunodepleted extracts, but free TBP is sufficient for U6. In fact, 0.38M-TFIIIB is unable to support U6 transcription in this context.[46] Recombinant hBRF has only a slightly inhibitory effect upon the ability of TBP to support U6 expression.[46] This suggests that 0.38M-TFIIIB contains one or more components besides TBP and hBRF that prevent it from functioning at the U6 promoter. hBRF may not be involved in U6 transcription, since Mital et al[46] found that immunodepletion of hBRF inhibits VA_I but does not affect U6. However, Wang and Roeder[178] observed inhibition of U6 and 7SK expression following immunodepletion with anti-hBRF antibodies. The reason for this discrepancy is uncertain. One possibility is that type III promoters use a modified form of hBRF or a distinct BRF-like factor that is recognized by the anti-hBRF antibodies of Mital et al but not by those of Wang and Roeder.

Teichmann and Seifart[114,177] used chromatography on EMD-DEAE-Fractogel (EDF) to separate two distinct forms of human TFIIIB. One form coeluted with TBP and hBRF in 300 mM KCl, efficiently supported VA_I but not U6 transcription, and was named hTFIIIB-β.[114] The other form eluted from EDF in 200 mM KCl, was free of TBP and hBRF, active for U6 but not VA_I expression, and was named hTFIIIB-α.[114,177] These two activities also behave differently on Sephacryl-S300 HR and Cibacron Blue.[114] hTFIIIB-α appears to be a subcomplex of hTFIIIB-β, and may be generated by dissociation of the latter.[114] It is therefore possible that hTFIIIB-α is related to 0.48M-TFIIIB. The most highly purified fractions of hTFIIIB-α and -β revealed multiple bands on silver staining. Following SDS-PAGE of hTFIIIB-α, optimal U6 transcription was reconstituted using renatured proteins of ~90 kD, ~60 kD and ~25 kD.[114] However, hTFIIIB-α has a size of ~60 kD, as estimated using gradient centrifugation or gel filtration.[114]

In a system reconstituted with human factors, a yeast B" fraction was found to be able to substitute for hTFIIIB-α in supporting expression of a human U6 gene.[177] The B" fraction contained both TFIIIE and the 90 kD subunit of yeast TFIIIB.[177] An antiserum raised against the 90 kD yeast subunit was able to deplete hTFIIIB-α of an activity required for transcription of human U6.[177] Human polypeptides of 90, 67 and 65 kD were retained by this antiserum.[177] An additional component that was not bound by this antiserum was required to reconstitute hTFIIIB-α activity.[177] This component could be substituted for using partially purified yeast TFIIIE.[177] The authors concluded that hTFIIIB-α may be related functionally to a combination of yeast TFIIIE and the 90 kD B" subunit of yeast TFIIIB.[177] In contrast, yeast fractions were unable to substitute for hTFIIIB-β in supporting tRNA synthesis in a human system.[177]

TFIIIC

TFIIIC is one of the largest and most complex transcription factors to have been studied. Like TFIIIB, it was first purified from yeast, where it is also referred to as τ factor. Partially purified yeast TFIIIC sediments in glycerol gradients as a large macromolecule of at least 300 kD.[179,180] A similar size estimate was obtained by gel filtration, suggesting that the protein is relatively isometric.[180] Yeast TFIIIC appears by scanning transmission electron microscopy (STEM) as two tightly associated globular domains, each of ~300 kD and ~10 nm in diameter.[181,182] Quantitative gel retardation analyses suggest that yeast TFIIIC has either a single DNA-binding site, or that multiple sites must be equivalent and independent.[183,184] An apparent dissociation constant (K_{app}) of 10^{-10} M was measured for binding to the $tRNA^{Glu3}$ gene.[184]

Yeast TFIIIC has been purified by a combination of ion-exchange and affinity chromatography on specific tDNA or B-block columns.[125,185-188] There is substantial agreement between different laboratories concerning its polypeptide composition. Major polypeptides of approximately 138, 131, and 95 kD copurify with TFIIIC transcriptional and DNA-binding activity, and additional polypeptides of approximately 91, 60, and 55 kD are also consistently observed.[133,186-189] Gabrielsen et al[186] found that species of approximately 138, 131, 95 and 60 kD remained associated with a tRNA gene following preparative scale nondenaturing gel electrophoresis of affinity purified TFIIIC. Antibodies raised against the 138, 131 or 95 kD species did not cross-react and each of these antisera could inhibit transcription, immunodeplete an extract of TFIIIC activity, and retard the migration of a TFIIIC/tDNA complex in a nondenaturing gel.[186,187] The 138, 131, 95, 91, and 55 kD polypeptides can all be crosslinked specifically to tDNA.[133,186,189] The lack of cross-reactivity between antibodies raised against the separated components,[186,187,189] and their clearly distinct spatial distributions within a TFIIIC/tDNA complex, as revealed by photocrosslinking,[133,189] argue against the possibility that some of these species are degradation products of others. Since yeast TFIIIC is often referred to as τ, its subunits have been named τ_{138}, τ_{131}, τ_{95}, τ_{91}, τ_{60}, and τ_{55}.

Marzouki et al[190] found that limited protease treatment generated a smaller form of TFIIIC, as judged by its mobility in nondenaturing gels. DNA-binding affinity was not reduced, but the protease-resistant TFIIIC form bound only the B-block of a $tRNA^{Glu3}$ gene.[190] The authors suggested that TFIIIC contains separate domains, perhaps linked by a flexible hinge, which can interact with the two variably spaced ICR sequence blocks; the protease-resistant B-block-binding domain, termed τ_B, is the major quantitative determinant of binding, but interaction of the rest of the molecule, termed τ_A, with the A-block is also required for transcription.[190] These conclusions are supported by direct visualization studies using STEM, in which most TFIIIC/tDNA complexes appear dumbbell shaped, with individual protein domains bound separately to the A- and B-blocks.[181] In a small proportion of complexes the entire TFIIIC molecule was bound to a single site centered upon the B-block.[181] On U6 genes, which have a large interblock spacing, one-half of the dumbbell occupies the B-block whereas the other half was observed to bind nonspecifically to DNA approximately 70 bp upstream.[182] Gabrielsen et al[186] found that τ_B is recognized by an anti- τ_{138} antibody but not an anti-τ_{95} antibody, indicating that τ_{138} participates in the τ_B domain. TFIIIC fractions consisting primarily of just τ_{138} and τ_{95} still retain a substantial proportion of DNA binding activity,[191] indicating that these polypeptides are the major quantitative determinants of tDNA recognition. This conclusion is strongly supported by the fact that only τ_{138} is accessible to photocrosslinking from the B-block, the primary binding site for TFIIIC.[133] A polypeptide of approximately this size that specifically binds to tDNA has also been detected in a partially purified TFIIIC fraction by Southwestern blotting.[192] In contrast, τ_{131}, τ_{95}, and τ_{55} can be crosslinked to N_3RdU residues in and around the A-block,[133] and are therefore likely to comprise the τ_A domain. τ_{91} is specifically accessible to photoprobes located downstream of the ICR.[189]

Genes have been cloned for the τ_{138}, τ_{131}, and τ_{95} subunits of TFIIIC from *S. cerevisiae*.[188,193-195] Ruth et al[165] also reported the cloning of genes encoding τ_{91} and τ_{55}, although this work has yet to be published. TFC1, which encodes τ_{95}, was isolated by screening genomic DNA with oligonucleotides corresponding to the sequences of tryptic peptides.[188] It was also isolated independently by probing an expression library with a polyclonal antiserum raised against purified τ_{95}.[194] It is a single copy gene that is essential for growth.[188] It is expressed at relatively low levels, such that its transcript is approximately 100 times less abundant than the phosphoglycerate kinase mRNA.[194] The predicted gene product contains 649 amino acids and shows no extensive homology to other known proteins.[188,194] A region with strong homology to a helix-turn-helix DNA-binding domain begins at amino acid 247.[188] However, τ_{95} alone does not bind DNA,[188,194] which suggests that τ_A may contain a composite DNA-binding domain involving more than one polypeptide. Residues 510-620 are highly acidic, with a net charge of -31.[188] (Fig. 4.8)

The gene for the τ_{138} component of τ_B was cloned by screening a genomic library with probes corresponding to peptide sequences obtained from tryptic digests.[193] This single-copy gene has been named TFC3 and is essential for viability, as shown by gene disruption.[193] An elegant epitope-tagging approach demonstrated the presence of one copy of the TFC3 gene product per TFIIIC/tDNA complex.[193] TFC3 encodes an 1160 amino acid protein with a calculated mass of 138 kD, that is highly basic and has a predicted pI of 10.16.[193] There is no extensive homology to other proteins, although two regions resemble HMG boxes, one at the N-terminus and one near the C-terminus.[193] The HMG box is a DNA-binding motif of approximately 80 amino acids that was initially discovered in the abundant high-mobility group chromosomal proteins.[196,197] It is found in a number of transcription factors, including the class I factor UBF,[198,199] the testis-determining factor SRY,[200] and the lymphoid-specific factor LEF-1.[201] A common feature of HMG box-containing proteins is the ability to bend or loop DNA.[196] For example, LEF-1 is estimated to bend DNA by as much as 130°.[202] The presence of potential HMG boxes in

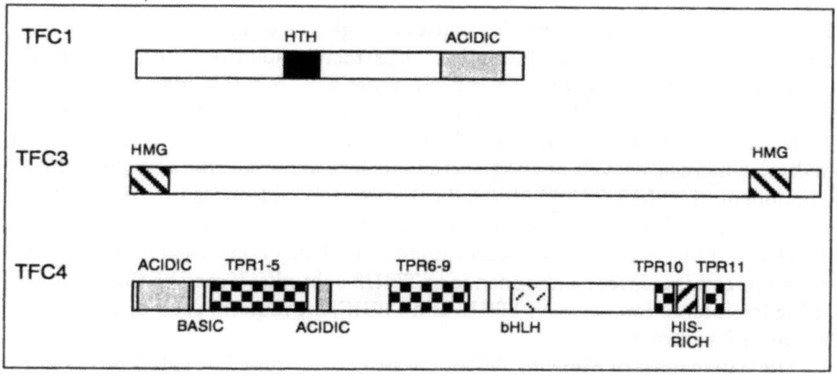

Fig. 4.8. Sequence features of the τ_{95}, τ_{138} and τ_{131} subunits of *S. cerevisiae* TFIIIC are encoded by the TFC1, TFC3 and TFC4 genes, respectively. Regions that are rich in acidic or basic amino acids are drawn shaded. HTH denotes a region with homology to a helix-turn-helix motif. HMG indicates the position of putative HMG boxes. TPR refers to tetratricopeptide repeats and bHLH denotes a region with homology to a basic helix-loop-helix motif.

TFIIIC may account for its ability to bend class III promoters.[180,203] A single Glu to Gly substitution at residue 349 of τ_{138} reduces the affinity of TFIIIC for tDNA.[184] The level of TFIIIC and the synthesis of tRNA and 5S rRNA are reduced in vivo as a consequence of this mutation.[184]

The TFC4 gene, which encodes the τ_{131} subunit of yeast TFIIIC, was cloned on the basis of protein microsequence data.[195] It is unique and essential for cell viability.[195,204] Its conceptual translation product is 1025 amino acids long and is rich in aromatic residues.[195] The N-terminal region has many acidic residues and is essential for viability.[160] A putative helix-loop-helix (HLH) dimerization motif is located around residue 600.[195] This HLH homology is preceded by a basic sequence that resembles the DNA-binding domain of many HLH-containing factors, such as MyoD and c-myc.[195] Several deletions within the basic and HLH homology regions can be tolerated.[160] The TFC4 sequence also encodes 11 copies of the 34 amino acid tetratricopeptide (TPR) motif.[195] TPR units are believed to form amphipathic α-helical regions that are punctuated by proline-induced turns.[205] Repeat copies of such units are thought to associate to form a "snap helix" secondary structure, which may serve as a protein-protein interaction domain.[206,207] With the exception of the ninth, deletion of any of the TPR repeats in TFC4 is lethal.[160] Gel retardation assays using an epitope-tagged TFC4 gene product demonstrated that a TFIIIC/tDNA complex contains a single copy of this polypeptide.[195]

TFC4 was cloned independently using a strategy based on the ability of target genes to suppress a mutation in the A-block of a tRNA gene.[208] Transcription from the mutated tRNA promoter allowed readthrough of an amber suppressor tRNA located immediately downstream of the mutant gene.[208] It was the coexpression of the amber suppressor that was directly selected for.[208] This scheme allowed the isolation of eight independent dominant mutant strains that suppressed the A-block mutation.[208] One such strain was found to overexpress the BRF1 gene.[139] Another strain, named PCF1-1, was found to result from a point mutation in the

TFC4 gene.[204] The PCF1-1 mutation causes a single His to Tyr substitution of residue 190, within the second TPR repeat.[204] Extracts made from the PCF1-1 strain transcribed a variety of tRNA, 5S and 7SL genes at 4- to 7-fold higher levels than wild-type extracts.[204,208,209] The mutation was shown to facilitate recruitment of TFIIIB into a transcription complex.[204] In addition, PCF1-1 extracts were also found to display increased levels of yBRF and an increase in the specific activity of the B" component of TFIIIB.[25,204,209] Two-hybrid assays suggest that the second TPR region of τ131 inhibits the interaction between τ131 and B".[165] This can explain how the PCF1-1 mutation stimulates recruitment of TFIIIB. It is less clear how this point mutation can increase the abundance of TFIIIB subunits; it may be that by facilitating the interaction between TFIIIB and TFIIIC, the mutation stabilizes yBRF or enhances its extractability.

The native mass of *Xenopus* TFIIIC has been determined by gel filtration to be approximately 400 kD, assuming a globular conformation.[210] Keller et al[210] found that B-block DNA affinity-purified preparations of the *Xenopus* factor contained predominant bands of 200, 160, 85, 75, and 31 kD, and that the 85 kD subunit could be crosslinked specifically to B-block sequences. In contrast, Cohen and Reynolds[211] found a 55 kD polypeptide to be the predominant silver-stained species in *Xenopus* PC-C fractions and in fractions prepared by B-block DNA affinity chromatography. They cloned the cDNA encoding this polypeptide by screening a *Xenopus* ovary expression library with a multimerized B-block sequence.[211] The predicted gene product, designated YB3, contains 305 amino acids and binds specifically to B-block sequences.[211] Western blotting showed YB3 to be present in PC-C but not in PC-A or PC-B fractions.[211] YB3 RNA is present in all adult tissues tested; it is abundant in ovaries, gastrulae and later stage embryos and is rare at the morula and blastula stages.[211] YB3 is a member of the Y-box family.[211] This family contains a variety of nucleic acid binding proteins with similar primary structures and which share an 80 amino acid domain with strong homology to CS 7.4, the major cold shock protein of *E. coli*.[212] The first member of this family was the CCAAT box-binding factor YB-1,[213] which is approximately 85% homologous to YB3.[211] Cohen and Reynolds[211] pointed out that although YB3 displays several of the properties expected of TFIIIC, its extreme abundance in various TFIIIC fractions argues against its being a subunit of TFIIIC, since other subunits would be expected to be present in equimolar amounts. It may, therefore, be an alternative B-block binding factor, perhaps with a regulatory role.[211]

Human TFIIIC was first resolved into two components, TFIIIC1 and TFIIIC2, using either Mono Q or oligonucleotide-affinity columns.[214,215] TFIIIC1 elutes from phosphocellulose between 280 and 390 mM KCl.[132] It can therefore be found in both PC-B and PC-C fractions. As a consequence it has often eluded detection when crude fractions were used in complementation assays. Whereas both components are required to reconstitute transcription of VA_I, 5S and tRNA genes,[132,176,214-218] 7SK and U6 transcription requires TFIIIC1 but not TFIIIC2.[176,217,218] Both TFIIIC1 and TFIIIC2 are inactivated by heat treatment at 47°C for 15 minutes.[176] Sedimentation analysis suggested masses of up to 200 kD for TFIIIC1 and of 400-500 kD for TFIIIC2, assuming that each is globular.[215] The subunit composition of TFIIIC1 has yet to be determined. It may exist in two forms, with slightly different chromatographic and functional properties.[132]

Following glycerol gradient sedimentation of highly purified TFIIIC2, Yoshinaga et al[219] localized VA_I binding activity to fractions containing 5 polypeptides, of 240, 110, 100, 80, and 60 kD (termed α, β, γ, δ, ε, respectively). These fractions also displayed transcriptional activity in the presence of TFIIIB, TFIIIC1, and pol III.[219] The same 5 bands were obtained with approximately equal stoichiometry when TFIIIC2 was affinity-purified using a B-block oligonucleotide.[216] Kovelman and Roeder[220] also found that 5 polypeptides copurify with TFIIIC2 activity, and the sizes in this case (220, 110, 102, 90, and 63 kD) agree closely with those determined by Yoshinaga et al.[219] These subunits remained associated following immunoprecipitation and washing in 1M NaCl,[217] thereby demonstrating the stability of the TFIIIC2 complex.

TFIIIC2 binds to the B-block region of VA_I and $tRNA^{MetI}$ genes.[132,215] TFIIIC1 increases the intensity of the B-block footprint generated by TFIIIC2 and extends it both 3' and 5' to include the A-block.[132,215] It remains to be determined whether TFIIIC1 binds directly to the A-block or induces rearrangements in TFIIIC2 which allow the latter to do so. The TFIIIC1 fractions of Wang and Roeder also protect the termination region of VA_I and $tRNA^{MetI}$ genes.[132] Deletion of this downstream region was found to prevent TFIIIC1 from extending and enhancing the B-block footprint.[132] Thus, sequences in the vicinity of the terminator may be required for cooperative interactions between TFIIIC1 and TFIIIC2. However, Oettel et al[218] reported that the terminator-binding activity can be separated chromatographically from TFIIIC1. These workers referred to the factor that binds to termination regions as $TFIIIC_0$.[218] Fractions containing this activity generated DNase I footprints around the 3' ends of 5S, VA_I and $tRNA^{MetI}$ genes.[218] $TFIIIC_0$ fractions could substitute for TFIIIC1 in supporting 5S, tRNA or U6 transcription, but, unlike TFIIIC1, did not enhance DNA binding by TFIIIC2.[218]

Quantitative gel retardation analysis indicates that TFIIIC2 binds nonspecific DNA with relatively low affinity (Kd, 6×10^{-4} M) and binds to the B-block of VA_I with high affinity (Kd, 2×10^{-11} M).[221] A single class of independent binding sites is observed.[221] TFIIIC2 has a protease-resistant DNA-binding domain, reminiscent of yeast τB.[222] Several studies found that the 240-220 kD α polypeptide is labeled specifically by UV cross-linking to bromodeoxyuridine-substituted VA_I probes.[219-221] Digestion with polioviral 3C protease has been used to dissect the subunit interactions within TFIIIC2.[223] The 110 kD β subunit associates with an N-terminal 83 kD fragment of the 240 kD α subunit (residues 1-732); this subcomplex retains DNA-binding activity, but is unable to support transcription.[223] The 100 kD γ subunit and the 60 kD ε subunit interact with a C-terminal ~125 kD fragment of the α subunit.[223] The 80 kD δ subunit is degraded by 3C protease.[223]

Hoeffler et al[224] resolved two DNA-binding forms of HeLa cell TFIIIC2 on nondenaturing gels. The slower migrating form has been named TFIIIC2a whereas the high mobility form is referred to as TFIIIC2b.[225] The two forms produce identical footprints on VA_I and have similar DNA-binding affinities (Kd, 2.2×10^{-11} M for TFIIIC2a and 1.5×10^{-11} M for TFIIIC2b).[220,224] However, after chromatographic separation, transcriptional activity was associated only with TFIIIC2a.[220,224] Silver staining revealed differences in the composition of the two forms, such that the TFIIICβ 110 kD subunit was missing from TFIIIC2b whereas an additional band of 77 kD was present instead.[220,225] (Fig. 4.9) Sinn et al[225] found that antibodies against TFIIICβ did not recognize the 77 kD component of TFIIIC2b. This suggests that the 77 kD polypeptide is not a proteolytic product of TFIIICβ, although it is possible

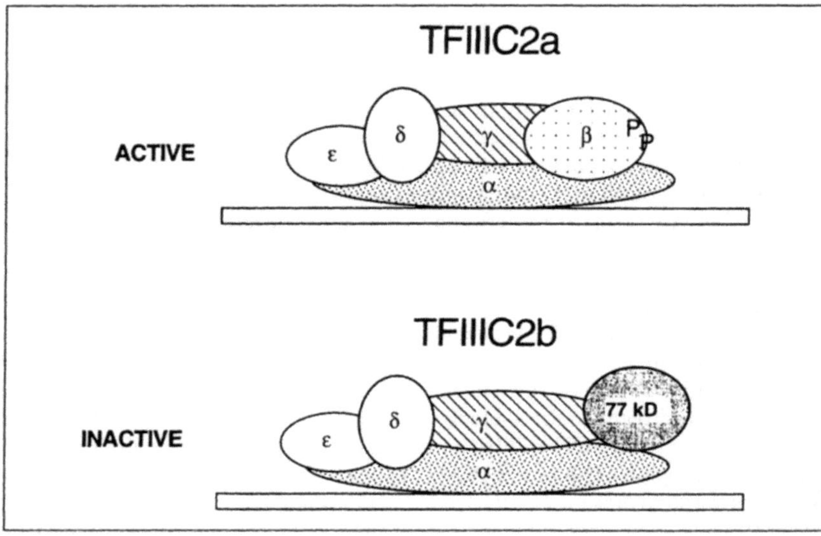

Fig. 4.9. Apparent subunit composition of two distinct forms of TFIIIC2. Both forms contain the α, γ, δ and ε subunits. The β subunit is only found in the TFIIIC2a form, whereas a 77 kD polypeptide is associated with only the TFIIIC2b form. TFIIIC2a is transcriptionally active, whereas TFIIIC2b is not.

that the epitopes recognized by the antibody have been degraded. Indeed, Shen et al[223] found that an antiserum against TFIIICβ suppressed all B-block-binding activity, suggesting that all TFIIIC2 molecules contain some form of TFIIICβ. This is consistent with the observation that a subcomplex comprising the N-terminal domain of TFIIICα bound to TFIIICβ can bind to DNA, whereas the TFIIICα subunit alone cannot.[223] Treatment with acid phosphatase converts TFIIIC2a into the inactive TFIIIC2b.[224] The differences observed on silver staining are therefore likely to reflect phosphorylation-induced changes either in mobility or in polypeptide composition. Thus, the transcriptional activity of TFIIIC can be modulated by phosphorylation. Metabolic labeling with ^{32}Pi demonstrated that all five subunits of TFIIIC2 are phosphorylated in HeLa cells.[223] TFIIIC from mouse[11] or *Xenopus*[210] also migrates as two or more species in nondenaturing gels. As in the human system, treatment of mouse TFIIIC with acid phosphatase has been shown to cause an increase in the high mobility form relative to the slowly migrating species.[11]

cDNAs encoding the largest subunit of TFIIIC2 (TFIIICα) have been cloned from both human and rat.[216,217] This was achieved by sequencing peptides from affinity-purified TFIIIC2 and using the sequence information to design primers for screening cDNA libraries.[216,217] The TFIIICα gene is single copy, and encodes a product with a calculated mass of 243 kD and a predicted pI of 6.9.[216,217] The rat and human sequences are 74% identical overall, with the C-terminal third of the polypeptide being less conserved than the rest.[217] Several regions of unusual charge density are conserved between the human and rat polypeptides.[217] (Fig. 4.10) Data base searches revealed little homology to other known genes and no recognized

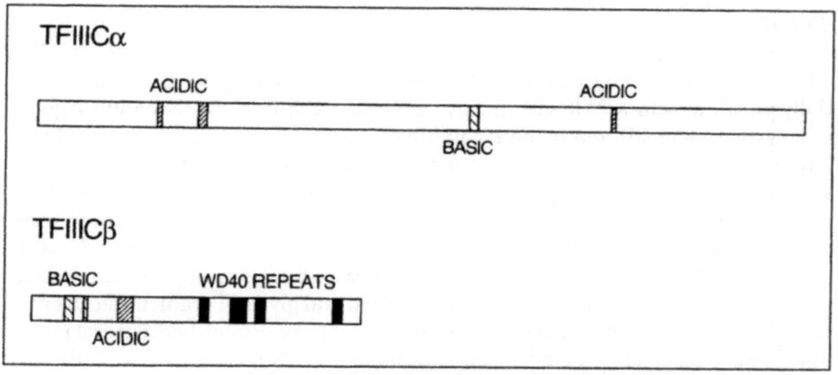

Fig. 4.10. Sequence features of the 220 kD TFIIICα and 110 kD TFIIICβ subunits of human TFIIIC2. Regions that are rich in acidic or basic residues are represented by cross-hatching. WD40 repeats are indicated.

DNA-binding motif.[216,217] The TFIIICα subunit alone exhibits no specific B-block-binding activity.[223] However, residues 1-732 can bind DNA specifically when associated with the β subunit.[223] TFIIICα is only 10% similar to τ_{138}, the B-block-binding subunit of yeast, and the conserved residues are scattered throughout the sequence.[216,217] Nevertheless, antisera against TFIIICα specifically depleted TFIIIC2 from extracts and recognized a TFIIIC2/B-block complex in band shift assays.[216,217] Whereas antisera raised against internal regions of TFIIICα supershift the TFIIIC2/B-block complex,[216,223] antibodies that recognize the N-terminus of TFIIICα abolish B-block binding.[217] This is consistent with the finding that the N-terminal region of TFIIICα is involved in binding DNA.[223]

Microsequencing of purified TFIIIC2 has also provided a route to the cloning of cDNAs encoding the 110 kD TFIIICβ subunit, which is found in TFIIIC2a but not in TFIIIC2b.[225] The cDNA encodes a 911 amino acid protein with a predicted mass of 100.7 kD and a calculated pI of 7.0.[225] The sequence reveals several short stretches of highly charged residues and five putative WD40 repeats.[225] (Fig. 4.10) The WD40 motif has been implicated in protein-protein interactions.[226,227] Apart from these short motifs, no significant homology was found with other proteins in the data base, including the subunits of yeast TFIIIC.[225] An antibody raised against recombinant TFIIICβ supershifted the complex formed between TFIIIC2 and the VA_I gene in a nondenaturing gel.[225] Immunoprecipitation with this antibody reduced transcription to undetectable levels.[225] Both TFIIIC2a and TFIIIC1 were required to restore transcription in the immunodepleted extract.[225] This suggests that TFIIIC1 associates tightly with TFIIIC2a in nuclear extracts.

In contrast to the findings of the Berk and Roeder laboratories, Seifart's group[228,229] found major polypeptides of 250, 220, 120, 110, 55 and 25 kD after purifying human TFIIIC. The 110 and 55 kD polypeptides were reported to bind DNA in Southwestern blotting and UV-crosslinking experiments.[228-230] DNA binding by the 55 kD polypeptide was found to require divalent metal cations.[230] This subunit is the same size as the *Xenopus* B-block binding protein YB3.[211]

TFIIIE

An additional general factor has been identified in yeast.[169] This factor is called TFIIIE and is required for efficient transcription of 5S and tRNA genes using partially purified factors.[169] It will also stimulate transcription of tRNA and U6 genes in a system reconstituted using recombinant TFIIIB plus TFIIIC and pol III that have been purified to near homogeneity.[165] TFIIIE copurifies with TFIIIB under some conditions, and is present in both the B' and B" fractions defined by Kassavetis et al.[126] However, it does not appear to be an essential component, since U6 transcription can be reconstituted in vitro using just recombinant TFIIIB and highly purified pol III.[163] TFIIIE is inactivated by protease and has an apparent molecular weight of 30-60 kD, as estimated by gel filtration and glycerol gradient sedimentation.[169] The role of TFIIIE in transcription has yet to be established. Several possibilities have been suggested:[165] it may stimulate TFIIIB recruitment, stabilize the transcription complex, or catalyze conformational rearrangements in TFIIIB.

TFIIID

Ottonello et al[13] resolved an activity, termed TFIIID, that is required for transcription of tRNA and 5S genes in silkworm. Neither silkworm TFIIIC, TFIIIB nor TFIIID is able on its own to interact stably with the $tRNA^{AlaC}$ gene, but combinations of TFIIIB and TFIIID or TFIIIC and TFIIID can form stable template-bound complexes.[13,231,232] A TFIIID activity has not been reported in other systems. It has been suggested that silkworm TFIIIC and TFIIID represent a subdivision of the components associated with TFIIIC in other organisms.[231,233] The inability of silkworm TFIIIC to bind independently to a $tRNA^{AlaC}$ gene could then be explained by the absence in this fraction of a subunit required for DNA-binding. Silkworm TFIIIC and TFIIID might also bear some equivalence to human TFIIIC1 and TFIIIC2.

La

Patients suffering from the autoimmune diseases systemic lupus erythematosus and Sjogren's syndrome frequently produce antibodies against an abundant 50 kD phosphoprotein called La (Lupus antigen). La is confined to the nucleus.[234,235] It binds transiently in a 1:1 molar ratio to the U-rich 3' ends of all nascent class III transcripts in mammals, frogs, flies and yeast.[234-241] It has also been shown to protect nascent B1 RNA from post-transcriptional 3' processing.[242]

HeLa extracts that have been immunodepleted of La are severely compromised in their ability to transcribe class III genes.[237,239] The small number of pol III transcripts synthesized in the absence of La are 1 to 5 U residues shorter at their 3' ends than those made in its presence.[243] Addition of purified La to the immunodepleted extracts restores RNA length and a low level of transcription.[243] Furthermore, pure recombinant La can stimulate the release of transcripts from immobilized class III templates and thereby facilitate multiple reinitiations.[242,244] Such effects are specific, since La does not affect T7 RNA polymerase and other DNA- or RNA-binding proteins do not exhibit this activity.[242,244] These results have led to the suggestion that La is a general class III factor that is required for completion of transcription and RNA release prior to recycling of pol III. Furthermore, Maraia[244] found that recombinant La can stimulate single round transcription assays and proposed that it contributes to initiation as well as termination. Phosphorylation at residue 366 by casein kinase II inhibits the ability of La to stimulate RNA synthesis from transcription complexes assembled on the VA_I promoter.[245] Phosphorylation of La has no effect upon its ability to bind RNA, which indicates that its effect on transcription is not simply a product of RNA binding.[245]

Active human class III transcription complexes can be isolated on glycerol gradients, and shown to contain TFIIIB, TFIIIC, and pol III activities, but less than 1% of the total extract protein.[246] The rate of sedimentation is slightly reduced in such complexes isolated from extracts depleted of La protein,[243] suggesting that this may also be an integral component. However, La, like pol III, can be preferentially stripped from transcription complexes using heparin and sarkosyl.[244]

Human, bovine, mouse, frog, fly and yeast genes encoding La have been cloned.[235,247-249] The La sequence displays good homology to the ~80 amino acid RNA recognition motif (RRM) that is found in many RNA binding proteins, including some present in U1 and U2 snRNPs.[250] The most highly conserved region of La lies adjacent and N-terminal to the RRM and this is also required for binding to pol III transcripts.[235] All La molecules examined have a highly charged C terminus that is not involved in specific RNA recognition.[235] This region is necessary for La to activate transcription.[245] La was reported to have ATPase activity capable of melting a synthetic DNA-RNA hybrid in a reaction requiring ATP hydrolysis.[251] These properties are strongly reminiscent of the rho termination factor of *E. coli*, which uses energy derived from ATP hydrolysis to unwind the DNA-RNA duplex at the 3' terminus of a nascent transcript, thereby facilitating RNA release.[252] This is consistent with the idea that La is a termination factor, which prevents the sequestration of stalled pol III molecules on incomplete transcripts by catalyzing transcript release. However, Maraia et al[242] were unable to detect any helicase activity with pure recombinant La and found that ATP depletion or inclusion of nonhydrolyzable ATP analogs had no effect on the ability of La to release RNA from transcription complexes.

Although La is present from yeast to man, a role in pol III transcription has so far only been detected in mammals. Indeed, work in lower systems appears to argue against a requirement for La. Using purified *Xenopus* pol III, Campbell and Setzer[253] obtained efficient termination in the apparent absence of La and observed no effect of adding recombinant La. Although La is present in *Saccharomyces cerevisiae*, strains with a null allele of the gene encoding La are fully viable.[235,254] Extracts prepared from such strains display no defect in tRNA synthesis and addition of recombinant La does not stimulate transcription.[254] The lack of requirement for La in yeast could be explained by redundancy. Once the genome of *S. cerevisiae* was fully sequenced, however, it became clear that there is only one protein that is homologous to vertebrate La throughout its length.[254] Two additional loci encode proteins that share a conserved domain with all La proteins.[254] However, gene disruption experiments demonstrated that neither of these loci is essential for viability nor functionally redundant with La.[254] It therefore seems that La is not a class III transcription factor in *S. cerevisiae*. Indeed, yeast La has been shown to be involved in the post-transcriptional maturation of tRNAs, facilitating the endonucleolytic removal of the 3' trailer sequence.[254] Other post-transcriptional roles have also been proposed for La, and both frog and human La have been implicated in the nuclear import and nuclear retention of pol III transcripts.[255-257] A general function as a chaperone of pol III transcripts has been postulated.[254] It remains entirely possible that mammalian La has evolved an additional role in transcription. However, only in vitro evidence has been provided to date that this is the case. Categorical proof will require the demonstration that La contributes to pol III transcription in vivo.

Gene-Specific Factors

TFIIIA

5S gene transcription requires the presence of an additional, gene-specific factor called TFIIIA.[6,10] TFIIIA has been the subject of considerable study for a number of reasons.[258,259] In addition to its ability to bind to 5S genes, it can also bind in a sequence-specific fashion to 5S rRNA.[260] Another unusual feature of TFIIIA is that it binds to single-stranded M13 DNA over 30 times more avidly than to its double-stranded binding site in a 5S gene.[261] It has also been reported to show DNA-dependent ATPase activity.[262] It was the first eukaryotic transcription factor to be purified to homogeneity[10] and also the first eukaryotic transcription factor to have its gene cloned.[263] Its purification was greatly facilitated by its considerable abundance in early *Xenopus* oocytes, where it occurs in a complex with 5S rRNA.[10] The cDNA was cloned on the basis of peptide sequence data.[263]

The predominant form of TFIIIA in *Xenopus* somatic cells is about 2 kD larger than the 38.5 kD form found in oocytes, and is referred to as TFIIIA' or S-TFIIIA.[264-268] S-TFIIIA is indistinguishable functionally from the smaller form that is abundant in oocytes.[268] A distinct TFIIIA-related polypeptide of 42 kD has also been reported,[269] but the significance of this protein for 5S transcription has been questioned.[270] S-TFIIIA is encoded by the same gene as the 38.5 kD form, but has 22 additional amino acids at its N-terminus.[268] The principal transcriptional start of the TFIIIA gene in somatic cells is located 284 bp upstream from the start site utilized in oocytes (referred to as +1).[268,271] The +1 site is also used in somatic cells, with the result that these cells contain both forms of TFIIIA.[271] In addition to these two pol II start sites, pol III also transcribes the TFIIIA gene, but the resultant transcripts are not polyadenylated and probably not spliced, and their functional significance is unclear.[271]

TFIIIA is extremely asymmetric both when free in solution[272] and when bound to DNA.[273,274] The 4 kD C-terminal arm can be removed by papain digestion without affecting the affinity of DNA binding or the extent of the footprint produced.[274-276] In contrast, deletion of just 29 residues from the N-terminus abolishes the ability to bind to 5S DNA.[277] The predicted protein sequence of TFIIIA is 344 amino acids long and, apart from 90 C-terminal and 10 N-terminal residues, is composed of 9 tandem 27 amino acid repeats, each characterized by pairs of cysteine and histidine residues at precisely repeated positions.[263,278] Hydrophobic amino acids are clustered at corresponding positions of each repeat.[278] (Fig. 4.11) Most of these repeats are encoded by separate exons.[279] Miller et al[278] proposed that the nine repeats form small, flexibly linked, compact and independent domains, each enclosing a central zinc ion coordinated to the invariant cysteines and histidines within a loop-like structure that directly contacts DNA, termed a "finger". Zinc ions are essential for the DNA binding activity of TFIIIA.[280] Extended X-ray absorption fine structure analysis confirmed that the coordination sphere of the zinc sites in TFIIIA consists of two cysteine and two histidine residues.[281] TFIIIA was the founder member of a large family of DNA-binding proteins that utilize zinc fingers (reviewed in refs. 282-285). Berg[286] proposed that an individual

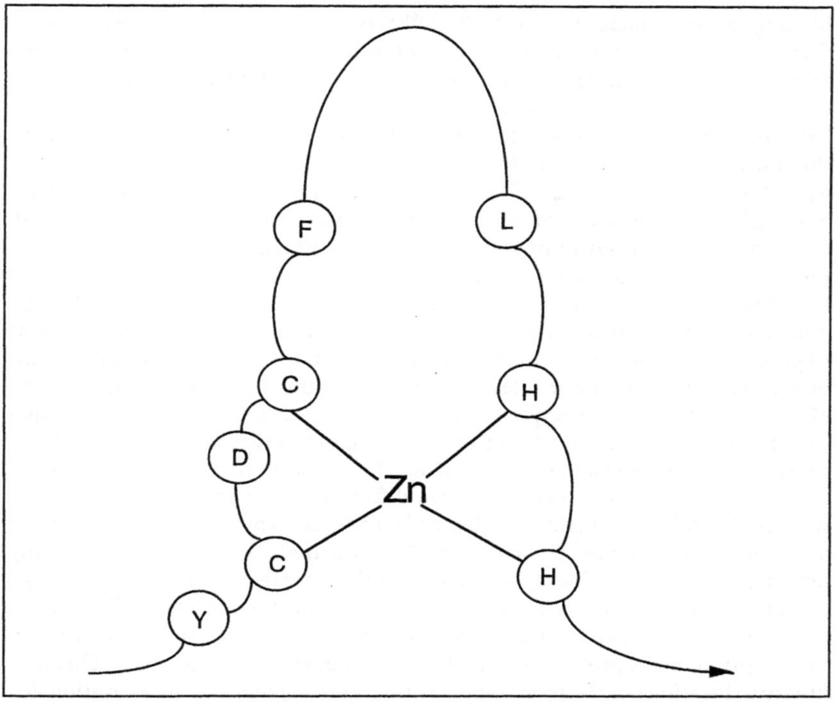

Fig. 4.11. Schematic illustration of the residues conserved between the finger motifs of *Xenopus* TFIIIA. The cysteine and histidine residues that coordinate a zinc ion are indicated. Abbreviations are: Y, tyrosine; C, cysteine; D, aspartate; F, phenylalanine; L, leucine; H, histidine.

finger domain consists of an antiparallel β-sheet packed against an α-helix, and that the latter lies in the major groove of DNA. NMR studies with single-finger peptides provided support for this model.[287,288] The crystal structure of a three-finger polypeptide of the mouse Zif 268 protein bound to DNA has confirmed that the α-helix of each finger lies in the major groove, with the N-terminal exposed face of the helices contacting adjacent 3 bp subsites.[289]

TFIIIA is highly modular, with clusters of zinc fingers binding to distinct regions of the 5S gene ICR over a distance of about 50 bp.[290-293] This modularity is well illustrated by the fact that progressive deletion into the DNA-binding domain generates TFIIIA fragments that can still give partial footprints on 5S genes. Thus, 32.5 or 27 kD fragments produced by deletion from the C-terminus of the protein only give the 3' two-thirds of the footprint observed with full-length TFIIIA.[274,276,290] The DNA-binding domain is divided into three distinct regions, two of which are compact and composed of fingers 1-3 and 7-9, and one of which is extended and consists of fingers 4-6.[291-294] Fingers 1-3 bind to the C-block of the 5S promoter, finger 5 binds to the intermediate element, fingers 7-9 bind to the A-block, and fingers 4 and 6 span these promoter regions.[291-298] A fragment containing only fingers 1-3 can bind independently to the C-block and contribute 95% of the total binding energy of intact TFIIIA.[291,293,295-300] Structural disruption of any individual finger by the substitution of asparagine for the first of the ligand-binding histidines reduces the affinity of TFIIIA for 5S DNA by 2.6- to 27-fold, depending on which finger is mutated.[301] Mutation of finger 3 has the most detrimental effect upon binding to the 5S ICR, but all the fingers contribute to the binding energy.[301] Nevertheless, disruption of any one of fingers 1-6 has no discernible effect upon the ability of the protein to support 5S transcription in vitro.[301] This implies that DNA-binding by TFIIIA is not a limiting step in the transcription of 5S genes. In contrast, analogous substitutions in fingers 7-9 result in partial or complete loss of 5S expression.[301] Disruption of finger 9 causes a severe loss of transcription, even though DNA-binding is only reduced by 2.7-fold.[301] These mutations have similar, although more complex, effects upon expression in vivo.[302] The data suggest that whereas fingers 1-6 are required just for binding to the ICR, fingers 7-9 perform an additional structural or catalytic role in the 5S transcription pathway. Removal of the 4 kD region C-terminal to finger 9 also inhibits transcriptional activity, although DNA binding is unaffected.[274] Point-mutagenesis of this C-terminal region defined a stretch of 14 to 18 residues that is critical for transcription.[303] The sequence of this region is highly conserved between different frog species.[303] It may form a structurally isolated unit, since it is flanked on both sides by proline residues, which often form flexible hinges or sharp turns.[303] The activity of this motif is unusually sensitive to location.[303] The eight amino acids that separate this region from finger 9 can be replaced without effect, but their removal inactivates TFIIIA.[303] Furthermore, insertions between finger 9 and the activating motif cause a decrease in transcription that is proportional to the number of residues inserted.[303] This sensitivity to location contrasts with the position-independence of the activation domains of many factors that regulate pol II.

In addition to its role as a transcription factor, TFIIIA also serves important functions in the storage and transport of 5S rRNA.[234,304-307] The molecular mechanisms which allow TFIIIA to bind to both DNA and RNA have attracted considerable attention.[259] Rhodes and Klug[308] suggested that the DNA helix of the 5S gene ICR might adopt an A-form conformation and thereby resemble the geometry of RNA rather than a typical B-form DNA duplex. In this way, TFIIIA could recognize similar structures in its DNA and RNA binding sites. Crystal structure analysis of a C-block DNA fragment found a geometry closer to A-form than to B-form.[309] However, CD and NMR measurements indicated that DNA fragments corresponding to either the C-block or the entire ICR adopt a structure in solution that differs from the A-form[310,311] and is irregular and intermediate between the classical A- and B-form duplexes, with a wide major groove.[312] Importantly, CD analysis demonstrated that the binding of TFIIIA does not induce any gross structural transition in the geometry of the ICR.[310]

TFIIIA may recognize RNA by a fundamentally different mechanism from that employed in binding DNA. The fact that the helices of zinc fingers are too wide to fit into the major groove of an A-form RNA duplex suggests that RNA binding may involve contacts with the phosphate backbone rather than direct recognition of specific nucleotides.[298] In contrast to DNA recognition, the major structural information required for specific RNA binding is provided principally by the secondary and tertiary structure of 5S RNA and not by its primary sequence.[298,313-317] The DNA sequences required for TFIIIA recognition are not included in the primary RNA contact domain.[318,319] Point mutations that prevent 5S RNA from binding

TFIIIA cluster into loop A and helices II and V, which constitute the central core of the structure.[298,318-320] However, almost all the RNA molecule is needed to maintain the optimal shape for TFIIIA binding. Most of the mutations within the helices that have a deleterious effect upon TFIIIA recognition also disrupt base pairing.[298,320] The relative orientation of helices II and V, as well as their structural integrity, is important for binding.[298] RNA footprint analysis using cobra snake venom CV1 endonuclease reveals that RNA helices I, III and IV are also contacted by TFIIIA, even though point mutations in these regions have no significant effect upon complex formation.[298] Zinc fingers 1-3 interact with helix IV, fingers 4-7 with helices II and V and loops A and E, and fingers 8-9 are positioned over helix III and loop B.[298,320] As such, the nine fingers are located over homologous primary sequences in both 5S RNA and 5S DNA. However, whereas fingers 1-3 contribute most of the affinity of DNA binding, fingers 4-7 are of greatest importance in binding RNA.[298,300,320,321] Therefore, the RNA-binding and DNA-binding functions of TFIIIA are separable and distinct.[296,298,320,321] Indeed, fingers 4-7 have a higher affinity than intact TFIIIA for 5S RNA, suggesting that the RNA-binding function is compromised to some extent to allow the polypeptide to perform its other roles.[320] It is not clear what structural features distinguish the DNA-binding fingers from the RNA-binding ones. Mutagenesis and phage display imply that the residues used by fingers 4 and 6 for RNA recognition are the same as those used by fingers 1 and 2 for contacting DNA.[300] For example, alanine substitution of four residues predicted to lie in the DNA recognition helix of finger 4 can reduce RNA binding by 77-fold.[300] However, fingers 1-3 are joined by a linker sequence that is common to many DNA-binding finger proteins but is not found in fingers 4-9.[299,321] Substitutions in the linkers between fingers 1-3 can abolish specific binding of TFIIIA to DNA, suggesting that the linker sequence may dictate the positioning or flexibility of the fingers.[299,321]

TFIIIA has an important role in the nuclear export of 5S RNA.[234] Two alternative and independent pathways exist for transporting 5S RNA out of the nucleus in *Xenopus* oocytes; one involves the formation of a 5S RNP complex with ribosomal protein L5 and the other is as a 7S RNP complex with TFIIIA.[234] A mutant 5S RNA that is unable to interact with TFIIIA and L5 is retained in the nucleus.[234] The 7S RNP in *Xenopus* oocytes contains one molecule of TFIIIA and one molecule of oocyte-type 5S RNA.[260,322] In contrast, somatic 5S RNA binds preferentially to L5 to form the 5S RNP.[323] Whereas the 5S RNP shuttles between the nucleus and the cytoplasm, the 7S RNP is retained in the cytoplasm.[324] (Fig. 4.12) Thus, in contrast to the two export pathways, 5S RNA only returns to the nucleus for ribosomal assembly when complexed with L5.[324] TFIIIA cannot enter the nucleus if it is bound to 5S RNA.[324] The nuclear localization signal of TFIIIA is located within the finger region and may be masked in the 7S RNP.[324] Return of TFIIIA to the nucleus therefore requires the exchange of 5S RNA from the 7S to the 5S RNP. During early oogenesis, 5S RNA is made in excess over other ribosomal components and is sequestered in the cytoplasm in complex with TFIIIA. As the oocytes develop, L5 levels increase and TFIIIA levels decrease, allowing the stored 5S RNA to exchange into 5S RNPs and thereby re-enter the nucleus for assembly into ribosomes.[325]

Transport of TFIIIA to the cytoplasm as an RNP complex can deplete the nucleus of this factor, so that it loses the ability to transcribe a subsequently injected 5S gene.[234] This provides a negative feedback loop for regulating the production of 5S RNA. A simpler version of this feedback inhibition is also observed in vitro, where

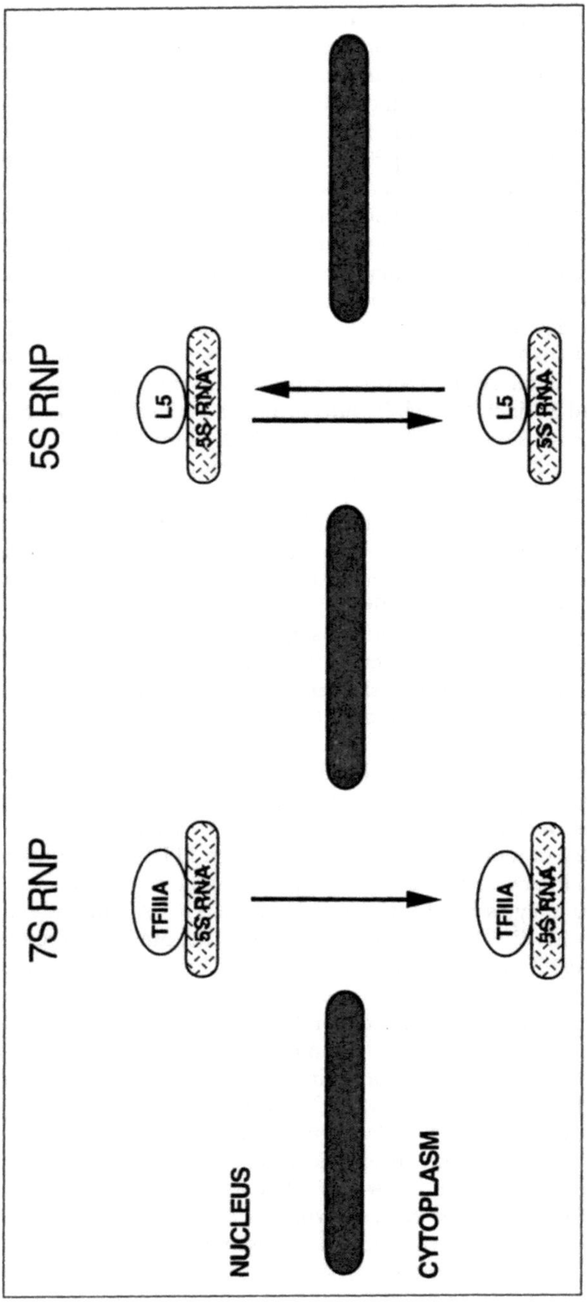

Fig. 4.12. Nucleocytoplasmic transport of 5S rRNA and TFIIIA in *Xenopus*. 5S rRNA can leave the nucleus when complexed with TFIIIA to form a 7S RNP. 5S rRNA can travel both into and out of the nucleus in complex with ribosomal protein L5 as a 5S RNP.

transcription of 5S genes can be inhibited by the presence of excess exogenous 5S RNA.[260,302,326] Following mutation of finger 6, which plays a primary role in RNA binding but only a subsidiary role in DNA binding, TFIIIA directs 5S transcription that is less susceptible to feedback inhibition.[302] As a consequence, the finger 6 mutant activates 5S expression more efficiently than wild-type TFIIIA in microinjected *Xenopus* embryos.[302]

In addition to the 7S RNP, previtellogenic oocytes of amphibians accumulate a 42S storage particle that consists of oocyte-specific 5S rRNA, various tRNAs, and polypeptides of 50 kD and 43 kD, present in a molar ratio of 1:3:2:1.[327] Whereas the 50 kD polypeptide is a divergent form of elongation factor EF-1α,[328] the 43 kD protein (p43) is related to TFIIIA.[329] Like TFIIIA, p43 has nine zinc fingers, and seven of these are precisely the same size as their counterparts in TFIIIA.[329] The 33% amino acid identity between the proteins consists mainly of the conserved residues characteristic of zinc fingers and there is complete divergence at the N- and C-termini of the proteins.[329] The gene for p43 is only expressed in oocytes.[329] Unlike TFIIIA, p43 binds exclusively to 5S RNA and not to DNA.[329] RNase protection patterns suggest that p43 and TFIIIA bind to the same general regions of 5S RNA and in the same orientation.[330] However, the primary role in RNA binding is performed by the N-terminal fingers of p43, in contrast to the case with TFIIIA, where the N-terminal fingers make only a subsidiary contribution.[296]

TFIIIA has also been purified from HeLa cells.[331,332] Moorefield and Roeder[332] identified a 42 kD polypeptide that cofractionates with TFIIIA activity in both transcription and DNA-binding assays. This polypeptide retained its ability to direct specific 5S gene transcription following elution and renaturation from SDS-polyacrylamide gels.[332] Antibodies raised against *Xenopus* TFIIIA were shown to cross-react with this 42 kD polypeptide.[332] Peptide sequences obtained from this factor have been used to isolate a cDNA that encodes a nine finger protein with 60% identity to the finger region of *Xenopus* TFIIIA.[332] The footprint produced by the 42 kD human TFIIIA on a human 5S gene closely resembles that produced by *Xenopus* TFIIIA on its homologous gene.[332] Human TFIIIA also bound to a *Xenopus* 5S gene, but had little affinity for a yeast 5S gene.[332] The abundance of TFIIIA was estimated to be approximately 4000 molecules per HeLa cell.[332] In contrast, Seifart et al[331] identified a 35 kD polypeptide from HeLa cells as being able to bind 5S genes and stimulate their transcription. This polypeptide was not recognized by antisera against *Xenopus* TFIIIA[230,331] and its identity is unclear.

TFIIIA from *Bombyx* copurifies with a polypeptide of 33.5 kD.[333] Fractions containing this protein give footprints on the noncoding strand of the silkworm 5S gene that extend from +45 to +96.[333] The extent of this protection is strikingly similar to the footprint obtained using TFIIIA from *Xenopus*.[333] *Xenopus* TFIIIA will support 5S transcription in a silkworm system.[333]

S. cerevisiae TFIIIA appears as a cluster of bands of around 50 kD when photocrosslinked to a 5S gene.[189] It has been purified to apparent homogeneity and found to be associated with polypeptides of 48 and 51 kD.[334] The gene for yeast TFIIIA (TFC2 or PZF1) was discovered immediately upstream of the gene for the ABC23 subunit of pol III, such that the putative translational initiation codons are separated by only 233 bp.[335,336] These genes are transcribed divergently.[335,336] Such proximity raises the obvious possibility of coregulation. TFC2 is a single copy gene that is essential for growth.[335] Conceptual translation predicts a 429 amino acid protein of 50 kD.[335,336] Like the *Xenopus* protein, yeast TFIIIA contains nine zinc

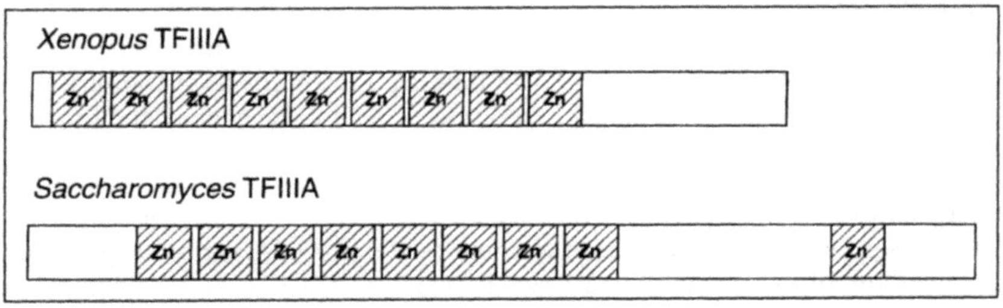

Fig. 4.13. Schematic comparison of the locations of the nine zinc fingers within the primary structures of TFIIIA from *Xenopus laevis* and from *Saccharomyces cerevisiae.*

fingers with invariant cysteine and histidine pairs.[335,336] (Fig. 4.13) A truncated polypeptide containing just fingers 1-3 of yeast TFIIIA is sufficient to bind specifically to 5S DNA, but will not support transcription.[337] Transcription is not impaired by removal of the N- or C-terminal domains that flank the zinc fingers, but is reduced by deletion of finger 9 and is abolished by deletion of finger 1.[337] The 81 amino acid region between fingers 8 and 9 is highly basic[335] and is essential for transcription.[337]

Although TFIIIA contains nine zinc fingers both in *Saccharomyces* and *Xenopus*, it has evolved surprisingly rapidly. Even among frog species there is considerable divergence, with TFIIIA from *X. laevis* showing only 84% identity to the *X. borealis* protein and 63% identity to the *Rana catesbeiana* protein.[338] Nevertheless, the spacing between the zinc-binding residues of homologous fingers is highly conserved among frog TFIIIA proteins.[338] These spacings are very different in *S. cerevisiae* TFIIIA, which overall is only ~20% identical to *X. laevis* TFIIIA, with most of the shared residues occurring in the finger motifs.[335,336] Whereas the fingers are contiguous in amphibian TFIIIAs, finger 9 in the yeast protein is separated from the others by 81 amino acids.[335,336] The N- and C-terminal domains flanking the nine zinc fingers are the most divergent in both size and sequence when frog TFIIIAs are compared either with each other or with yeast TFIIIA.[335,336,338] Hydrodynamic studies suggest that yeast TFIIIA, like the *Xenopus* protein, is an asymmetric polypeptide.[334] The DNase I footprint of yeast TFIIIA on a yeast 5S gene extends from +66 to +95,[339] whereas that of *Xenopus* TFIIIA on a *Xenopus* 5S gene reaches from +45 to +96.[10] *Xenopus* TFIIIA binds to a yeast 5S gene with 100- to 1000-fold lower affinity than does the homologous yeast factor, consistent with the divergence of the DNA-binding domains.[334] Similarly, yeast TFIIIA has little or no affinity for a *Xenopus* 5S gene.[340] Like its *Xenopus* counterpart, yeast TFIIIA requires zinc in order to bind DNA.[340] As in *Xenopus*, yeast TFIIIA binds 5S rRNA in competition with the 5S gene.[16]

Camier et al[341] used an elegant genetic approach to demonstrate that the synthesis of 5S rRNA is the only essential function of TFIIIA in *S. cerevisiae*. This was achieved by using the H1 promoter to drive TFIIIA-independent transcription of a downstream 5S gene.[341] The H1-5S fusion construct on a multicopy plasmid allowed yeast to survive in the absence of TFIIIA.[341] This established that 5S transcription is the only essential function of TFIIIA. However, the mutant cells grew more slowly than wild-type,[341] which raises the possibility that TFIIIA performs additional non-essential functions that facilitate normal growth. Yeast TFIIIA binds

5S rRNA and may, therefore, be involved in its nucleocytoplasmic transport, like its amphibian counterpart. Alternatively, the RNA produced by the H1-5S hybrid may be less active than normal 5S rRNA and this may account for the growth defect.[341]

PSE-Binding Factor(s)

Several distinct PSE-binding factors have been reported. The observation that the U6 and U2 PSEs are functionally interchangeable[107,342,343] suggested that the same PSE-binding factor may be employed by both class II and class III snRNA genes. A factor termed PSE1, that is related to the autoantigen Ku, has been reported to bind to the PSE of U1 and to stimulate its transcription,[344,345] but is not required for U6 transcription.[346,347] Instead, the efficiency of U6 transcription was found to correlate with the affinity of various wild-type and mutant PSE sequences for a distinct factor called PBP.[86,346,348] PBP copurifies through a number of chromatographic procedures with an activity required for transcription of both U6 and U1.[86,346] It may also be required for expression of the $tRNA^{Sec}$ gene, which utilizes a PSE motif.[349] PBP is a heat-labile factor found in the PC-C fraction.[86,347] Glycerol gradient analysis indicated a native mass of ~90 kD for PBP.[346] It generates a DNase I footprint that extends from -78 to -42 of the mouse U6 promoter and encompasses the PSE.[346] It has a high affinity for this PSE (Kd, 1.3×10^{-11} M) and a relatively low affinity for nonspecific DNA (Kd, 1.2×10^{-5} M).[347] PBP binds to PSEs quite slowly, but once formed, the resultant complex is very stable.[347] The DNA-binding domain is relatively resistant to protease treatment, especially when bound to DNA.[347]

Roeder and colleagues have described a factor called PTF which has similar PSE-binding and chromatographic properties to PBP.[176,350,351] PTF purified from HeLa cells is a complex of four subunits designated PTFα (180 kD), PTFβ (55 kD), PTFγ (45 kD) and PTFδ (44 kD).[176,351] Approximately equimolar amounts of these four polypeptides copurify chromatographically and also coimmunoprecipitate in 0.4 M KCl and 0.1% NP-40, which shows that they form a stable complex.[351] Indeed, these four subunits remain associated during sedimentation through a glycerol gradient containing 4 M urea.[176] Preparative-scale native gel electrophoresis was used to demonstrate that all four subunits are also present when PTF binds to the PSE.[176] PTFα can be crosslinked directly to PSE DNA.[176] Glycerol gradient sedimentation suggests a native mass for PTF of ~200 kD, assuming a globular conformation.[176] However, gel filtration indicated a natural size of ~500 kD,[176] which may mean that native PTF is not globular.

Microsequencing of highly purified PTF provided sequence information that was used to clone cDNAs encoding human PTF_β, PTFγ and PTFδ.[351,352] PTF_β is composed of 411 amino acids and has a calculated mass of 47 kD.[352] It is extremely acidic, with a pI of 4.9.[352,353] At pH 7.4, the subunit is predicted to have a net charge of -22.[353] The C-terminal region contains potential leucine zipper and zinc finger motifs.[352,353] This subunit can be photocrosslinked to PSE DNA in a sequence-specific manner in the context of the natural complex, but not as an isolated recombinant polypeptide.[353] The C-terminal 250 residues are 28% identical to a *C. elegans* protein of unknown function.[352] In band shift assays, antiserum raised against recombinant PTF_β specifically disrupts a complex composed of PTF and Oct-1 bound to the PSE of the 7SK gene.[352]

PTFγ is a 368 amino acid polypeptide with a calculated mass of 43 kD.[351] It is predicted to be predominantly α-helical, and is highly charged (36% of all residues) and basic (pI 9.5).[351] The N-terminal region has a high content of aromatic residues, whereas the C-terminal half is very hydrophilic.[351] $PTF_δ$ contains 334 amino acids and has a calculated mass of 36 kD.[351] It is both acidic (pI 5.7) and rich in proline (14%).[351] Neither PTFγ nor $PTF_δ$ bind independently to DNA,[351] consistent with the crosslinking data which suggest that the two larger subunits are responsible for contacting DNA.[176,353] However, antisera against recombinant PTFγ or $PTF_δ$ supershift the complex formed in a native gel between PTF, Oct-1 and the 7SK PSE, whereas preimmune sera are without effect.[351] This clearly establishes the presence of PTFγ and $PTF_δ$ in DNA-bound PTF.

Hernandez and colleagues have described a PSE-binding complex which they named $SNAP_C$.[107,354] Like PTF, this has a native mass of ~200 kD when analyzed on glycerol gradients.[107] Highly purified SNAPc was found to contain polypeptides of 190, 50, 45 and 43 kD,[354,355] sizes which correspond closely to those of the PTF subunits. Microsequencing of purified SNAPc provided information that was used to clone cDNAs encoding the 50 kD (SNAP50), 43 kD (SNAP43) and 45 kD (SNAP45) subunits.[353-355] These cDNAs were found to match those obtained for $PTF_β$, PTFγ and $PTF_δ$, respectively.[351,355] It is therefore highly likely that PTF and SNAPc are the same.

Immunodepletion of extracts using antibodies raised against recombinant SNAP50/$PTF_β$, SNAP43/PTFγ or SNAP45/$PTF_δ$ was found to repress specifically transcription from the 7SK and U6 promoters, whereas the VA_I and AdML promoters were unaffected.[351-355] Immunodepletion with these antibodies also inhibited expression from the U1 and U2 promoters.[351-355] Transcription could be restored by the addition of highly purified PTF/SNAPc.[351-353,355] These data demonstrate that SNAP50/$PTF_β$, SNAP43/PTFγ and SNAP45/$PTF_δ$ are part of a complex that is required specifically by both pol II- and pol III-transcribed U snRNA genes.

PTF/SNAPc interacts directly with TBP.[107,351,354] Indeed, TBP copurifies extensively with PTF/SNAPc and is coimmunoprecipitated using antibodies against $PTF_β$/SNAP50 or PTFγ/SNAP43.[107,351,352,354] Both PTFγ/SNAP43 and $PTF_δ$/SNAP45 bind directly to the conserved C-terminal domain of TBP, whereas $PTF_β$/SNAP50 does not.[351,354,355] These results demonstrate that TBP associates with PTF/SNAPc to form a complex. However, the amount of TBP that copurifies with SNAPc is variable.[355] Furthermore, the interaction between TBP and PTF/$SNAP_c$ can be disrupted using 0.4 M KCl or anti-TBP antibodies,[107,351,352] which suggests that TBP is less tightly bound than in complexes such as SL1 and TFIID. PTF/SNAPc also differs from other TBP-containing complexes in being compatible with transcription by more than one polymerase, since it is required by U1 and U2, as well as U6 and 7SK.[107,176,351,354] hBRF can also be coimmunoprecipitated with PTF, and this interaction shows similar salt sensitivity to the association between TBP and PTF.[352] Antisera against PTF immunoprecipitate TBP and hBRF in equimolar stoichiometries.[352] These results suggest that PTF interacts with a form of TFIIIB, rather than free TBP.

It seems likely that a single PSE-binding factor activates pol III transcription of U6 and 7SK genes and that this factor has received different names from different laboratories. PBP, PTF and SNAPc have similar chromatographic properties and each bind preferentially to the mouse U6 PSE.[107,346,350] PBP and PTF have both been shown to be heat-labile.[176,347] PTF and SNAPc have the same native mass, very similar polypeptide components and have been shown by cloning to share the $PTF_β$/

SNAP50, PTFγ/SNAP43 and PTFδ/SNAP45 subunits.[107,176,351,354,355] However, the native mass of ~90 kD determined for PBP by glycerol gradient sedimentation[346] is significantly less than the value of ~200 kD found for PTF and SNAPc using the same technique.[107,176] It seems likely that the size estimate for PBP was erroneous, but until the composition of PBP has been documented one cannot exclude the possibility that it is distinct from PTF/SNAPc. In contrast, the evidence is compelling that PTF/SNAPc is a *bona fide* PSE-binding protein that supports transcription of 7SK and U snRNA genes. Expression of these genes is lost if extracts are depleted using a PSE oligonucleotide or antibodies against recombinant subunits; furthermore, transcription of the same genes can be fully restored in the depleted extracts by addition of highly purified PTF/SNAPc.[176,351,353,355] These data suggest that PTF/SNAPc is the only PSE-binding factor that is required for transcription of 7SK and U6 snRNA genes.

Oct-1

Octamer motifs are found upstream of vertebrate U6 and 7SK genes, as well as in other type III promoters. Murphy et al[356] showed that purified Oct-1 and Oct-2 can bind to the 7SK promoter and stimulate its transcription by more than 10-fold. This provided the first demonstration that a purified class II factor can also activate transcription by pol III. More recently, recombinant Oct-1 has been shown to bind to the DSE of the human U6 gene and to stimulate transcription of constructs containing 7SK or U6 proximal promoter sequences fused to a single octamer motif.[350,357,358] Partially purified *Xenopus* Oct-1 was found to bind to the DSE of the *Xenopus* U6 gene.[359]

Oct-1 is a ubiquitous factor present in all tissues tested, whereas Oct-2 expression is more restricted.[360] Oct-1 and Oct-2 are closely related members of the POU domain family of transcription factors (see refs. 361 and 362 for reviews). This family is characterized by a bipartite DNA-binding domain consisting of a 75-82 amino acid POU-specific (POU_S) domain followed by a short linker and then a 60 amino acid homeodomain (POU_H).[363,364] Both the POU_S and POU_H domains of Oct-1 adopt a helix-turn-helix conformation in solution.[364-366] POU_H is nearly identical to other homeodomains, whereas POU_S is much more similar to the DNA-binding domains of the bacteriophage λ and 434 repressor and Cro proteins.[364] Each domain contributes specific contacts to an octamer DNA site.[364,367] The POU-specific domain binds the 5' end of the octamer and the homeodomain binds the 3' end.[364,368] Recognition occurs in the major groove and bends the DNA.[364,369,370] The DNA-binding domain of Oct-2 is 87% identical to that of Oct-1, consistent with their shared DNA-binding specificity.[363,364] Outside of the POU domains, Oct-1 and Oct-2 both have glutamine-rich activation domains and Oct-2 also has an activation domain that is rich in serine, threonine and proline.[371-373] (Fig. 4.14) However, the POU domain alone of Oct-1 is sufficient to activate transcription of 7SK or U6.[350,358] It achieves this by stimulating the binding of PTF/SNAPc to the PSE.[350,358] The pituitary transcription factor Pit-1 is 50% homologous to Oct-1 in its POU domain and can bind to octamer sequences, but is unable to activate U6 transcription.[358] This differential ability can be attributed to a single residue difference near the N-terminal end of the POU-specific domains of Oct-1 and Pit-1.[358] 7SK transcription is stimulated more strongly by full-length Oct-1 than by its POU domain alone, suggesting that the glutamine-rich domain may also contribute to full activation.[350] Indeed, a short segment from this domain can stimulate U6 transcription when

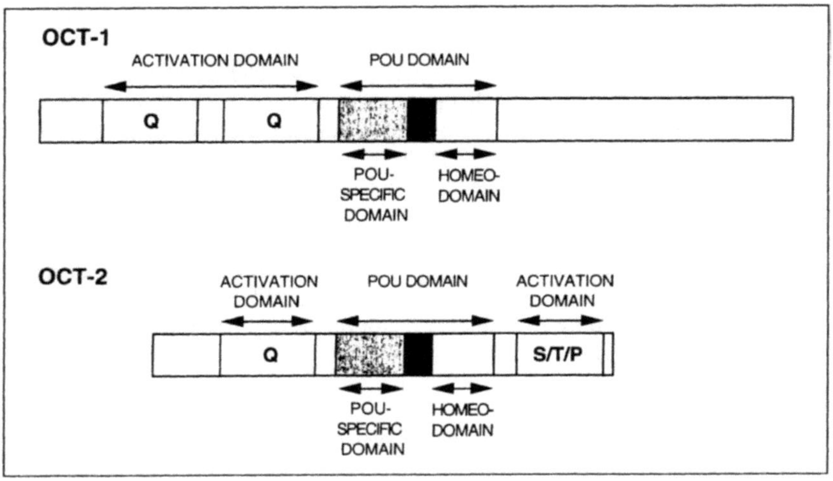

Fig. 4.14. Sequence features of the human Oct-1 and Oct-2 polypeptides. Both contain a DNA-binding POU domain that comprises a POU-specific domain separated from a homeodomain by a linker region. Both also contain glutamine-rich activation domains (Q). In addition, Oct-2 has an activation domain that is rich in serine, threonine and proline residues (S/T/P).

fused to a heterologous DNA-binding domain.[374] It remains to be determined whether the activation domain, like the POU domain, contacts PTF/SNAPc, or whether it interacts with another component of the transcription apparatus; the latter possibility might allow Oct-1 to stimulate multiple steps in the initiation process.

References

1. Jaehning JA, Woods PS, Roeder RG. Purification, properties, and subunit structure of DNA-dependent RNA polymerase III from uninfected and adenovirus-2-infected KB cells. J Biol Chem 1977; 252:8762-8771.
2. Weil PA, Segall J, Harris B et al. Faithful transcription of eukaryotic genes by RNA polymerase III in systems reconstituted with purified DNA templates. J Biol Chem 1979; 254:6163-6173.
3. Cozzarelli NR, Gerrard SP, Schlissel M et al. Purified RNA polymerase III accurately and efficiently terminates transcription of 5S RNA genes. Cell 1983; 34:829-835.
4. Hammond CI, Holland MJ. Purification of yeast RNA polymerases using heparin agarose affinity chromatography. Transcriptional properties of the purified enzymes on defined templates. J Biol Chem 1983; 258:3230-3241.
5. Parker CS, Roeder RG. Selective and accurate transcription of the *Xenopus laevis* 5S RNA genes in isolated chromatin by purified RNA polymerase III. Proc Natl Acad Sci USA 1977; 74:44-48.
6. Segall J, Matsui T, Roeder RG. Multiple factors are required for the accurate transcription of purified genes by RNA polymerase III. J Biol Chem 1980; 255:11986-11991.
7. Singh K, Carey M, Saragosti S et al. Expression of enhanced levels of small RNA polymerase III transcripts encoded by the B2 repeats in Simian virus 40-transformed mouse cells. Nature 1985; 314:553-556.

8. Carey MF, Singh K, Botchan M et al. Induction of specific transcription by RNA polymerase III in transformed cells. Mol Cell Biol 1986; 6:3068-3076.
9. Kickhoefer VA, Searles RP, Kedersha NL et al. Vault ribonucleoprotein particles from rat and bullfrog contain a related small RNA that is transcribed by RNA polymerase III. J Biol Chem 1993; 268:7868-7873.
10. Engelke DR, Ng S-Y, Shastry BS et al. Specific interaction of a purified transcription factor with an internal control region of 5S RNA genes. Cell 1980; 19:717-728.
11. White RJ, Stott D, Rigby PWJ. Regulation of RNA polymerase III transcription in response to Simian virus 40 transformation. EMBO J 1990; 9:3713-3721.
12. Shastry BS, Ng S-Y, Roeder RG. Multiple factors involved in the transcription of class III genes in *Xenopus laevis*. J Biol Chem 1982; 257:12979-12986.
13. Ottonello S, Rivier DH, Doolittle GM et al. The properties of a new polymerase III transcription factor reveal that transcription complexes can assemble by more than one pathway. EMBO J 1987; 6:1921-1927.
14. Burke DJ, Schaack J, Sharp S et al. Partial purification of *Drosophila* Kc cell RNA polymerase III transcription components: evidence for shared 5S RNA and tRNA gene factors. J Biol Chem 1983; 258:15224-15231.
15. Burke DJ, Soll D. Functional analysis of fractionated *Drosophila* Kc cell tRNA gene transcription components. J Biol Chem 1985; 260:816-823.
16. Klekamp MS, Weil PA. Specific transcription of homologous class III genes in yeast soluble cell-free extracts. J Biol Chem 1982; 257:8432-8441.
17. Taylor MJ, Segall J. Characterization of factors and DNA sequences required for accurate transcription of the *Saccharomyces cerevisiae* 5S RNA gene. J Biol Chem 1985; 260:4531-4540.
18. Johnson DL, Fan RS, Treinies ML. Analysis of the molecular mechanisms for the species-specific transcription of *Drosophila* and human tRNA gene transcription components. J Biol Chem 1991; 266:16037-16043.
19. Gill G. Complexes with a common core. Current Biology 1992; 2:565-567.
20. Green MR. Transcriptional transgressions. Nature 1992; 357:364-365.
21. Sharp PA. TATA-binding protein is a classless factor. Cell 1992; 68:819-821.
22. White RJ, Jackson SP. The TATA-binding protein: a central role in transcription by RNA polymerases I, II and III. Trends Genet 1992; 8:284-288.
23. Hernandez N. TBP, a universal eukaryotic transcription factor? Genes Dev 1993; 7:1291-1308.
24. Rigby PWJ. Three in one and one in three: it all depends on TBP. Cell 1993; 72:7-10.
25. Willis IM. RNA polymerase III. Genes, factors and transcriptional specificity. Eur J Biochem 1993; 212:1-11.
26. Struhl K. Duality of TBP, the universal transcription factor. Science 1994; 263:1103-1104.
27. Mattaj IW, Dathan NA, Parry HD et al. Changing the RNA polymerase specificity of U snRNA gene promoters. Cell 1988; 55:435-442.
28. Murphy S, Moorefield B, Pieler T. Common mechanisms of promoter recognition by RNA polymerases II and III. Trends Genet 1989; 5:122-126.
29. Simmen KA, Mattaj IW. Complex requirements for RNA polymerase III transcription of the *Xenopus* U6 promoter. Nucleic Acids Res 1990; 18:5649-5657.
30. Margottin F, Dujardin G, Gerard M et al. Participation of the TATA factor in transcription of the yeast U6 gene by RNA polymerase C. Science 1991; 251:424-426.
31. Lobo SM, Lister J, Sullivan ML et al. The cloned RNA polymerase II transcription factor IID selects RNA polymerase III to transcribe the human U6 gene in vitro. Genes Dev 1991; 5:1477-1489.
32. Simmen KA, Bernues J, Parry HD et al. TFIID is required for in vitro transcription of the human U6 gene by RNA polymerase III. EMBO J 1991; 10:1853-1862.

33. White RJ, Jackson SP, Rigby PWJ. A role for the TATA-box-binding protein component of the transcription factor IID complex as a general RNA polymerase III transcription factor. Proc Natl Acad Sci USA 1992; 89:1949-1953.
34. White RJ, Rigby PWJ, Jackson SP. The TATA-binding protein is a general transcription factor for RNA polymerase III. J Cell Science 1992; 16 (Supp): 1-7.
35. Radebaugh CA, Matthews JL, Geiss GK et al. TATA box-binding protein (TBP) is a constituent of the polymerase I-specific transcription initiation factor TIF-IB (SL1) bound to the rRNA promoter and shows differential sensitivity to TBP-directed reagents in polymerase I, II, and III transcription factors. Mol Cell Biol 1994; 14:597-605.
36. Sharma S, Gopinathan KP. Role of TATATAA element in the regulation of $tRNA^{Gly1}$ gene expression in *Bombyx mori* is position dependent. J Mol Biol 1996; 262:396-406.
37. McBryant SJ, Meier E, Leresche A et al. TATA-box DNA binding activity and subunit composition of RNA polymerase III transcription factor IIIB from *Xenopus laevis*. Mol Cell Biol 1996; 16:4639-4647.
38. Huet J, Sentenac A. The TATA-binding protein participates in TFIIIB assembly on tRNA genes. Nucleic Acids Res 1993; 20:6451-6454.
39. Lobo SM, Tanaka M, Sullivan ML et al. A TBP complex essential for transcription from TATA-less but not TATA-containing RNA polymerase III promoters is part of the TFIIIB fraction. Cell 1992; 71:1029-1040.
40. Pruzan R, Chatterjee PK, Flint SJ. Specific transcription from the adenovirus E2E promoter by RNA polymerase III requires a subpopulation of TFIID. Nucleic Acids Res 1992; 20:5705-5712.
41. Simmen KA, Bernues J, Lewis JD et al. Cofractionation of the TATA-binding protein with the RNA polymerase III transcription factor TFIIIB. Nucleic Acids Res 1992; 20:5889-5898.
42. Taggart AKP, Fisher TS, Pugh BF. The TATA-binding protein and associated factors are components of pol III transcription factor TFIIIB. Cell 1992; 71:1015-1028.
43. White RJ, Jackson SP. Mechanism of TATA-binding protein recruitment to a TATA-less class III promoter. Cell 1992; 71:1041-1053.
44. Meyers RE, Sharp PA. TATA-binding protein and associated factors in polymerase II and polymerase III transcription. Mol Cell Biol 1993; 13:7953-7960.
45. Lescure A, Lutz Y, Eberhard D et al. The N-terminal domain of the human TATA-binding protein plays a role in transcription from TATA-containing RNA polymerase II and III promoters. EMBO J 1994; 13:1166-1175.
46. Mital R, Kobayashi R, Hernandez N. RNA polymerase III transcription from the human U6 and adenovirus type 2 VAI promoters has different requirements for human BRF, a subunit of human TFIIIB. Mol Cell Biol 1996; 16:7031-7042.
47. Cormack BP, Struhl K. The TATA-binding protein is required for transcription by all three nuclear RNA polymerases in yeast cells. Cell 1992; 69:685-696.
48. Schultz MC, Reeder RH, Hahn S. Variants of the TATA-binding protein can distinguish subsets of RNA polymerase I, II, and III promoters. Cell 1992; 69:697-702.
49. Poon D, Knittle RA, Sabelko KA et al. Genetic and biochemical analyses of yeast TATA-binding protein mutants. J Biol Chem 1993; 268:5005-5013.
50. Heard DJ, Kiss T, Filipowicz W. Both *Arabidopsis* TATA binding protein (TBP) isoforms are functionally identical in RNA polymerase II and III transcription in plant cells: evidence for gene-specific changes in DNA binding specificity of TBP. EMBO J 1993; 12:3519-3528.
51. Whitehall SK, Kassavetis GA, Geiduschek EP. The symmetry of the yeast U6 RNA gene's TATA box and the orientation of the TATA-binding protein in yeast TFIIIB. Genes Dev 1995; 9:2974-2985.

52. Joazeiro CAP, Kassavetis GA, Geiduschek EP. Alternative outcomes in assembly of promoter complexes: the roles of TBP and a flexible linker in placing TFIIIB on tRNA genes. Genes Dev 1996; 10:725-739.
53. Trivedi A, Vilalta A, Gopalan S et al. TATA-binding protein is limiting for both TATA-containing and TATA-lacking RNA polymerase III promoters in *Drosophila* cells. Mol Cell Biol 1996; 16:6909-6916.
54. Buratowski S, Hahn S, Sharp PA et al. Function of a yeast TATA element-binding protein in a mammalian transcription system. Nature 1988; 334:37-42.
55. Cavallini B, Huet J, Plassat JL et al. A yeast activity can substitute for the HeLa TATA box factor. Nature 1988; 344:77-80.
56. Horikoshi M, Wang CK, Fujii H et al. Purification of a yeast TATA box-binding protein that exhibits human transcription factor IID activity. Proc Natl Acad Sci USA 1989; 86:4843-4847.
57. Cavallini B, Faus I, Matthes H et al. Cloning of the gene encoding the yeast protein BTF1, which can substitute for the human TATA box-binding factor. Proc Natl Acad Sci USA 1989; 86:9803-9807.
58. Hahn S, Buratowski S, Sharp PA et al. Isolation of the gene encoding the yeast TATA binding protein TFIID: a gene identical to the *SPT15* suppressor of TY element insertions. Cell 1989; 58:1173-1181.
59. Horikoshi M, Wang KC, Fujii H et al. Cloning and structure of a yeast gene encoding a general transcription initiation factor TFIID that binds to the TATA box. Nature 1989; 341:299-303.
60. Schmidt MC, Kao CC, Pei R et al. Yeast TATA-box transcription factor gene. Proc Natl Acad Sci USA 1989; 86:7785-7789.
61. Hoffmann A, Horikoshi M, Wang CK et al. Cloning of the *Schizosaccharomyces pombe* TFIID gene reveals a strong conservation of functional domains present in *Saccharomyces cerevisiae* TFIID. Genes Dev 1990; 4:1141-1148.
62. Fikes JD, Becker DM, Winston F et al. Striking conservation of TFIID in *Schizosaccharomyces pombe* and *Saccharomyces cerevisiae*. Nature 1990; 346:291-294.
63. Eisenmann DM, Arndt KM, Ricupero SL et al. SPT3 interacts with TFIID to allow normal transcription in *Saccharomyces cerevisiae*. Genes Dev 1992; 6:1319-1331.
64. Hoffman A, Sinn E, Yamamoto T et al. Highly conserved core domain and unique N terminus with presumptive regulatory motifs in a human TATA factor (TFIID). Nature 1990; 346:387-390.
65. Kao CC, Lieberman PM, Schmidt MC et al. Cloning of a transcriptionally active human TATA binding factor. Science 1990; 248:1646-1650.
66. Peterson MG, Tanese N, Pugh BF et al. Functional domains and upstream activation properties of cloned human TATA binding protein. Science 1990; 248:1625-1630.
67. Tamura T, Sumita K, Fujino I et al. Striking homology of the 'variable' N-terminal as well as the 'conserved core' domains of the mouse and human TATA-factors (TFIID). Nucleic Acids Res 1991; 19:3861-3865.
68. Hashimoto S, Fujita H, Hasegawa S et al. Conserved structural motifs within the N-terminal domain of TFIIDτ from *Xenopus*, mouse and human. Nucleic Acids Res 1992; 20:1783.
69. Hoey T, Dynlacht BD, Peterson MG et al. Isolation and characterization of the *Drosophila* gene encoding the TATA box binding protein, TFIID. Cell 1990; 61:1179-1186.
70. Muhich ML, Iida CT, Horikoshi M et al. cDNA clone encoding *Drosophila* transcription factor TFIID. Proc Natl Acad Sci USA 1990; 87:9148-9152.
71. Gasch A, Hoffmann A, Horikoshi M et al. *Arabidopsis thaliana* contains two genes for TFIID. Nature 1990; 346:390-394.

72. Haass MM, Feix G. Two different cDNAs encoding TFIID proteins of maize. FEBS Lett 1992; 301:294-298.
73. McAndrew MB, Read M, Sims PFG et al. Characterisation of the gene encoding an unusually divergent TATA-binding protein (TBP) from the extremely A + T-rich human malaria parasite *Plasmodium falciparum*. Gene 1993; 124:165-171.
74. Wong JM, Liu F, Bateman E. Cloning and expression of the *Acanthamoeba castellanii* gene encoding transcription factor TFIID. Gene 1992; 117:91-97.
75. Rowlands T, Baumann P, Jackson SP. The TATA-binding protein: a general transcription factor in eukaryotes and archaebacteria. Science 1994; 264:1326-1329.
76. Crowley TE, Hoey T, Liu J-K et al. A new factor related to TATA-binding protein has highly restricted expression pattern in *Drosophila*. Nature 1993; 361:557-561.
77. Mittal V, Hernandez N. Role for the amino-terminal region of human TBP in U6 snRNA transcription. Science 1997; 275:1136-1140.
78. Horikoshi M, Yamamoto T, Ohkuma Y et al. Analysis of structure-function relationships of yeast TATA box binding factor TFIID. Cell 1990; 61:1171-1178.
79. Poon D, Schroeder S, Wang CK et al. The conserved carboxy-terminal domain of *Saccharomyces cerevisiae* TFIID is sufficient to support normal cell growth. Mol Cell Biol 1991; 11:4809-4821.
80. Kim TK, Roeder RG. Involvement of the basic repeat domain of TATA-binding protein (TBP) in transcription by RNA polymerases I, II, and III. J Biol Chem 1994; 269:4891-4894.
81. Cormack BP, Strubin M, Ponticelli AS et al. Functional differences between yeast and human TFIID are localized to the highly conserved region. Cell 1991; 65:341-348.
82. Gill G, Tjian R. A highly conserved domain of TFIID displays species specificity in vivo. Cell 1991; 65:333-340.
83. Reddy P, Hahn S. Dominant negative mutations in yeast TFIID define a bipartite DNA-binding region. Cell 1991; 65:349-357.
84. Zhou QA, Schmidt MC, Berk AJ. Requirement for acidic amino acid residues immediately N-terminal to the conserved domain of *Saccharomyces cerevisiae* TFIID. EMBO J 1991; 10:1843-1852.
85. Lieberman PM, Schmidt MC, Kao CC et al. Two distinct domains in the yeast transcription factor IID and evidence for a TATA box-induced conformational change. Mol Cell Biol 1991; 11:63-74.
86. Bernues J, Simmen KA, Lewis JD et al. Common and unique transcription factor requirements of human U1 and U6 snRNA genes. EMBO J 1993; 12:3573-3585.
87. Zhou Q, Boyer TG, Berk AJ. Factors (TAFs) required for activated transcription interact with TATA box-binding protein conserved core domain. Genes Dev 1993; 7:180-187.
88. Rudloff U, Eberhard D, Grummt I. The conserved core domain of the human TATA binding protein is sufficient to assemble the multisubunit RNA polymerase I-specific transcription factor SL1. Proc Natl Acad Sci USA 1994; 91:8229-8233.
89. Kuddus R, Schmidt MC. Effect of the non-conserved N-terminus on the DNA binding activity of the yeast TATA binding protein. Nucleic Acids Res 1993; 21:1789-1796.
90. Cormack BP, Struhl K. Regional codon randomization: defining a TATA-binding protein surface required for RNA polymerase III transcription. Science 1993; 262:244-248.
91. Arndt KM, Ricupero SL, Eisenmann DM et al. Biochemical and genetic characterization of a yeast TFIID mutant that alters transcription in vivo and DNA binding in vitro. Mol Cell Biol 1992; 12:2372-2382.
92. Strubin M, Struhl K. Yeast and human TFIID with altered DNA-binding specificity for TATA elements. Cell 1992; 68:721-730.

93. Yamamoto T, Horikoshi M, Wang J et al. A bipartite DNA binding domain composed of direct repeats in the TATA box binding factor TFIID. Proc Natl Acad Sci USA 1992; 89:2844-2848.
94. Lee DK, Horikoshi M, Roeder RG. Interaction of TFIID in the minor groove of the TATA element. Cell 1991; 67:1241-1250.
95. Starr DB, Hawley DK. TFIID binds in the minor groove of the TATA box. Cell 1991; 67:1231-1240.
96. Horikoshi M, Bertuccioli C, Takada R et al. Transcription factor TFIID induces DNA bending upon binding to the TATA element. Proc Natl Acad Sci USA 1992; 89:1060-1064.
97. Schmidt MC, Zhou Q, Berk AJ. Sp1 activates transcription without enhancing DNA-binding activity of the TATA box factor. Mol Cell Biol 1989; 9:3299-3307.
98. Hoopes BC, LeBlanc JF, Hawley DK. Kinetic analysis of yeast TFIID-TATA box complex formation suggests a multi-step pathway. J Biol Chem 1992; 267: 11539-11547.
99. Nikolov DB, Hu S-H, Lin J et al. Crystal structure of TFIID TATA-box binding protein. Nature 1992; 360:40-46.
100. Kim Y, Geiger JH, Hahn S et al. Crystal structure of a yeast TBP/TATA-box complex. Nature 1993; 365:512-520.
101. Kim JL, Nikolov DB, Burley SK. Co-crystal structure of TBP recognizing the minor groove of a TATA element. Nature 1993; 365:520-527.
102. Singer VL, Wobbe CR, Struhl K. A wide variety of DNA sequences can functionally replace a yeast TATA element for transcriptional activation. Genes Dev 1990; 4:636-645.
103. Timmers HT, Sharp PA. The mammalian TFIID protein is present in two functionally distinct complexes. Genes Dev 1991; 5:1946-1956.
104. Brou C, Chaudhary S, Davidson I et al. Distinct TFIID complexes mediate the effect of different transcriptional activators. EMBO J 1993; 12:489-499.
105. Chiang C-M, Ge H, Wang Z et al. Unique TATA-binding protein-containing complexes and cofactors involved in transcription by RNA polymerases II and III. EMBO J 1993; 12:2749-2762.
106. Poon D, Weil PA. Immunopurification of yeast TATA-binding protein and associated factors. J Biol Chem 1993; 268:15325-15328.
107. Sadowski CL, Henry RW, Lobo SM et al. Targeting TBP to a non-TATA box *cis*-regulatory element: a TBP-containing complex activates transcription from snRNA promoters through the PSE. Genes Dev 1993; 7:1535-1548.
108. Pugh BF, Tjian R. Diverse transcriptional functions of the multisubunit eukaryotic TFIID complex. J Biol Chem 1992; 267:679-682.
109. Tansey WP, Herr W. TAFs: guilt by association? Cell 1997; 88:729-732.
110. Comai L, Tanese N, Tjian R. The TATA-binding protein and associated factors are integral components of the RNA polymerase I transcription factor, SL1. Cell 1992; 68:965-976.
111. Eberhard D, Tora L, Egly J-M et al. A TBP-containing multiprotein complex (TIF-IB) mediates transcription specificity of murine RNA polymerase I. Nucleic Acids Res 1993; 21:4180-4186.
112. Zomerdijk JCBM, Beckmann H, Comai L et al. Assembly of transcriptionally active RNA polymerase I initiation factor SL1 from recombinant subunits. Science 1994; 266:2015-2018.
113. Matsui T, Segall J, Weil PA et al. Multiple factors required for accurate initiation of transcription by purified RNA polymerase II. J Biol Chem 1980; 255:11992-11996.
114. Teichmann M, Seifart KH. Physical separation of two different forms of human TFIIIB active in the transcription of the U6 or the VAI gene in vitro. EMBO J 1995; 14:5974-5983.

115. Timmers HT, Meyers RE, Sharp PA. Composition of transcription factor B-TFIID. Proc Natl Acad Sci USA 1992; 89:8140-8144.
116. Kassavetis GA, Joazeiro CAP, Pisano M et al. The role of the TATA-binding protein in the assembly and function of the multisubunit yeast RNA polymerase III transcription factor, TFIIIB. Cell 1992; 71:1055-1064.
117. White RJ, Gottlieb TM, Downes CS et al. Mitotic regulation of a TATA-binding-protein-containing complex. Mol Cell Biol 1995; 15:1983-1992.
118. Mitchell MT, Hobson GM, Benfield PA. TATA box-mediated polymerase III transcription in vitro. J Biol Chem 1992; 267:1995-2005.
119. Mitchell MT, Benfield PA. TATA box-mediated in vitro transcription by RNA polymerase III. J Biol Chem 1993; 268:1141-1150.
120. Wang Y, Stumph WE. RNA polymerase II/III transcription specificity determined by TATA box orientation. Proc Natl Acad Sci USA 1995; 92:8606-8610.
121. Beckmann H, Chen J-L, O'Brien T et al. Coactivator and promoter-selective properties of RNA polymerase I TAFs. Science 1995; 270:1506-1509.
122. Kassavetis GA, Braun BR, Nguyen LH et al. *S. cerevisiae* TFIIIB is the transcription initiation factor proper of RNA polymerase III, while TFIIIA and TFIIIC are assembly factors. Cell 1990; 60:235-245.
123. Klekamp MS, Weil PA. Partial purification and characterization of the *Saccharomyces cerevisiae* transcription factor TFIIIB. J Biol Chem 1986; 261:2819-2827.
124. Klekamp MS, Weil PA. Properties of yeast class III gene transcription factor TFIIIB. J Biol Chem 1987; 262:7878-7883.
125. Kassavetis GA, Riggs DL, Negri R et al. Transcription factor IIIB generates extended DNA interactions in RNA polymerase III transcription complexes on tRNA genes. Mol Cell Biol 1989; 9:2551-2566.
126. Kassavetis GA, Bartholomew B, Blanco JA et al. Two essential components of the *Saccharomyces cerevisiae* transcription factor TFIIIB: transcription and DNA-binding properties. Proc Natl Acad Sci USA 1991; 88:7308-7312.
127. Huet J, Conesa C, Manaud N et al. Interactions between yeast TFIIIB components. Nucleic Acids Res 1994; 22:3433-3439.
128. Fuhrman SA, Engelke DR, Geiduschek EP. HeLa cell RNA polymerase III transcription factors. J Biol Chem 1984; 259:1934-1943.
129. Huibregtse JM, Evans CF, Engelke DR. Comparison of tRNA gene transcription complexes formed in vitro and in nuclei. Mol Cell Biol 1987; 7:3212-3220.
130. Waldschmidt R, Jahn D, Seifart KH. Purification of transcription factor IIIB from HeLa cells. J Biol Chem 1988; 263:13350-13356.
131. Giardina CA, Wu CW. The identification of two antagonistic activities in a *Xenopus* oocyte extract that can modulate the in vitro transcription of RNA polymerase III genes. J Biol Chem 1990; 265:9121-9130.
132. Wang Z, Roeder RG. TFIIIC1 acts through a downstream region to stabilize TFIIIC2 binding to RNA polymerase III promoters. Mol Cell Biol 1996; 16:6841-6850.
133. Bartholomew B, Kassavetis GA, Braun BR et al. The subunit structure of *Saccharomyces cerevisiae* transcription factor IIIC probed with a novel photocrosslinking reagent. EMBO J 1990; 9:2197-2205.
134. Bartholomew B, Braun BR, Kassavetis GA et al. Probing close DNA contacts of RNA polymerase III transcription complexes with the photoactive nucleoside 4-thiodeoxythymidine. J Biol Chem 1994; 269:18090-18095.
135. Persinger J, Bartholomew B. Mapping the contacts of yeast TFIIIB and RNA polymerase III at various distances from the major groove of DNA by DNA photoaffinity labeling. J Biol Chem 1996; 271:33039-33046.

136. Bartholomew B, Kassavetis GA, Geiduschek EP. Two components of *Saccharomyces cerevisiae* transcription factor IIIB (TFIIIB) are stereospecifically located upstream of a tRNA gene and interact with the second-largest subunit of TFIIIC. Mol Cell Biol 1991; 11:5181-5189.
137. Buratowski S, Zhou H. A suppressor of TBP mutations encodes an RNA polymerase III transcription factor with homology to TFIIB. Cell 1992; 71:221-230.
138. Colbert T, Hahn S. A yeast TFIIB-related factor involved in RNA polymerase III transcription. Genes Dev 1992; 6:1940-1949.
139. Lopez-de-Leon A, Librizzi M, Tuglia K et al. PCF4 encodes an RNA polymerase III transcription factor with homology to TFIIB. Cell 1992; 71:211-220.
140. Orphanides G, Lagrange T, Reinberg D. The general transcription factors of RNA polymerase II. Genes Dev 1996; 10:2657-2683.
141. Roeder RG. The role of general initiation factors in transcription by RNA polymerase II. Trends Biochem Sci 1996; 21:327-334.
142. Malik S, Hisatake K, Sumimoto H et al. Sequence of general transcription factor TFIIB and relationships to other initiation factors. Proc Natl Acad Sci USA 1991; 88:9553-9557.
143. Ha I, Lane WS, Reinberg D. Cloning of a human gene encoding the general transcription initiation factor IIB. Nature 1991; 352:689-695.
144. Hisatake K, Malik S, Roeder RG et al. Conserved structural motifs between *Xenopus* and human TFIIB. Nucleic Acids Res 1991; 19:6639.
145. Yamashita S, Wada K, Horikoshi M et al. Isolation and characterization of a cDNA encoding *Drosophila* transcription factor TFIIB. Proc Natl Acad Sci USA 1992; 89:2839-2843.
146. Wampler SL, Kadonaga JT. Functional analysis of *Drosophila* transcription factor IIB. Genes Dev 1992; 6:1542-1552.
147. Pinto I, Ware DE, Hampsey M. The yeast *SUA7* gene encodes a homolog of human transcription factor TFIIB and is required for normal start site selection in vivo. Cell 1992; 68:977-988.
148. Bagby S, Kim S, Maldonado E et al. Solution structure of the C-terminal core domain of human TFIIB: similarity to cyclin A and interaction with TATA-binding protein. Cell 1995; 82:857-867.
149. Nikolov DB, Chen H, Halay ED et al. Crystal structure of a TFIIB-TBP-TATA-element ternary complex. Nature 1995; 377:119-128.
150. Zhu W, Zheng Q, Colangelo CM et al. The N-terminal domain of TFIIB from *Pyrococcus furiosis* forms a zinc ribbon. Nature Struct Biol 1996; 3:122-124.
151. Gibson TJ, Thompson JD, Blocker A et al. Evidence for a protein domain superfamily shared by the cyclins, TFIIB and RB/p107. Nucleic Acids Res 1994; 22:946-952.
152. Barberis A, Muller CW, Harrison SC et al. Delineation of two functional regions of transcription factor TFIIB. Proc Natl Acad Sci USA 1993; 90:5628-5632.
153. Buratowski S, Zhou H. Functional domains of transcription factor TFIIB. Proc Natl Acad Sci USA 1993; 90:5633-5637.
154. Ha I, Roberts S, Maldonado E et al. Multiple functional domains of human transcription factor IIB: distinct interactions with two general transcription factors and RNA polymerase II. Genes Dev 1993; 7:1021-1032.
155. Hisatake K, Roeder RG, Horikoshi M. Functional dissection of TFIIB domains required for TFIIB-TFIID-promoter complex formation and basal transcription activity. Nature 1993; 363:744-747.
156. Khoo B, Brophy B, Jackson SP. Conserved functional domains of the RNA polymerase III general transcription factor BRF. Genes Dev 1994; 8:2879-2890.

157. Colgan J, Wampler S, Manley JL. Interaction between a transcriptional activator and transcription factor IIB in vivo. Nature 1993; 362:549-553.
158. Burton ZF, Killeen M, Sopta M et al. RAP30/74: a general initiation factor that binds to RNA polymerase II. Mol Cell Biol 1988; 8:1602-1613.
159. White RJ, Khoo BC-E, Inostroza JA et al. The TBP-binding repressor Dr1 differentially regulates RNA polymerases I, II and III. Science 1994; 266:448-450.
160. Chaussivert N, Conesa C, Shaaban S et al. Complex interactions between yeast TFIIIB and TFIIIC. J Biol Chem 1995; 270:15353-15358.
161. Malik S, Lee DK, Roeder RG. Potential RNA polymerase II-induced interactions of transcription factor TFIIB. Mol Cell Biol 1993; 13:6253-6259.
162. Roberts SGE, Green MR. Activator-induced conformational change in general transcription factor TFIIB. Nature 1994; 371:717-720.
163. Kassavetis GA, Nguyen ST, Kobayashi R et al. Cloning, expression, and function of *TFC5*, the gene encoding the B" component of the *Saccharomyces cerevisiae* RNA polymerase III transcription factor TFIIIB. Proc Natl Acad Sci USA 1995; 92:9786-9790.
164. Roberts S, Miller SJ, Lane WS et al. Cloning and functional characterization of the gene encoding the TFIIIB90 subunit of RNA polymerase III transcription factor TFIIIB. J Biol Chem 1996; 271:14903-14909.
165. Ruth J, Conesa C, Dieci G et al. A suppressor of mutations in the class III transcription system encodes a component of yeast TFIIIB. EMBO J 1996; 15:1941-1949.
166. Aasland R, Stewart AF, Gibson T. The SANT domain: a putative DNA-binding domain in the SWI-SNF and ADA complexes, the transcriptional corepressor N-CoR and TFIIIB. Trends Biochem Sci 1996; 21:87-88.
167. Kumar A, Kassavetis GA, Geiduschek EP et al. Functional dissection of the B" component of RNA polymerase III transcription factor IIIB: a scaffolding protein with multiple roles in assembly and initiation of transcription. Mol Cell Biol 1997; 17:1868-1880.
168. Bartholomew B, Durkovich D, Kassavetis GA et al. Orientation and topography of RNA polymerase III in transcription complexes. Mol Cell Biol 1993; 13:942-952.
169. Dieci G, Duimio L, Coda-Zabetta F et al. A novel RNA polymerase III transcription factor fraction that is not required for template commitment. J Biol Chem 1993; 268:11199-11207.
170. Dieci G, Duimio L, Peracchia G et al. Selective inactivation of two components of the multiprotein transcription factor TFIIIB in cycloheximide growth-arrested yeast cells. J Biol Chem 1995; 270:13476-13482.
171. Chao DM, Young RA. Activation without a vital ingredient. Nature 1996; 383:119-120.
172. Huang W, Wong JM, Bateman E. TATA elements direct bi-directional transcription by RNA polymerases II and III. Nucleic Acids Res 1996; 24:1158-1163.
173. Wang Y, Jensen RC, Stumph WE. Role of TATA box sequence and orientation in determining RNA polymerase II/III transcription specificity. Nucleic Acids Res 1996; 24:3100-3106.
174. Roberts S, Colbert T, Hahn S. TFIIIC determines RNA polymerase III specificity at the TATA-containing yeast U6 promoter. Genes Dev 1995; 9:832-842.
175. Larminie CGC, White RJ. Identification of a putative BRF homologue in the genome of *Caenorhabditis elegans*. DNA Sequence 1998; in press.
176. Yoon J-B, Murphy S, Bai L et al. Proximal sequence element-binding transcription factor (PTF) is a multisubunit complex required for transcription of both RNA polymerase II- and RNA polymerase III-dependent small nuclear RNA genes. Mol Cell Biol 1995; 15:2019-2027.

177. Teichmann M, Dieci G, Huet J et al. Functional interchangeability of TFIIIB components from yeast and human cells in vitro. EMBO J 1997; 16:4708-4716.
178. Wang Z, Roeder RG. Structure and function of a human transcription factor TFIIIB subunit that is evolutionarily conserved and contains both TFIIB- and high-mobility-group protein 2-related domains. Proc Natl Acad Sci USA 1995; 92:7026-7030.
179. Ruet A, Camier S, Smagowicz W et al. Isolation of a class C transcription factor which forms a stable complex with tRNA genes. EMBO J 1984; 3:343-350.
180. Stillman DJ, Better M, Geiduschek EP. Electron-microscopic examination of the binding of a large RNA polymerase III transcription factor to a tRNA gene. J Mol Biol 1985; 185:451-455.
181. Schultz P, Marzouki N, Marck C et al. The two DNA-binding domains of yeast transcription factor τ as observed by scanning transmission electron microscopy. EMBO J 1989; 8:3815-3824.
182. Burnol A-F, Margottin F, Schultz P et al. Basal promoter and enhancer element of yeast U6 snRNA gene. J Mol Biol 1993; 233:644-658.
183. Baker RE, Gabrielsen O, Hall BD. Effects of $tRNA^{Tyr}$ point mutations on the binding of yeast RNA polymerase III transcription factor C. J Biol Chem 1986; 261:5275-5282.
184. Lefebvre O, Ruth J, Sentenac A. A mutation in the largest subunit of yeast TFIIIC affects tRNA and 5S RNA synthesis. J Biol Chem 1994; 269:23374-23381.
185. Camier S, Gabrielsen O, Baker R et al. A split binding site for transcription factor τ on the $tRNA^{Glu3}$ gene. EMBO J 1985; 4:491-500.
186. Gabrielsen OS, Marzouki N, Ruet A et al. Two polypeptide chains in yeast transcription factor τ interact with DNA. J Biol Chem 1989; 264:7505-7511.
187. Parsons MC, Weil PA. Purification and characterization of *Saccharomyces cerevisiae* transcription factor TFIIIC. J Biol Chem 1990; 265:5095-5103.
188. Swanson RN, Conesa C, Lefebvre O et al. Isolation of *TFC1*, a gene encoding one of two DNA-binding subunits of yeast transcription factor τ (TFIIIC). Proc Natl Acad Sci USA 1991; 88:4887-4891.
189. Braun BR, Bartholomew B, Kassavetis GA et al. Topography of transcription factor complexes on the *Saccharomyces cerevisiae* 5S RNA gene. J Mol Biol 1992; 228:1063-1077.
190. Marzouki M, Camier S, Piwet A et al. Selective proteolysis defines two DNA binding domains in yeast transcription factor τ. Nature 1986; 323:176-178.
191. Gabrielsen OS, Hornes E, Korsnes L et al. Magnetic DNA affinity purification of yeast transcription factor tau—a new purification principle for the ultrarapid isolation of near homogeneous factor. Nucleic Acids Res 1989; 17:6253-6267.
192. Johnson DL, Wilson SL. Identification of a 150-kilodalton polypeptide that copurifies with yeast TFIIIC and binds specifically to tRNA genes. Mol Cell Biol 1989; 9:2018-2024.
193. Lefebvre O, Carles C, Conesa C et al. *TFC3*: gene encoding the B-block binding subunit of the yeast transcription factor IIIC. Proc Natl Acad Sci USA 1992; 89:10512-10516.
194. Parsons MC, Weil PA. Cloning of *TFC1*, the *Saccharomyces cerevisiae* gene encoding the 95-kDa subunit of transcription factor TFIIIC. J Biol Chem 1992; 267:2894-2901.
195. Marck C, Lefebvre O, Carles C et al. The TFIIIB-assembling subunit of yeast transcription factor TFIIIC has both tetratricopeptide repeats and basic helix-loop-helix motifs. Proc Natl Acad Sci USA 1993; 90:4027-4031.
196. Lilley DM. HMG has DNA wrapped up. Nature 1992; 357:282-283.
197. Ner SS. HMGs everywhere. Current Biology 1992; 2:208-210.

198. Jantzen H-M, Admon A, Bell SP et al. Nucleolar transcription factor hUBF contains a DNA-binding motif with homology to HMG proteins. Nature 1990; 344:830-836.
199. Bachvarov D, Moss T. The RNA polymerase I transcription factor xUBF contains 5 tandemly repeated HMG homology boxes. Nucleic Acids Res 1991; 19:2331-2335.
200. Sinclair AH, Berta P, Palmer MS et al. A gene from the human sex-determining region encodes a protein with homology to a conserved DNA-binding motif. Nature 1990; 346:240-244.
201. Travis A, Amsterdam A, Belanger C et al. LEF-1, a gene encoding a lymphoid-specific protein with an HMG domain, regulates T-cell receptor α enhancer function. Genes Dev 1991; 5:880-894.
202. Giese K, Cox J, Grosschedl R. The HMG domain of lymphoid enhancer factor I bends DNA and facilitates assembly of functional nucleoprotein structures. Cell 1992; 69:185-195.
203. Leveillard T, Kassavetis GA, Geiduschek EP. *Saccharomyces cerevisiae* transcription factors IIIB and IIIC bend the DNA of a $tRNA^{Gln}$ gene. J Biol Chem 1991; 266:5162-5168.
204. Rameau G, Puglia K, Crowe A et al. A mutation in the second largest subunit of TFIIIC increases a rate-limiting step in transcription by RNA polymerase III. Mol Cell Biol 1994; 14:822-830.
205. Sikorski RS, Boguski MS, Goebl M et al. A repeating amino acid motif in *CDC23* defines a family of proteins and a new relationship among genes required for mitosis and RNA synthesis. Cell 1990; 60:307-317.
206. Hirano T, Kinoshita N, Morikawa K et al. Snap helix with knob and hole: essential repeats in *S. pombe* nuclear protein *nuc2*$^+$. Cell 1990; 60:319-328.
207. Lamb JR, Tugendreich S, Hieter P. Tetratrico peptide repeat interactions: to TPR or not to TPR? Trends Biochem Sci 1995; 20:257-259.
208. Willis IM, Schmidt P, Soll D. A selection for mutants of the RNA polymerase III transcription apparatus: *PCF1* stimulates transcription of tRNA and 5S RNA genes. EMBO J 1989; 8:4281-4288.
209. Willis I, Oksman A, Lopez-De-Leon A. The PCF1-1 mutation increases the activity of the transcription factor (TF) IIIB fraction from *Saccharomyces cerevisiae.* Nucleic Acids Res 1992; 20:3725-3730.
210. Keller HJ, Romaniuk PJ, Gottesfeld JM. Interaction of *Xenopus* TFIIIC with the TFIIIA.5 S RNA gene complex. J Biol Chem 1992; 267:18190-18198.
211. Cohen I, Reynolds WF. The *Xenopus* YB3 protein binds the B box element of the class III promoter. Nucleic Acids Res 1991; 19:4753-4759.
212. Wolffe AP, Tafuri S, Ranjan M et al. The Y-box factors: a family of nucleic acid binding proteins conserved from *Escherichia coli* to man. New Biol 1992; 4:290-298.
213. Didier DK, Schiffenbauer J, Woulfe SL et al. Characterization of the cDNA encoding a protein binding to the major histocompatibility complex class II Y box. Proc Natl Acad Sci USA 1988; 85:7322-7326.
214. Dean N, Berk AJ. Separation of TFIIIC into two functional components by sequence specific DNA affinity chromatography. Nucleic Acids Res 1987; 15:9895-9907.
215. Yoshinaga SK, Boulanger PA, Berk AJ. Resolution of human transcription factor TFIIIC into two functional components. Proc Natl Acad Sci USA 1987; 84:3585-3589.
216. L'Etoile ND, Fahnestock ML, Shen Y et al. Human transcription factor IIIC box B binding subunit. Proc Natl Acad Sci USA 1994; 91:1652-1656.
217. Lagna G, Kovelman R, Sukegawa J et al. Cloning and characterization of an evolutionarily divergent DNA-binding subunit of mammalian TFIIIC. Mol Cell Biol 1994; 14:3053-3064.

218. Oettel S, Hartel F, Kober I et al. Human transcription factors $IIIC_2$, $IIIC_1$ and a novel component $IIIC_0$ fulfill different aspects of DNA binding to various pol III genes. Nucleic Acids Res 1997; 25:2440-2447.
219. Yoshinaga SK, L'Etoile ND, Berk AJ. Purification and characterization of transcription factor IIIC2. J Biol Chem 1989; 264:10726-10731.
220. Kovelman R, Roeder RG. Purification and characterization of two forms of human transcription factor IIIC. J Biol Chem 1992; 267:24446-24456.
221. Boulanger PA, Yoshinaga SK, Berk AJ. DNA-binding properties and characterization of human transcription factor TFIIIC2. J Biol Chem 1987; 262:15098-15105.
222. Boulanger PA, L' Etoile ND, Berk AJ. A DNA-binding domain of human transcription factor IIIC2. Nucleic Acids Res 1989; 17:7761-7770.
223. Shen Y, Igo M, Yalamanchili P et al. DNA binding domain and subunit interactions of transcription factor IIIC revealed by dissection with poliovirus 3C protease. Mol Cell Biol 1996; 16:4163-4171.
224. Hoeffler WK, Kovelman R, Roeder RG. Activation of transcription factor IIIC by the adenovirus E1A protein. Cell 1988; 53:907-920.
225. Sinn E, Wang Z, Kovelman R et al. Cloning and characterization of a TFIIIC2 subunit (TFIIICβ) whose presence correlates with activation of RNA polymerase III-mediated transcription by adenovirus E1A expression and serum factors. Genes Dev 1995; 9:675-685.
226. Komachi K, Redd M, Johnson A. The WD repeats of Tup1 interact with the homeo domain protein α 2. Genes Dev 1994; 8:2857-2867.
227. Tzamarias D, Struhl K. Functional dissection of the yeast Cyc8-Tup1 transcriptional co-repressor complex. Nature 1994; 369:758-761.
228. Schneider HR, Waldschmidt R, Jahn D et al. Purification of human transcription factor IIIC and its binding to the gene for ribosomal 5S RNA. Nucleic Acids Res 1989; 17:5003-5016.
229. Schneider HR, Waldschmidt R, Seifart KH. Human transcription factor IIIC contains a polypeptide of 55 kDa specifically binding to Pol III genes. Nucleic Acids Res 1990; 18:4743-4750.
230. Waldschmidt R, Schneider HR, Seifart KH. Human transcription factor IIIC binds to its cognate promoter sequences in a metal coordinated fashion. Nucleic Acids Res 1991; 19:1455-1459.
231. Young LS, Rivier DH, Sprague KU. Sequences far downstream from the classical tRNA promoter elements bind RNA polymerase III transcription factors. Mol Cell Biol 1991; 11:1382-1392.
232. Young LS, Ahnert N, Sprague KU. Silkworm TFIIIB binds both constitutive and silk gland-specific $tRNA^{Ala}$ promoters but protects only the constitutive promoter from DNase I cleavage. Mol Cell Biol 1996; 16:1256-1266.
233. Sprague KU. New twists in class III transcription. Curr Opin Cell Biol 1992; 4:475-479.
234. Guddat U, Bakken AH, Pieler T. Protein-mediated nuclear export of RNA: 5S rRNA containing small RNPs in *Xenopus* oocytes. Cell 1990; 60:619-628.
235. Yoo CJ, Wolin SL. La proteins from *Drosophila melanogaster* and *Saccharomyces cerevisiae*: a yeast homolog of the La autoantigen is dispensable for growth. Mol Cell Biol 1994; 14:5412-5424.
236. Rinke J, Steitz JA. Precursor molecules of both human 5S ribosomal RNA and transfer RNAs are bound by a cellular protein reactive with Anti-La Lupus antibodies. Cell 1982; 29:149-159.
237. Gottesfeld JM, Andrews DL, Hoch SA. Association of an RNA polymerase III transcription factor with a ribonucleoprotein complex recognized by autoimmune sera. Nucleic Acids Res 1984; 12:3185-3200.

238. Stefano JE. Purified lupus antigen La recognizes an oligouridylate stretch common to the 3' termini of RNA polymerase III transcripts. Cell 1984; 36:145-154.
239. Gottlieb E, Steitz JA. The RNA binding protein La influences both the accuracy and the efficiency of RNA polymerase III transcription in vitro. EMBO J 1989; 8:841-850.
240. Kramerov DA, Tillib SV, Shumyatsky GP et al. The most abundant nascent poly(A)$^+$ RNAs are transcribed by RNA polymerase III in murine tumor cells. Nucleic Acids Res 1990; 18:4499-4506.
241. Schwemmle M, Clemens MJ, Hilse K et al. Localization of Epstein-Barr virus-encoded RNAs EBER-1 and EBER-2 in interphase and mitotic Burkitt lymphoma cells. Proc Natl Acad Sci USA 1992; 89:10292-10296.
242. Maraia RJ, Kenan DJ, Keene JD. Eukaryotic transcription termination factor La mediates transcript release and facilitates reinitiation by RNA polymerase III. Mol Cell Biol 1994; 14:2147-2158.
243. Gottlieb E, Steitz JA. Function of the mammalian La protein: evidence for its action in transcription termination by RNA polymerase III. EMBO J 1989; 8:851-861.
244. Maraia RJ. Transcription termination factor La is also an initiation factor for RNA polymerase III. Proc Natl Acad Sci USA 1996; 93:3383-3387.
245. Fan H, Sakulich AL, Goodier JL et al. Phosphorylation of the human La antigen on serine 366 can regulate recycling of RNA polymerase III transcription complexes. Cell 1997; 88:707-715.
246. Jahn D, Wingender E, Seifart KH. Transcription complexes for various class III genes differ in parameters of formation and stability towards salt. J Mol Biol 1987; 193:303-313.
247. Chambers JC, Keene JD. Isolation and analysis of cDNA clones expressing human lupus La antigen. Proc Natl Acad Sci USA 1985; 82:2115-2119.
248. Chambers JC, Kenan D, Martin BJ et al. Genomic structure and amino acid sequence domains of the human La autoantigen. J Biol Chem 1988; 263:18043-18051.
249. Chan EKL, Sullivan KF, Tan EM. Ribonucleoprotein SS-B/La belongs to a protein family with consensus sequences for RNA-binding. Nucleic Acids Res 1989; 17:2233-2244.
250. Query CC, Bentley RC, Keene JD. A common RNA recognition motif identified within a defined U1 RNA binding domain of the 70K U1 snRNP protein. Cell 1989; 57:89-101.
251. Bachmann M, Pfeifer K, Schroder HC et al. Characterization of the autoantigen La as a nucleic acid-dependent ATPase/dATPase with melting properties. Cell 1990; 60:85-93.
252. Brennan CA, Dombroski AJ, Platt T. Transcription termination factor rho is an RNA-DNA helicase. Cell 1987; 48:945-952.
253. Campbell FE, Setzer DR. Transcription termination by RNA polymerase III: uncoupling of polymerase release from termination signal recognition. Mol Cell Biol 1992; 12:2260-2272.
254. Yoo CJ, Wolin SL. The yeast La protein is required for the 3' endonucleolytic cleavage that matures tRNA precursors. Cell 1997; 89:393-402.
255. Boelens WC, Palacios I, Mattaj IW. Nuclear retention of RNA as a mechanism for localization. RNA 1995; 1:273-283.
256. Simons FHM, Rutjes SA, Van Venrooij WJ et al. the interactions with Ro60 and La differentially affect nuclear export of hY1 RNA. RNA 1996; 2:264-273.
257. Grimm C, Lund E, Dahlberg JE. In vivo selection of RNAs that localize in the nucleus. EMBO J 1997; 16:793-806.
258. Shastry BS. Xenopus transcription factor IIIA (XTFIIIA): after a decade of research. Prog Biophys Mol Biol 1991; 56:135-144.

259. Pieler T, Theunissen O. TFIIIA: nine fingers-three hands? Trends Biochem Sci 1993; 18:226-230.
260. Pelham HRB, Brown DD. A specific transcription factor that can bind either the 5S RNA gene or 5S RNA. Proc Natl Acad Sci USA 1980; 77:4170-4174.
261. Hanas JS, Bogenhagen DF, Wu C-W. Binding of *Xenopus* transcription factor A to 5S RNA and to single-stranded DNA. Nucleic Acids Res 1984; 12:2745-2758.
262. Hazuda DJ, Wu C-W. DNA-activated ATPase activity associated with *Xenopus* transcription factor A. J Biol Chem 1986; 261:12202-12208.
263. Ginsberg AM, King BO, Roeder RG. *Xenopus* 5S gene transcription factor, TFIIIA: characterization of a cDNA clone and measurement of RNA levels throughout development. Cell 1984; 39:479-489.
264. Pelham HRB, Wormington WM, Brown DD. Related 5S RNA transcription factors in *Xenopus* oocytes and somatic cells. Proc Natl Acad Sci USA 1981; 78:1760-1764.
265. Shastry BS, Honda BM, Roeder RG. Altered levels of a 5S gene-specific transcription factor (TFIIIA) during oogenesis and embryonic development of *Xenopus laevis*. J Biol Chem 1984; 259:11373-11382.
266. Andrews MT, Brown DD. Transient activation of oocyte 5S RNA genes in *Xenopus* embryos by raising the level of the trans-acting factor TFIIIA. Cell 1987; 51:445-453.
267. Darby MK, Andrews TM, Brown DD. Transcription complexes that program *Xenopus* 5S RNA genes are stable in vivo. Proc Natl Acad Sci USA 1988; 85:5516-5520.
268. Kim SH, Darby MK, Joho KE et al. The characterization of the TFIIIA synthesized in somatic cells of *Xenopus laevis*. Genes Dev 1990; 4:1602-1610.
269. Blanco J, Millstein L, Razik MA et al. Two TFIIIA activities regulate expression of the *Xenopus* 5S RNA gene families. Genes Dev 1989; 3:1602-1612.
270. Brown DD. Is there a *Xenopus* transcription factor that can substitute for TFIIIA? Re: Two TFIIIA activities regulate expression of the *Xenopus* 5S RNA gene families. Genes Dev 1991; 5:1737-1738.
271. Martinez E, Lagna G, Roeder RG. Overlapping transcription by RNA polymerases II and III of the *Xenopus* TFIIIA gene in somatic cells. J Biol Chem 1994; 269:25692-25698.
272. Bieker JJ, Roeder RG. Physical properties and DNA-binding stoichiometry of a 5S gene-specific transcription factor. J Biol Chem 1984; 259:6158-6164.
273. Sakonju S, Brown DD. Contact points between a positive transcription factor and the *Xenopus* 5S RNA gene. Cell 1982; 31:395-405.
274. Smith DR, Jackson IJ, Brown DD. Domains of the positive transcription factor specific for the *Xenopus* 5S RNA gene. Cell 1984; 37:645-652.
275. Hayes J, Tullius TD, Wolffe AP. A protein-protein interaction is essential for stable complex formation on a 5S RNA gene. J Biol Chem 1989; 264:6009-6012.
276. Xing YY, Worcel A. The C-terminal domain of transcription factor IIIA interacts differently with different 5S RNA genes. Mol Cell Biol 1989; 9:499-514.
277. Fiser-Littell KM, Duke AL, Yanchick JS et al. Deletion of the N-terminal region of *Xenopus* transcription factor IIIA inhibits specific binding to the 5S RNA gene. J Biol Chem 1988; 263:1607-1610.
278. Miller J, McLachlan AD, Klug A. Repetitive zinc-binding domains in the protein transcription factor IIIA from *Xenopus* oocytes. EMBO J 1985; 4:1609-1614.
279. Tso JY, Van Den Berg DJ, Korn LJ. Structure of the gene for *Xenopus* transcription factor TFIIIA. Nucleic Acids Res 1986; 14:2187-2200.
280. Hanas JS, Hazuda DJ, Bogenhagen DF et al. *Xenopus* transcription factor A requires zinc for binding to the 5S RNA gene. J Biol Chem 1983; 258:14120-14125.

281. Diakun GP, Fairall L, Klug A. EXAFS study of the zinc-binding sites in the protein transcription factor IIIA. Nature 1986; 324:698-699.
282. Enver T. A pulling out of fingers. Nature 1985; 317:385-386.
283. Vincent A. TFIIIA and homologous genes. The 'finger' proteins. Nucleic Acids Res 1986; 14:4385-4391.
284. Evans RM, Hollenberg SM. Zinc fingers: gilt by association. Cell 1988; 52:1-3.
285. Berg JM. Zinc finger domains: hypotheses and current knowledge. Annu Rev Biophys Chem 1990; 19:405-421.
286. Berg JM. Proposed structure for the zinc-binding domains from transcription factor IIIA and related proteins. Proc Natl Acad Sci USA 1988; 85:99-102.
287. Parraga G, Horvath SJ, Eisen A et al. Zinc-dependent structure of a single-finger domain of yeast ADR1. Science 1988; 241:1489-1492.
288. Lee MSL, Gippert GP, Soman KV et al. Three-dimensional solution structure of a single zinc finger DNA-binding domain. Science 1989; 245:635-637.
289. Pavletich NP, Pabo CO. Zinc finger-DNA recognition: crystal structure of a Zif268-DNA complex at 2.1 A. Science 1991; 252:809-817.
290. Vrana KE, Churchill MEA, Tullius TD et al. Mapping functional regions of transcription factor TFIIIA. Mol Cell Biol 1988; 8:1684-1696.
291. Clemens KR, Liao X, Wolf V et al. Definition of the binding sites of individual zinc fingers in the transcription factor IIIA-5S RNA gene complex. Proc Natl Acad Sci USA 1992; 89:10822-10826.
292. Fairall L, Rhodes D. A new approach to the analysis of DNase I footprinting data and its application to the TFIIIA/5S DNA complex. Nucleic Acids Res 1992; 20:4727-4731.
293. Hayes JJ, Clemens KR. Locations of contacts between individual zinc fingers of *Xenopus laevis* transcription factor IIIA and the internal control region of a 5S RNA gene. Biochemistry 1992; 31:11600-11605.
294. Hayes JJ, Tullius TD. Structure of the TFIIIA-5S DNA complex. J Mol Biol 1992; 227:407-417.
295. Christensen JH, Hansen PK, Lillelund O et al. Sequence-specific binding of the N-terminal three-finger fragment of *Xenopus* transcription factor IIIA to the internal control region of a 5S RNA gene. FEBS Lett 1991; 281:181-184.
296. Darby MK, Joho KE. Differential binding of zinc fingers from *Xenopus* TFIIIA and p43 to 5S RNA and the 5S RNA gene. Mol Cell Biol 1992; 12:3155-3164.
297. Liao XB, Clemens KR, Tennant L et al. Specific interaction of the first three zinc fingers of TFIIIA with the internal control region of the *Xenopus* 5 S RNA gene. J Mol Biol 1992; 223:857-871.
298. Theunissen O, Rudt F, Guddat U et al. RNA and DNA Binding zinc fingers in *Xenopus* TFIIIA. Cell 1992; 71:679-690.
299. Choo Y, Klug A. A role in DNA binding for the linker sequences of the first three zinc fingers of TFIIIA. Nucleic Acids Res 1993; 21:3341-3346.
300. Friesen WJ, Darby MK. Phage display of RNA binding zinc fingers from transcription factor IIIA. J Biol Chem 1997; 272:10994-10997.
301. Del Rio S, Setzer DR. The role of zinc fingers in transcriptional activation by transcription factor IIIA. Proc Natl Acad Sci USA 1993; 90:168-172.
302. Rollins MB, Del Rio S, Galey AL et al. Role of TFIIIA zinc fingers in vivo: analysis of single-finger function in developing *Xenopus* embryos. Mol Cell Biol 1993; 13:4776-4783.
303. Mao X, Darby MK. A position-dependent transcription-activating domain in TFIIIA. Mol Cell Biol 1993; 13:7496-7506.
304. Ford P. Non-coordinated accumulation and synthesis of 5S ribonucleic acid by ovaries of *Xenopus laevis*. Nature 1971; 233:561-564.

305. Denis H, Mairy M. Biochemical studies on oogenesis. I. Intracellular RNA distribution in small oocytes of *Xenopus laevis*. Eur J Biochem 1972; 25:524-534.
306. Picard B, Wegnez M. Isolation of a 7S particle from *Xenopus laevis* oocytes: a 5S RNA-protein complex. Proc Natl Acad Sci USA 1979; 76:241-245.
307. Tafuri SR, Wolffe AP. Dual roles for transcription and translation factors in the RNA storage particles of *Xenopus* oocytes. Trends Cell Biol 1993; 3:94-98.
308. Rhodes D, Klug A. An underlying repeat in some transcriptional control sequences corresponding to half a double helical turn of DNA. Cell 1986; 46:123-132.
309. McCall M, Brown T, Hunter WN et al. The crystal structure of d(GGATGGGAG) forms an essential part of the binding site for transcription factor IIIA. Nature 1986; 322:661-664.
310. Gottesfeld JM, Blanco J, Tennant LL. The 5S gene internal control region is B-form both free in solution and in a complex with TFIIIA. Nature 1987; 329:460-462.
311. Aboul-ela F, Varani G, Walker GT et al. The TFIIIA recognition fragment d(GGATGGGAG)·d(CTCCCATCC) is B-form in solution. Nucleic Acids Res 1988; 16:3559-3572.
312. Fairall L, Martin S, Rhodes D. The DNA binding site of the *Xenopus* transcription factor IIIA has a non-B-form structure. EMBO J 1989; 8:1809-1817.
313. Romaniuk PJ, Leal de Stevenson I, Wong HH. Defining the binding site of *Xenopus* transcription factor IIIA on 5S RNA using truncated and chimeric 5S RNA molecules. Nucleic Acids Res 1987; 15:2737-2755.
314. Christiansen J, Brown R, Sproat B et al. *Xenopus* transcription factor IIIA binds primarily at junctions between double helical stems and internal loops in oocyte 5S RNA. EMBO J 1987; 6:453-460.
315. Baudin F, Romaniuk PJ. A difference in the importance of bulged nucleotides and their parent base pairs in the binding of transcription factor IIIA to *Xenopus* 5S RNA and 5S RNA genes. Nucleic Acids Res 1989; 17:2043-2056.
316. Romaniuk PJ. The role of highly conserved single-stranded nucleotides of *Xenopus* 5S RNA. Biochemistry 1989; 28:1388-1395.
317. You QM, Romaniuk PJ. The effects of disrupting 5S RNA helical structures on the binding of *Xenopus* transcription factor IIIA. Nucleic Acids Res 1990; 18:5055-5062.
318. Sands MS, Bogenhagen DF. TFIIIA binds to different domains of 5S RNA and the *Xenopus borealis* 5S RNA gene. Mol Cell Biol 1987; 7:3985-3993.
319. You QM, Veldhoen N, Baudin F et al. Mutations in 5S DNA and 5S RNA have different effects on the binding of *Xenopus* transcription factor IIIA. Biochemistry 1991; 30:2495-2500.
320. McBryant SJ, Veldhoen N, Gedulin B et al. Interaction of the RNA binding fingers of *Xenopus* transcription factor IIIA with specific regions of 5 S ribosomal RNA. J Mol Biol 1995; 248:44-57.
321. Clemens KR, Wolf V, McBryant SJ et al. Molecular basis for specific recognition of both RNA and DNA by a zinc finger protein. Science 1993; 260:530-533.
322. Honda BM, Roeder RG. Association of a 5S gene transcription factor with 5S RNA and altered levels of the factor during cell differentiation. Cell 1980; 22:119-126.
323. Allison LA, North MT, Neville LA. Differential binding of oocyte-type and somatic-type 5S rRNA to TFIIIA and ribosomal protein L5 in *Xenopus* oocytes: specialization for storage versus mobilization. Dev Biol 1995; 168:284-295.
324. Rudt F, Pieler T. Cytoplasmic retention and nuclear import of 5S ribosomal RNA containing RNPs. EMBO J 1996; 15:1383-1391.
325. Allison LA, Romaniuk PJ, Bakken AH. RNA-protein interactions of stored 5S RNA with TFIIIA and ribosomal protein L5 during *Xenopus* oogenesis. Dev Biol 1991; 144:129-144.

326. Woodland HR. Stable gene expression in vitro. Nature 1982; 297:457-458.
327. Picard B, Le Maire M, Wegnez M et al. Biochemical research on oogenesis. Composition of the 42S storage particles of *Xenopus laevis* oocytes. Eur J Biochem 1980; 109:665-669.
328. Mattaj IW, Coppard NJ, Brown RS et al. 42S p48-the most abundant protein in previtellogenic *Xenopus* oocytes-resembles elongation factor 1α structurally and functionally. EMBO J 1987; 6:2409-2413.
329. Joho KE, Darby MK, Crawford ET et al. A finger protein structurally similar to TFIIIA that binds exclusively to 5S RNA in *Xenopus*. Cell 1990; 61:293-300.
330. Sands MS, Bogenhagen DF. Two zinc finger proteins from *Xenopus laevis* bind the same region of 5S RNA but with different nuclease protection patterns. Nucleic Acids Res 1991; 19:1797-1803.
331. Seifart KH, Wang L, Waldschmidt R et al. Purification of human transcription factor IIIA and its interaction with a chemically synthesized gene encoding human 5S rRNA. J Biol Chem 1989; 264:1702-1709.
332. Moorefield B, Roeder RG. Purification and characterization of human transcription factor IIIA. J Biol Chem 1994; 269:20857-20865.
333. Smith TPL, Young LS, Bender LB et al. Silkworm TFIIIA requires additional class III factors for committment to transcription complex assembly on a 5S RNA gene. Nucleic Acids Res 1995; 23:1244-1251.
334. Wang CK, Weil PA. Purification and characterization of *Saccharomyces cerevisiae* transcription factor IIIA. J Biol Chem 1989; 264:1092-1099.
335. Archambault J, Milne CA, Schappert KT et al. The deduced sequence of the transcription factor TFIIIA from *Saccharomyces cerevisiae* reveals extensive divergence from *Xenopus* TFIIIA. J Biol Chem 1992; 267:3282-3288.
336. Woychik NA, Young RA. Genes encoding transcription factor IIIA and the RNA polymerase common subunit RPB6 are divergently transcribed in *Saccharomyces cerevisiae*. Proc Natl Acad Sci USA 1992; 89:3999-4003.
337. Milne CA, Segall J. Mapping regions of yeast transcription factor IIIA required for DNA binding, interaction with transcription factor IIIC, and transcription activity. J Biol Chem 1993; 268:11364-11371.
338. Gaskins CJ, Hanas JS. Sequence variation in transcription factor IIIA. Nucleic Acids Res 1990; 18:2117-2123.
339. Braun BR, Riggs DL, Kassavetis GA et al. Multiple states of protein-DNA interaction in the assembly of transcription complexes on *Saccharomyces cerevisiae* 5S ribosomal RNA genes. Proc Natl Acad Sci USA 1989; 86:2530-2534.
340. Struksnes K, Forus A, Gabrielsen OS et al. Yeast TFIIIA + TFIIIC/τ-factor, but not yeast TFIIIA alone, interacts with the *Xenopus* 5S rRNA gene. Nucleic Acids Res 1991; 19:565-571.
341. Camier S, Dechampesme A-M, Sentenac A. The only essential function of TFIIIA in yeast is the transcription of 5S rRNA genes. Proc Natl Acad Sci USA 1995; 92:9338-9342.
342. Lobo SM, Hernandez N. A 7 bp mutation converts a human RNA polymerase II snRNA promoter into an RNA polymerase III promoter. Genes Dev 1989; 58:55-67.
343. Li J-M, Haberman RP, Marzluff WF. Common factors direct transcription through the proximal sequence elements (PSEs) of the embryonic sea urchin U1, U2, and U6 genes despite minimal sequence similarity among the PSEs. Mol Cell Biol 1996; 16:1275-1281.
344. Gunderson SI, Knuth MW, Burgess RR. The human U1 promoter correctly initiates transcription in vitro and is activated by PSE1. Genes Dev 1990; 4:2048-2060.
345. Knuth MW, Gunderson SI, Thompson NE et al. Purification and characterization of proximal sequence element-binding protein 1, a transcription activating pro-

tein related to Ku and TREF that binds the proximal sequence element of the human U1 promoter. J Biol Chem 1990; 265:17911-17920.
346. Waldschmidt R, Wanandi I, Seifart KH. Identification of transcription factors required for the expression of mammalian U6 genes in vitro. EMBO J 1991; 10:2595-2603.
347. Wanandi I, Waldschmidt R, Seifart KH. Mammalian transcription factor PBP. J Biol Chem 1993; 268:6629-6640.
348. Simmen KA, Waldschmidt R, Bernues J et al. Proximal sequence element factor binding and species specificity in vertebrate U6 snRNA promoters. J Mol Biol 1992; 223:873-884.
349. Meißner W, Wanandi I, Carbon P et al. Transcription factors required for the expression of *Xenopus laevis* selenocysteine tRNA in vitro. Nucleic Acids Res 1994; 22:553-559.
350. Murphy S, Yoon JB, Gerster T et al. Oct-1 and Oct-2 potentiate functional interactions of a transcription factor with the proximal sequence element of small nuclear RNA genes. Mol Cell Biol 1992; 12:3247-3261.
351. Yoon J-B, Roeder RG. Cloning of two proximal sequence element-binding transcription factor subunits (γ and δ) that are required for transcription of small nuclear RNA genes by RNA polymerases II and III and interact with the TATA-binding protein. Mol Cell Biol 1996; 16:1-9.
352. Bai L, Wang Z, Yoon J-B et al. Cloning and characterization of the β subunit of human proximal sequence element-binding transcription factor and its involvement in transcription of small nuclear RNA genes by RNA polymerases II and III. Mol Cell Biol 1996; 16:5419-5426.
353. Henry RW, Ma B, Sadowski CL et al. Cloning and characterization of SNAP50, a subunit of the snRNA-activating protein complex $SNAP_c$. EMBO J 1996; 15:7129-7136.
354. Henry RW, Sadowski CL, Kobayashi R et al. A TBP-TAF complex required for transcription of human snRNA genes by RNA polymerases II and III. Nature 1995; 374:653-656.
355. Sadowski CL, Henry RW, Kobayashi R et al. The SNAP45 subunit of the small nuclear RNA (snRNA) activating protein complex is required for RNA polymerase II and III snRNA gene transcription and interacts with the TATA box binding protein. Proc Natl Acad Sci USA 1996; 93:4289-4293.
356. Murphy S, Pierani A, Scheidereit C et al. Purified octamer binding transcription factors stimulate RNA polymerase III-mediated transcription of the 7SK RNA gene. Cell 1989; 59:1071-1080.
357. Danzeiser DA, Urso O, Kunkel GR. Functional characterization of elements in a human U6 small nuclear RNA gene distal control region. Mol Cell Biol 1993; 13:4670-4678.
358. Mittal V, Cleary MA, Herr W et al. The Oct-1 POU-specific domain can stimulate small nuclear RNA gene transcription by stabilizing the basal transcription complex $SNAP_c$. Mol Cell Biol 1996; 16:1955-1965.
359. Lescure A, Tebb G, Mattaj IW et al. A factor with Sp1 DNA-binding specificity stimulates *Xenopus* U6 snRNA in vivo transcription by RNA polymerase III. J Mol Biol 1992; 228:387-394.
360. Scholer HR, Hatzopoulos AK, Balling R et al. A family of octamer-specific proteins present during mouse embryogenesis: evidence for germline-specific expression of an Oct factor. EMBO J 1989; 8:2543-2550.
361. Rosenfeld MG. POU-domain transcription factors: pou-er-ful developmental regulators. Genes Dev 1991; 5:897-907.
362. Ruvkun G, Finney M. Regulation of transcription and cell identity by POU domain proteins. Cell 1991; 64:475-478.

363. Herr W, Sturm RA, Clerc RG et al. The POU domain: a large conserved region in the mammalian *pit-1, oct-1, oct-2,* and *Caenorhabiditis elegans unc-86* gene products. Genes Dev 1988; 2:1513-1516.
364. Herr W, Cleary MA. The POU domain: versatility in transcriptional regulation by a flexible two-in-one DNA-binding domain. Genes Dev 1995; 9:1679-1693.
365. Assa-Munt N, Mortishire-Smith RJ, Aurora R et al. The solution structure of the Oct-1 POU-specific domain reveals a striking similarity to the bacteriophage λ repressor DNA-binding domain. Cell 1993; 73:193-205.
366. Dekker N, Cox M, Boelens R et al. Solution structure of the POU-specific DNA-binding domain of Oct-1. Nature 1993; 362:852-855.
367. Verrijzer CP, Kal AJ, van der Vliet PC. The oct-1 homeo domain contacts only part of the octamer sequence and full oct-1 DNA-binding activity requires the POU-specific domain. Genes Dev 1990; 4:1964-1974.
368. Verrijzer CP, Alkema MJ, van Weperen WW et al. The DNA binding specificity of the bipartite POU domain and its subdomains. EMBO J 1992; 11:4993-5003.
369. Verrijzer CP, van Oosterhout JAWM, van Weperen WW et al. POU proteins bend DNA via the POU-specific domain. EMBO J 1991; 10:3007-3014.
370. Aurora R, Herr W. Segments of the POU domain influence one another's DNA-binding specificity. Mol Cell Biol 1992; 12:455-467.
371. Gerster T, Balmacedo CG, Roeder RG. The cell type-specific octamer transcription factor OTF-2 has two domains required for the activation of transcription. EMBO J 1990; 9:1635-1643.
372. Muller-Immergluck MM, Schaffner W, Matthias P. Transcription factor Oct-2A contains functionally redundant activating domains and works selectively from a promoter but not from a remote enhancer position in non-lymphoid (HeLa) cells. EMBO J 1990; 9:1625-1634.
373. Tanaka M, Herr W. Differential transcriptional activation by Oct-1 and Oct-2: interdependent activation domains induce Oct-2 phosphorylation. Cell 1990; 60:375-386.
374. Das G, Hinkley CS, Herr W. Basal promoter elements as a selective determinant of transcriptional activator function. Nature 1995; 374:657-660.

CHAPTER 5

Transcription Complex Formation on Class III Genes

The formation of transcription complexes, composed of factors bound to DNA, was initially investigated by means of the template exclusion assay. This approach monitors the ability of a gene that is preincubated with limiting amounts of factor(s) to exclude transcription of a second gene added subsequently; preferential transcription of the first gene indicates the stable interaction of a limiting component during the preincubation, thereby precluding its association with the second gene. Using such an assay with HeLa cell extracts, TFIIIC was shown to be able to bind alone to VA_I and $tRNA^{MetI}$ genes.[1] TFIIIB could only bind these genes after TFIIIC had bound.[1] In the case of a 5S gene, TFIIIC was found to bind after TFIIIA and before TFIIIB.[1,2] The same was shown to be true in yeast.[3] Carey et al[4] obtained similar results with an approach in which transcription complexes were separated from unbound factors by gel filtration through Sepharose 4B columns and then assayed for transcription. This method allowed the detection of a weak interaction between TFIIIC and 5S DNA in the absence of TFIIIA.[4] Using the template exclusion assay with separated TFIIIC1 and TFIIIC2, Dean and Berk[5] found that TFIIIC2 is the first factor to bind to VA_I or $tRNA^{Arg}$ genes. TFIIIC1 and TFIIIB can then interact in either order to form a preinitiation complex.[5] This is also likely to be true for 5S genes. TBP does not recognize TATA-less class III genes directly, nor does it interact stably with promoter-bound TFIIIC; instead, TBP recruitment occurs in association with TFIIIB.[6-8] Indeed, the sequence-specific DNA-binding function of TBP is not required for transcription of tRNA, 5S or VA_I genes.[9,10] However, once recruited as part of TFIIIB, TBP is positioned on the DNA in such a way that it can discriminate between different upstream sequences.[11] This may explain, at least in part, the general preference for A/T-richness in the 5'-flanking regions of class III genes. The final step in initiation complex formation involves the incorporation of pol III.[2,4,12,13] Each of the factors in the initiation complex appears to be required stoichiometrically, since they are removed from solution as a result of complex formation and the transcription rate is proportional to factor concentration when any of these proteins is made limiting.[1-3,5,8,12,14] The assembly pathways for initiation complex formation on type I and II promoters are schematized in Figures 5.1 and 5.2. Alternative assembly pathways may be compatible with expression in vivo.[15]

RNA Polymerase III Transcription, Second Edition, by Robert J. White.

Fig. 5.1. Flow chart indicating the order of interaction of factors and polymerase with a typical type II promoter such as that of a tRNA gene.

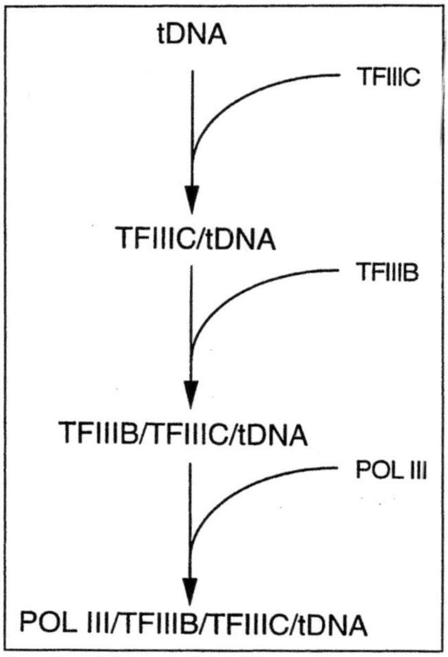

Fig. 5.2. Flow chart indicating the order of interaction of factors and polymerase with the promoter of a 5S RNA gene.

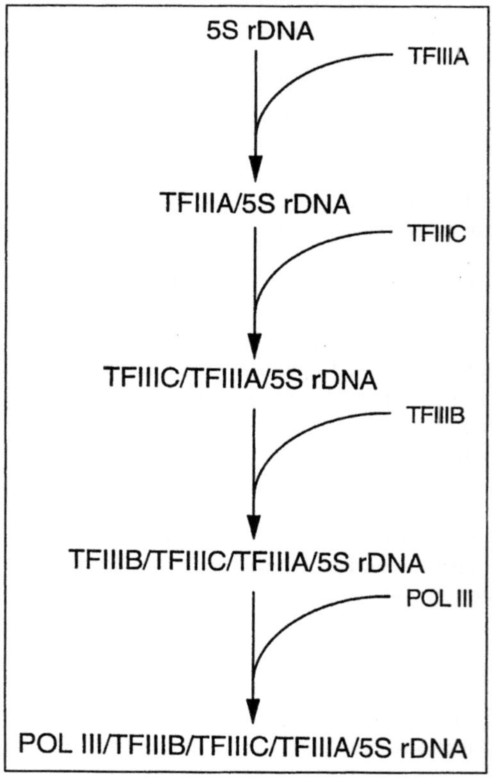

Because of their great length and flexible separation, class III transcription complexes are unlikely to be strictly linear structures. TFIIIA, TFIIIB and TFIIIC have all been shown to bend or loop DNA.[16-21] Thus, the factors themselves appear to be the primary determinants of the three-dimensional structure of the complexes.

Transcription Complex Assembly on Type II Promoters

Yeast

DNase I and DMS footprinting with purified protein indicates that yeast TFIIIC binds to tRNA genes within and around the A- and B-blocks.[22-32] Very similar protection patterns can be observed in vivo.[27] On some tRNA genes, the B-block becomes protected more rapidly than the A-block,[23,29] although this is not always seen.[25] TFIIIC protects both DNA strands from DNase I digestion.[25,26] It can contact both ICR blocks regardless of their relative helical orientation.[27,31,33] However, when excessive strain is imposed by abnormal interblock spacings, it is the A-block interaction that diminishes.[31,33-37] Footprints are generally weaker around the A- than the B-block.[23,26,29,31,32,37,38] Whereas a subclone containing the 3' half of the $tRNA^{Leu3}$ promoter was able to compete away the footprint from the entire gene, a subclone containing the 5' half competed little better than pBR322.[23] Baker et al[39] found that total deletion of A-block sequences from two tRNA genes reduced TFIIIC affinity only 2- to 5-fold, whereas single base changes within the B-block resulted in a 43- to 370-fold reduction. Thus mutations at the A-block have much less effect upon TFIIIC binding than they have upon promoter efficiency. This suggests that the role of the A-block in transcription involves more than just TFIIIC recruitment.

Footprinting and exonuclease digestion analyses of artificial yeast $tRNA^{Leu3}$ genes revealed that separations of 30-60 bp are optimal for TFIIIC binding.[31,33] The ICR blocks of yeast $tRNA^{Leu3}$ are separated by 74 bp due to the presence of an intron. Weak protection occurs within this region and alternates with unprotected sites, suggesting that the DNA may be folded or bent by protein binding.[23,27,33,34] Increasing the flexibility of the DNA by introducing single strand breaks in the interblock region does not enhance TFIIIC binding, which implies that the flexibility of the interaction is primarily due to the protein rather than the DNA molecule.[31] It is likely that τ_A, though constrained by its interaction with τ_B anchored at the B-block, is mobile enough to search for an optimal binding site. Electron microscopy has shown that the extended $tRNA^{Leu3}$ gene is sharply bent by TFIIIC, with the DNA entering and exiting on the same side of the complex.[16,35] TFIIIC also forms extended loops of DNA on the yeast U6 gene, which has an interblock spacing of ~200 bp.[37] The anomalous electrophoretic mobility of complexes containing TFIIIC and a $tRNA^{Gln}$ gene has provided evidence for a TFIIIC-induced bend of 90° centered near the transcriptional start site.[19] The ability of yeast TFIIIC to bend DNA may be provided by its τ_{138} subunit that contains two copies of an HMG homology.[40] HMG domains appear to play a major structural role in the bending or looping of DNA.[41]

The positions of the various subunits of yeast TFIIIC with respect to a tRNA promoter have been deduced by photocrosslinking.[42] (Fig. 5.3) The τ_{138} subunit of TFIIIC is primarily accessible to photoreactive residues located within and around the B-block.[42] The τ_{95} and τ_{55} subunits associate with the A-block region, on opposite faces of the DNA helix to one another and with the former somewhat upstream of the latter.[42] The τ_{91} subunit can be photocrosslinked to the downstream end of

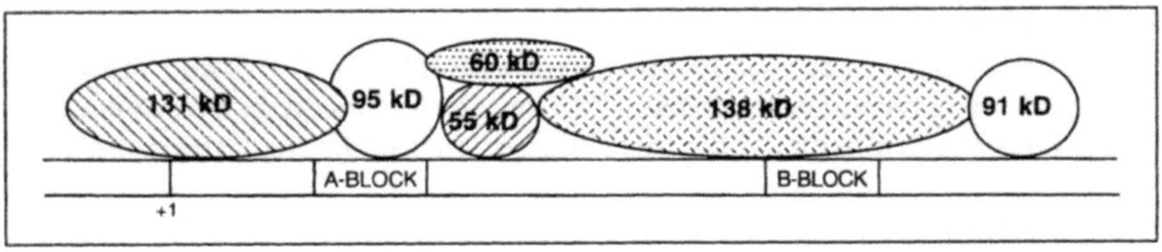

Fig. 5.3. Schematic illustration of the relative positions of the various subunits of *S. cerevisiae* TFIIIC when bound to a tRNA promoter.

class III genes.[43] Only the τ_{131} subunit of TFIIIC is accessible for crosslinking with photoreactive residues located upstream of the start site.[44,45] This subunit can also be crosslinked to residues downstream of the start site, but always together with other TFIIIC subunits.[42] The efficiency of crosslinking to τ_{131} is greatest for N_3RdU residues positioned near the initiation site, but specific interaction can still be detected as far upstream as -47, although it becomes very weak beyond -30.[44] It is also accessible to N_3RdU residues at the A-block (as are τ_{95} and τ_{55}) and in the interblock region at +47.[42] The proximity of this polypeptide to such a long stretch of DNA suggests that it is highly elongated and/or that the DNA is bent around it. Since DNase I footprints of TFIIIC do not extend beyond the start site,[30,32] the upstream extension of this protein is probably not tightly bound to DNA.

Binding of TFIIIC to yeast tRNA genes is half complete in less than a minute at 21°C.[30,31] In contrast, TFIIIB binding requires approximately 3 minutes to be half accomplished and nearly 20 minutes to reach completion.[30] With four different tRNA genes and a 5S RNA gene, addition of purified yeast TFIIIB has been shown to increase A-block protection and generate an upstream footprint extending over approximately 40 bp of 5' sequence.[30,32,46] That the A-block should make more contacts with a transcription complex than it does with TFIIIC alone is consistent with its essential role in transcription but minor role in TFIIIC binding. Upstream binding activity copurified precisely with TFIIIB transcriptional activity.[30] Furthermore, the rate of upstream binding is close to the rate of sequestration of TFIIIB into transcription complexes, as assayed either by transcriptional activity or by increased stability.[30] These data indicate that TFIIIB is responsible for the appearance of the upstream footprint.

Incorporation of TFIIIB into a complex can be prevented by linearization of the SUP4 $tRNA^{Tyr}$ gene at -31 or -28, but not at -135.[30] It can also be sterically impeded by the binding of other factors. Thus, the presence of prebound GCN4 protein within the TFIIIB footprint region inhibits transcription of a $tRNA^{Tyr}$ gene by 67-85%.[47] GCN4 had little effect when placed upstream of the TFIIIB binding site.[47] It was also unable to repress transcription if added after TFIIIB had already bound.[47] The residual transcription seen in the presence of prebound GCN4 was redirected to nearby downstream sites—a consequence of TFIIIB binding to alternative positions.[47] Once it had been relocated, DNA-bound TFIIIB remained at its alternative site even if GCN4 was subsequently stripped from the DNA.[47] This illustrates the tight but sequence-independent DNA-binding capability of TFIIIB. Binding of the bacterial Tet repressor 7 bp upstream of a $tRNA^{Glu}$ gene inhibits its expression in yeast cells, whereas binding of the same protein at -46 does not.[48] This suggests that TFIIIB recruitment can also be occluded in vivo.

TFIIIB binding induces a bend in the upstream DNA.[47] Both the B' and B" components contribute to this bending.[21] B' lies on the outside of the bend whereas the footprint induced by B" covers the opposite side of the helix on the inside of the curvature.[21] This bend is out of helical phase with the bend induced by TFIIIC.[19] DNA bending by TFIIIB may help explain why 5'-flanking sequences of type I and II promoters can often affect transcription levels despite there being a general lack of upstream sequence conservation. Any protein that bends DNA is likely to display some preference for sequences that confer static bending or flexibility.[47] Short A/T tracts can confer curvature upon a DNA molecule in solution and so may be particularly favored as upstream flanking sequences for class III genes.[47,49]

Since τ_{131} is the only subunit of TFIIIC that extends beyond the transcriptional start site, it is likely to be responsible for positioning TFIIIB.[44] Initial contact is made with the B' component of TFIIIB, and B" is only recruited if B' is already present.[50] Recombinant yBRF has been shown to bind directly to τ_{131}.[51] yBRF and τ_{131} have also been shown to interact by two-hybrid screening.[52] This interaction involves the N-terminal half of yBRF and the first 165 residues of τ_{131}.[52] Binding to yBRF was abolished by deletion of the N-terminal acidic region or the first TPR repeat of τ_{131}, whereas the other TPR repeats are not required.[52] However, the C-terminal part of τ_{131} seems to interfere with binding to yBRF.[52] The yBRF-binding domain may therefore be masked by intramolecular interactions, possibly between the first and second blocks of TPR repeats. Two-hybrid screening suggests that τ_{131} also interacts with B".[53] An internal deletion within the second TPR cluster of τ_{131} increases its association with B".[53] Another mutation within this TPR sequence accelerates the rate of formation of transcription complexes.[54,55] Binding sites for both yBRF and B" may therefore be masked within τ_{131} and become exposed through conformational changes during transcription complex assembly.

Librizzi et al[56] found that only 0.07-0.32% of available yBRF molecules give rise to full-length transcripts under conditions in which yBRF is limiting for tRNA gene transcription. This is the case with either native or recombinant yBRF.[56] They suggested that the function of yBRF may be limited by an unfavorable equilibrium of association with TFIIIC.[56] This could explain why yBRF is limiting for transcription both in whole cell extracts and in vivo.[57,58]

Addition of the B' fraction to a TFIIIC/tDNA complex results in the appearance of the 3' region of the upstream footprint (from +5 to -32), whereas addition of B" has no effect.[50] However, addition of B" to a B'/TFIIIC/tDNA complex generates the complete upstream footprint extending between -8 and -41.[50] The upstream footprint obtained with B' can also be generated using recombinant yBRF and recombinant TBP, but neither of these components individually extends the footprint seen with TFIIIC alone.[7] Assembly of the TFIIIB components can also be followed by gel retardation analysis. yBRF retards the migration of a TFIIIC/tDNA complex.[7] TBP will not bind directly to a TATA-less tRNA gene and has little or no effect upon the mobility of the TFIIIC/tDNA complex,[6,7] but retards the migration of a BRF/TFIIIC/tDNA complex.[7] Therefore, the order of interaction is yBRF, then TBP, and then B". yBRF contains two separable binding sites for TBP—one within its direct repeats and one in its C-terminal region.[51] TBP stabilizes the weak interaction between yBRF and the TFIIIC/tDNA complex, and is required for the recruitment of B".[7] In fact, it is likely that yBRF/TBP/B" bind to TFIIIC/tDNA simultaneously as a preexisting TFIIIB complex, rather than step-by-step as individual components. However, since the TFIIIB complex is relatively unstable in solution,[59] stepwise recruitment may also occur. Either way, the assembly pathway for the separated polypeptides nevertheless provides important information concerning polypeptide interactions within the complexes.

Both the yBRF and B" subunits of TFIIIB are accessible to photoreactive N_3RdU residues within the upstream footprinted region.[44] Whereas B" is labeled preferentially by N_3RdU located between -38 and -34, yBRF is labeled preferentially from -33 and -32.[44] Photocrosslinking of B" only occurs in the presence of B'.[50] Addition of yBRF alone to the TFIIIC/tDNA complex causes a slight increase in the crosslinking efficiency of τ_{131}, but yBRF itself is only weakly crosslinked in this complex.[7] Further addition of TBP produces a marked stimulation of crosslinking for both τ_{131} and yBRF; weak crosslinking of TBP to the region between -38 and -32 can also be seen in this complex.[7] Recruitment of B" strongly enhances crosslinking of yBRF, but diminishes crosslinking of both TBP and τ_{131}.[7] These changes in accessibility to specific photoactive N_3RdU residues suggest that a series of conformational rearrangements in the extant complexes accompany the successive assembly steps (Fig. 5.4). Protein footprinting suggests that two extensive regions of B" become buried, whereas a third region is unfolded when this polypeptide is recruited.[60] Kassavetis et al[7] proposed that complex formation may involve a cascade of conformational changes. yBRF binding may alter the conformation of τ_{131} such that it becomes more accessible to crosslinking from upstream sites. Interaction with τ_{131} may also unmask the high-affinity TBP-binding site located in the C-terminal half of yBRF.[51] TBP binding strongly stimulates crosslinking of τ_{131} and yBRF and generates the 3' part of the upstream footprint, perhaps by unmasking a cryptic DNA-binding site in yBRF. A similar explanation may account for the transformation to extreme stability that accompanies recruitment of B" to produce the complete upstream footprint. A number of precedents exist for such a scenario. For example, the σ^{70} factor of *E. coli* does not bind DNA independently, but does so following association with the core subunits of RNA polymerase.[61] Another example is p53, which has a cryptic DNA binding function that can be activated by a C-terminal deletion.[62] Extensive N- or C-terminal truncations of B" make no difference to the TFIIIB footprint.[60] This suggests that B' may be primarily or entirely responsible for the upstream footprint and that B" may generate the 5' extension in protection by inducing rearrangements.

The ability of upstream TATA boxes to influence transcription from type I and II promoters suggests that TBP is positioned within the complex so as to interact with sequences in the 5' flanking region. Although a TATA box is not required by type I and II promoters, TBP is nevertheless available to respond to such a sequence if present. TBP has a marked preference for A/T-rich sequences, but TFIIIB will still bind if the 5'-flanking region is entirely G/C-rich, albeit with reduced efficiency.[11] The resultant complex is transcriptionally active and resistant to heparin.[11] It has a similar electrophoretic mobility to TFIIIB bound to A/T-rich DNA, implying a comparable degree of DNA bending in each case.[11] Since TBP is likely to be responsible for most if not all TFIIIB-induced bending, this would suggest that TBP still binds to DNA even when directed to a G/C-rich site. When a TATAAA sequence upstream of a tRNA gene was substituted to TGTAAA, initiation at the +1 start site was inhibited.[11] The effect of this substitution could be suppressed by reconstituting TFIIIB with a mutant form of TBP that recognizes TGTAAA.[11] This demonstrates that the TBP subunit of TFIIIB binds directly to DNA during at least some stages of TFIIIC-mediated assembly. Therefore, TBP may be positioned over

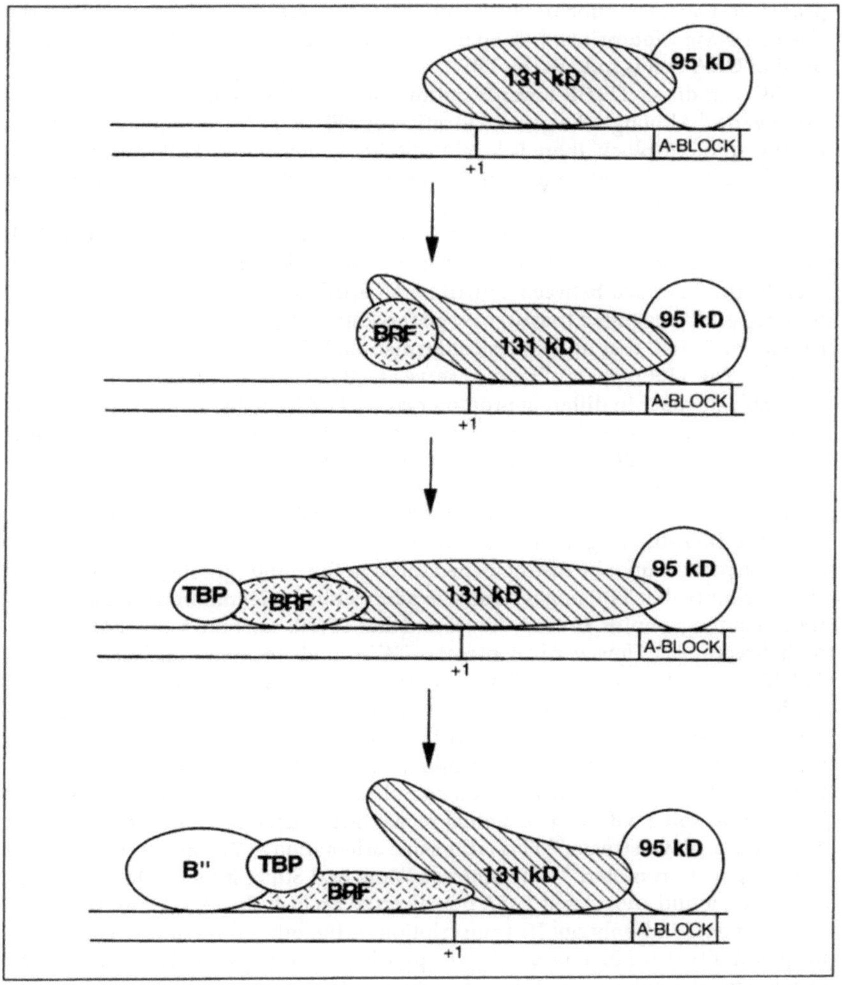

Fig. 5.4. Schematic model of TFIIIB assembly onto a TFIIIC/tDNA complex in *S. cerevisiae*. The figure only shows the factors associated with the 5' end of the promoter. It illustrates the order of interaction of the isolated TFIIIB subunits and the conformational changes that may be associated with the various binding steps.

the DNA in a similar way in TATA-less and TATA-containing promoters, even though it is recruited by very different mechanisms in these two situations. Alternatively, the interaction between TBP and 5'-flanking DNA may be transient. Crosslinking studies using photoreactive groups at variable distances from the helix suggested that TBP is located further from the major groove when assembled on a tRNA promoter than it is when bound to the AdML promoter.[63] Librizzi et al[56] showed that yBRF can increase by 3-fold the affinity of TBP for the TATA box of the AdML promoter. The interaction of yBRF and TBP is cooperative, but does not affect the size of DNase I footprints.[56] Cooperativity may result from yBRF stabilizing the bend in DNA that is induced by TBP.[56]

TBP can direct TFIIIB towards optimal sites within the upstream region of a tRNA gene.[11] Although the general location of TFIIIB is clearly dictated by TFIIIC relative to the A-block, there is local flexibility which allows TBP to select a preferred site.[11] Thus the alignment of TFIIIB relative to TFIIIC can vary as a function of upstream sequence.[11] The assembly is flexible enough to allow the positioning of TFIIIB at opposite faces of the DNA helix.[11] Alternative placements of TFIIIB can even occur on the same promoter, generating multiple transcription start sites.[11] Clearly, the interface between TFIIIB and TFIIIC is remarkably unconstrained. Joazeiro et al[11] suggested that τ_{131} may be the source of this elasticity. τ_{131} is the only subunit of TFIIIC that undergoes conformational changes during TFIIIB recruitment.[7,50] The TPR repeats of τ_{131} may serve as a flexible arm that can stretch across a variable distance in different promoter contexts. Alternatively, they may provide several alternative sites for the interface with TFIIIB. Whatever the mechanics, this flexibility allows TBP to scan the -30 region for the most thermodynamically favorable binding site.

Whitehall et al[64] investigated the orientation of TBP in promoter-bound TFIIIB. Since the DNA-binding domain of TBP is 2-fold symmetrical and since T:A and A:T base pairs cannot be distinguished through the minor groove, they addressed the possibility that TBP might be able to adopt either orientation within a class III transcription complex. Indeed, they found that TFIIIB can bind to the TATA box in both directions.[64] However, in contrast to TATA-mediated binding, TFIIIC recruits TFIIIB in one direction only.[64] This unidirectional assembly is likely to be dictated by the interaction of yBRF with τ_{131} and suggests that yBRF and B" associate with TBP in one specific orientation. TFIIIB was reconstituted with a mutant TBP carrying three substitutions in its C-terminal repeat in order to determine the orientation of TBP within a transcription complex.[64] These experiments showed that TBP is assembled with its C-terminal repeat facing upstream and its N-terminal repeat facing downstream.[64] This is the same orientation as TBP adopts when TFIID binds a pol II promoter.[65,66] In contrast, Wang and Stumph using *Drosophila* extracts that found an isolated TATA box directed exclusively pol II transcription in one direction and only pol III transcription in the other direction.[67] This raised the possibility that TBP is orientated in opposite directions in the TFIID and TFIIIB of multicellular eukaryotes. However, subsequent studies have demonstrated that this polymerase-specific directionality is the product of a particular TATA sequence and promoter context, and is not a general phenomenon.[68,69]

The role of TFIIIE in transcription complex formation is not clear. It does not bind stably to a tRNA gene, either alone or in the presence of TFIIIC, and is not involved in the recruitment of TFIIIC or TFIIIB.[70] It is, however, important in single-round transcription assays and so must participate in the initial formation and/or utilization of the initiation complex, rather than being restricted to a role in reinitiation.[70] As such, it may be analogous to class II factors such as TFIIE and TFIIF that are recruited into the preinitiation complex in association with or after the polymerase.[71,72]

Although yBRF binds directly to the C34 subunit of pol III,[51,73] stable recruitment of polymerase to the initiation complex is dependent upon the presence of both the B' and B" components of TFIIIB.[74] Pol III generates a DNase I footprint in the region surrounding the start site, from +21 to at least -3, and also reduces the degree of protection that occurs around the A-block.[32] The association is quite stable, and little or no dissociation of polymerase enzyme is detected during a 6-minute challenge with competitor.[75] Binding of pol III enhances photoaffinity labeling of yBRF[74] but reduces that of B".[63] The C160, C128 and C34 subunits of pol III in the preinitiation complex are all accessible to photoreactive residues situated upstream of the start site.[45,63,74] Of these, C34 projects furthest upstream, with weak crosslinking to -21, and is therefore available to interact with TFIIIB.[63,74] Indeed, C34 has been shown to interact with yBRF in vivo.[73] Furthermore, direct binding to the repeat region of yBRF has been demonstrated.[51] In contrast, C34 does not bind to the homologous region of TFIIB.[51] The selectivity of this interaction is likely to play a decisive part in determining polymerase specificity. C160, C128, C82, C53, C34 and ABC27 can all be crosslinked to sites downstream of +1 prior to initiation.[74]

Several factors contribute to the selection of the transcriptional start site. First of all, TFIIIC determines the general area in which TFIIIB can bind to DNA. However, the flexible interface between these two proteins allows TBP to scan the -30 region for an optimal binding site for TFIIIB.[11] Initiation then occurs 28-30 bp downstream of the 5' edge of this site.[11] Pol III itself has certain sequence preferences and will hunt within this small window for the best initiating nucleotide. Thus TFIIIB, TFIIIC and pol III all contribute to start site selection.

Although TFIIIC binding to the B-block is required for expression of the yeast U6 gene in vivo and in crude extracts, these components are dispensable for transcription with purified factors.[36,76-79] Joazeiro et al[79] demonstrated that yBRF, TBP and B" are all required for TFIIIC-independent U6 transcription.[79] Furthermore, once assembled onto the U6 TATA box in the absence of TFIIIC, TFIIIB displays the same stability, position and size of footprint as it does on a tRNA gene.[79] TBP alone can bind to the U6 TATA motif, but is much less resistant to heparin than the complete TFIIIB complex.[79] As well as conferring stability, yBRF and B" extend in both directions the footprint generated by TBP alone.[79]

Silkworm

Stable complex formation on the silkworm $tRNA^{AlaC}$ gene requires a combination of either TFIIIC and TFIIID or TFIIIB and TFIIID; none of these factors alone binds stably to this template.[80-82] It is possible that this requirement for more than one factor in order to stabilize binding is due to the degeneracy of the B-block of the $tRNA^{AlaC}$ promoter. As well as the A- and B-blocks, specific 3'-flanking sequences are required for binding by TFIIIC and TFIIID.[81] TFIIIC and TFIIID together generate a DNase I footprint that extends from -11 to +136 of the $tRNA^{AlaC}$ promoter.[81,82] Addition of TFIIIB extends the footprint as far as -34 and strengthens protection downstream of the start site.[82] Mutation of an A/T-rich element situated between -20 and -15 severely reduces both transcription and the TFIIIB footprint.[82] However, mutation of a TATAT sequence between -29 and -25 has little effect on transcription or binding.[82]

Under most conditions, the silkgland-specific $tRNA^{AlaSG}$ genes are transcribed much less actively than the constitutive $tRNA^{AlaC}$ genes.[83,84] The upstream flanking sequences of these genes are sufficient to confer their distinct transcriptional properties.[83] The lower activity of the $tRNA^{AlaSG}$ promoter correlates with a reduced ability to recruit TFIIIB.[82,84] Pol III is also recruited inefficiently, presumably due to the reduced binding of TFIIIB.[84] TFIIIB assembly occurs rapidly on both the $tRNA^{AlaC}$ and $tRNA^{AlaSG}$ genes, being essentially complete within 5 minutes (at 22°C); however, saturation of $tRNA^{AlaSG}$ promoters requires higher concentrations of TFIIIB.[82] Once TFIIIB has been assembled onto the $tRNA^{AlaSG}$ promoter, as judged by gel retardation, most of the resulting complexes are still unable to support transcription.[82] Furthermore, DNase I footprinting failed to detect any upstream protection of the $tRNA^{AlaSG}$ promoter by TFIIIB.[82] This might reflect the absence of a critical subunit or a failure to adopt an active conformation on the suboptimal upstream promoter sequences of $tRNA^{AlaSG}$ genes.

Vertebrates

As with the yeast factor, human TFIIIC binds predominantly to the B-block of tRNA and VA genes, as well as interacting weakly with the A-block.[4,85-88] Both TFIIIC1 and TFIIIC2 are necessary for transcription of VA, tRNA and 5S genes.[88-90] TFIIIC2 is the first to bind.[5] The A-block is dispensable for TFIIIC2 binding.[88,91] Interaction with the A-block requires the presence of TFIIIC1.[88,89] TFIIIC1 also enhances TFIIIC2 binding and extends the footprint both 3' and 5', thereby incorporating the A-block but leaving some interblock sequences unprotected.[88-90] Wang and Roeder[88] reported that the downstream termination region is required for this cooperative interaction between TFIIIC1 and TFIIIC2. Indeed, their TFIIIC1 fractions bound strongly to the termination region of tRNA and VA genes.[88] However, Oettel et al[90] obtained active TFIIIC1 fractions without terminator-binding activity and observed no effect of the termination region on the binding of TFIIIC1 or TFIIIC2. These workers reported a distinct fraction, which they called TFIIIC0, that gave a footprint over the termination sites of 5S, VA_I and $tRNA^{Met}$ genes.[90]

Studies to date have been unable to detect a footprint in the 5'-flanking region of vertebrate type II promoters that would correspond to the situation in yeast. This probably reflects weaker binding by TFIIIB in vertebrates. However, circumstantial evidence has been obtained which suggests that frog and human TFIIIB occupy a similar position to yeast TFIIIB, upstream of the gene.[92-94] Binding of lac repressor 32 bp upstream of a human $tRNA^{Ser}$ gene inhibits its expression in BSC40 cells.[94] In one study, the binding site for *E. coli* lac repressor was inserted at various positions upstream of a human $tRNA^{Ser}$ gene.[92] Transcription was completely repressed following prebinding of lac repressor to sites centered at -9, -15, -35 or -37, and was partially inhibited with sites at -43 and -46.[92] With the -43 and -46 sites, the residual initiation was redirected downstream from +1.[92] Lac repressor also inhibited expression of the -9 and -15 constructs when it was added after formation of the initiation complex or during ongoing transcription, but it had no effect on the other constructs when added after complex assembly.[92] This suggests that the human class III initiation complex extends to approximately -40, but that the initiation site remains accessible during transcription.[92] This correlates well with the situation in yeast, where TFIIIB provides strong protection up to around -40, but leaves the start site vacant.[30,32] However, in yeast once upstream binding of GCN4 has redirected initiation to sites downstream of the start, the aberrant initiation continues if GCN4 is subsequently removed.[47] This contrasts with the situation in humans, where release of prebound lac repressor restores initiation to +1.[92] This striking difference between yeast and humans is likely to reflect a less avid interaction between human TFIIIB and DNA.[92] Similar experiments have been carried out using *Xenopus* extracts.[93] In this case, binding of GCN4 to sites inserted between -22 and -47 severely represses transcription of the yeast SUP4 $tRNA^{Tyr}$ gene in a *Xenopus* extract.[93] This repression can be prevented if a factor present in PC-B is assembled onto the template before GCN4 is added.[93] Although the ability of GCN4 to block SUP4 transcription in *Xenopus* extracts parallels its effect in yeast, one significant difference was noted between the two systems: upstream binding of GCN4 redirected some residual initiation to aberrant downstream sites in yeast[47] but not in *Xenopus*.[93] Similar occlusion effects were obtained when a C-block was inserted at various positions upstream of the *Xenopus* $tRNA^{TyrD}$ gene and a recombinant polypeptide containing fingers 1-3 of TFIIIA was added.[93] In this case, transcription was severely inhibited by prior binding of the TFIIIA fragment to sites located between -17 and -43.[93] Order of addition experiments established that the blockage occurred after TFIIIC was bound but before pol III was recruited.[93] Again, there was no redirection of initiation observed in the *Xenopus* system.[93] Together, these results suggest that TFIIIB occupies a similar position in vertebrate transcription complexes to what has been documented in *S. cerevisiae.*

Human BRF has been shown to bind directly to the C39 subunit of human pol III, which is the homologue of yeast C34.[95] Two other pol III-specific subunits (C32 and C62) were also tested, but did not interact.[95] Conversely, recombinant C39 was found to bind directly to both hBRF and TBP.[95] It is therefore likely that at least some of the interactions between TFIIIB and pol III have been maintained through evolution.[95]

TFIIIB and TFIIIC fractions from humans are functionally incompatible with their counterparts from *Drosophila*.[96,97] The DNA-binding properties of TFIIIC are likely to be conserved between these species, and protein-protein interactions may be responsible for the incompatibility.[97] An alternative explanation that has yet to be excluded is that the TFIIIB and TFIIIC fractions from these two species contain different distributions of essential factors. Template commitment assays demonstrated that human TFIIIB is unable to form a stable complex with *Drosophila* TFIIIC.[97] In contrast, *Drosophila* TFIIIB can generate a stable complex with human TFIIIC, but this complex is transcriptionally inactive.[97] These data emphasize the importance of an accurate interface between TFIIIB and TFIIIC for the function of the transcription machinery.

Transcription Complex Assembly on Type I Promoters

Vertebrates

TFIIIA binding is the first step in transcription complex formation on type I promoters. *Xenopus* TFIIIA binds to 50 bp of the 5S ICR in a 1:1 molecular complex, with its N-terminus orientated towards the 3' end of the gene.[98-101] TFIIIA generates a strong DNase I footprint on somatic 5S genes that extends between approximately +45 and +96, with contacts primarily on the noncoding DNA strand.[99,101-108] Very similar footprints are produced on the major oocyte 5S gene and the 5S pseudogene, but protection of the trace oocyte 5S gene only extends from +64 to +96.[107] Full DNase I protection of 5S genes is obtained if TFIIIA is first treated with papain to remove the C-terminal 4 kD.[99,107,109] This treatment also has little effect upon the affinity of DNA-binding.[107,109] However, a 27 kD fragment produced by digestion of TFIIIA with trypsin only footprints from +64 to +96 and binds 4-fold less tightly.[99,107] This truncated footprint resembles that seen on a trace oocyte 5S gene with full-length TFIIIA.[107] Indeed, the 27 kD tryptic fragment gives the same footprint on the trace oocyte gene as does full-length TFIIIA, which suggests that the C-terminal third of the protein does not normally bind tightly to this gene.[107] This reduced region of interaction can account for the fact that the trace-oocyte gene has a 4-fold lower affinity for TFIIIA than does the somatic 5S gene.[104,107,110] In contrast, the somatic, major oocyte and pseudo 5S genes all bind TFIIIA with essentially the same affinity.[107,110] The footprint produced by human TFIIIA on a human 5S gene is almost indistinguishable from that produced by *Xenopus* TFIIIA on a *Xenopus* somatic 5S gene.[111]

The ability of truncated versions of TFIIIA to generate partial footprints illustrates its highly modular nature. Separate clusters of zinc fingers bind to the A-block, intermediate element and C-block regions of the 5S ICR.[101,108,112,113] Fingers 1-3 bind to the C-block, finger 5 binds to the intermediate element, fingers 7-9 bind to the A-block, and fingers 4 and 6 span these promoter regions.[112-114] (Fig. 5.5). The most energetically important contacts occur between nucleotides +80 to +92 and the N-terminal three zinc fingers.[99,101,108,112,113,115-118] TFIIIA binding can withstand extensive deletion into the ICR from the 5' side but not from the 3' side.[103] Studies involving methylation protection or interference also implicate contacts with the C-block region as being of primary importance.[104,105] Although binding to the A-block contributes little overall affinity to the interaction with DNA, it positions the activation domain of TFIIIA so that it can interact productively with the other components of the transcription apparatus.[99,109] A detailed analysis involving the individual amputation of each finger demonstrated that the third finger is the only one that is indispensable for sequence-specific DNA binding.[118] Nevertheless, an eight-ring polyamide that blocks the access of finger 4 inhibits TFIIIA binding in vitro and 5S gene expression in vivo.[119] Fingers 1-3 form a compact domain which contacts the C-block continuously from +81 to +91 in the major groove of the DNA duplex, primarily on the noncoding strand.[112-114] Fingers 7-9 interact continuously with the major groove of the A-block between +51 and +61 in a similar fashion,[112-114] although at this site binding is much weaker.[101,104,117] In both these cases TFIIIA appears to wrap around all faces of the helix in a manner that may resemble the complex formed between the three-finger protein Zif268 and DNA, as observed in its crystal structure.[120] Thus, approximately one turn of helix is protected at each end of the 5S ICR by binding of three fingers in the major groove. The intervening 20 bp is contacted by fingers 4-6, which form an extended domain that protects only one face of the DNA, leaving the opposite face exposed.[101,108,112,113] This long region of interaction is inconsistent with continued major groove binding, and a variety of methylation interference, missing nucleoside, hydroxyl radical footprinting and nuclease protection data indicate that the fingers cross the minor groove in this domain.[108,112-114] TFIIIA has been found to bend DNA,[17,18,20] and this may help to accommodate the central region of the ICR. Support for this comes from the detection of DNase I hypersensitive sites opposite to where fingers 4 and 6 bind.[112-114] Bending of the DNA near finger 6 may be necessitated by the unusually short linker regions that occur in this divergent module.[121]

Despite the large surface of sequence-specific interaction, the complex formed between *Xenopus* TFIIIA and 5S genes is relatively unstable.[1,2,12,122,123] The TFIIIA/5S gene complex has a half-life of 100 minutes in the presence of a competing 5S gene, but this drops to only 5-6 minutes when challenged with single-stranded M13 DNA.[123,124] This complex is stabilized by the recruitment of TFIIIC, such that the

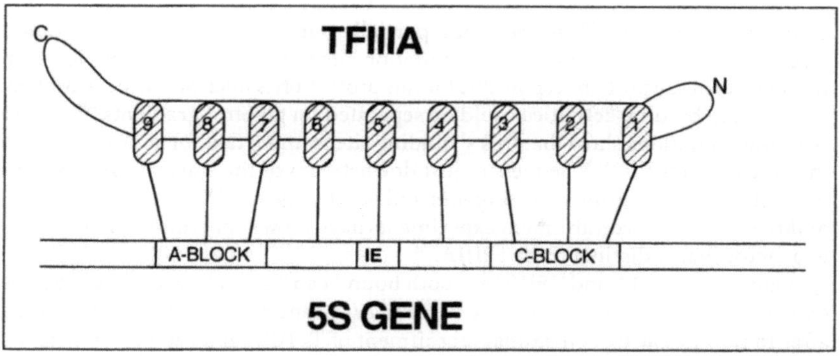

Fig. 5.5. Schematic diagram illustrating the orientation of *Xenopus* TFIIIA relative to a 5S promoter and the approximate region of the promoter which is contacted by each of the nine zinc fingers.

TFIIIC/TFIIIA/5S gene complex has a half-life of about 6 hours in the presence of 5S DNA competitor, although it remains unstable to single-stranded competitor.[1,2,12,109,124,125] TFIIIC1 may be responsible for strengthening the interaction between TFIIIA and DNA.[90] Stabilization of the TFIIIA/5S DNA complex by TFIIIC requires the C-terminal domain of TFIIIA and is not seen with the 34 kD fragment produced by papain digestion.[109] TFIIIC binding to the TFIIIA/DNA complex requires sequences in the A- and C-blocks, as well as elements located between the initiation site and the TFIIIA binding site.[124-127] The affinity of TFIIIC for the TFIIIA/somatic 5S DNA complex is 5- to 10-fold higher than its affinity for the TFIIIA/major oocyte 5S DNA complex.[124,125,128] Surprisingly, several studies found that TFIIIC recruitment to the TFIIIA/5S complex does not extend the footprint on *Xenopus* 5S DNA beyond that obtained with TFIIIA alone.[4,109,124,128,129] This could be explained if the interaction between TFIIIC and DNA in this complex is unstable or transient; the stabilization of TFIIIA binding produced by TFIIIC might then result from changes in protein conformation. However, Peck and coworkers have reported that a PC-C fraction can extend the TFIIIA footprint on a somatic 5S gene by an additional 25 bp towards the start site.[130,131] The resultant footprint displayed extensive protection between +20 and +95, with further alterations to the cleavage pattern between +1 and +20.[131] Oettel et al[90] also reported that TFIIIC2 can enlarge the footprint produced by human TFIIIA on a human 5S gene, such that protection reaches to +20. The human TFIIIC2/TFIIIA/5S DNA complex can be detected in a bandshift assay.[90] Once assembled in this complex, TFIIIC2 appears to retain the ability to bind a B-block, as addition of DNA fragments containing a B-block can supershift the TFIIIC2/TFIIIA/5S DNA complex.[90]

TFIIIC has also been observed to bind to 5S genes in the absence of TFIIIA, although this interaction is of relatively low affinity and is unlikely to represent a significant pathway for transcription complex assembly.[106,132] Schneider et al[132] reported a weak interaction between fractionated TFIIIC and the A-block region of the 5S promoter. However, the three C-terminal fingers of *Xenopus* TFIIIA are thought to occupy the major groove at the A-block and are therefore likely to prevent access of TFIIIC to the DNA helix in this region in a TFIIIC/TFIIIA/5S gene complex. Fradkin et al[106] found that a partially purified TFIIIC2 fraction generated a DNase I footprint around the initiation site between -10 and +18. However, the identity of the protein responsible for this protection is unclear, since the B-block and 5S gene binding activities could be separated on glycerol gradients.[106] Linker scanning mutations show that this 5' binding site is important for 5S transcription in some extracts.[106,125,133] Sequences just downstream of the start site are also important for transcription of *Drosophila* and yeast 5S genes,[134,135] and in the *Drosophila* case template commitment experiments have shown that this sequence binds to a factor that is distinct from TFIIIA.[135]

Only once TFIIIA and TFIIIC are both bound can TFIIIB be recruited onto the 5S gene.[2,4,12] Whereas the stable TFIIIC/TFIIIA/5S gene complex is fully assembled after 10 minutes at 30°C, maximal recruitment of TFIIIB requires 40 minutes.[2] As such, TFIIIB binding is the rate-limiting step for 5S transcription.[2] Pol III is then assembled onto the preformed TFIIIB/TFIIIC/TFIIIA/5S gene complex.[2,4,12] A fully assembled transcription complex protects approximately 180 bp of somatic 5S DNA from DNase I digestion, with the footprint extending from -24 to +159.[136] Protection around the start site takes longer to appear than that at the ICR, and may correspond to the relatively slow TFIIIB binding step.[136] GCN4 is able to repress transcription of an oocyte 5S gene when bound to a site inserted between -25 and -33.[93]

Silkworms

In silkworms, at least three factors are required to commit the 5S gene to transcription.[137] More than one combination of fractions is competent in this regard, which suggests that alternative pathways can be used to assemble the preinitiation complex.[137] A fraction containing TFIIIA protects the 5S gene noncoding strand between +45 and +96.[137] Template commitment requires the additional presence of TFIIIC and TFIIID or TFIIIB and TFIIID.[137] It is likely that the complex becomes committed only once a sufficient number of protein-DNA contacts are made.

Yeast

The DNase I footprint of a fully assembled transcription complex reconstituted on a yeast 5S gene covers approximately 165 bp.[43] A highly extended footprint is also observed on these genes in vivo.[138] The topography of the transcription complex that forms on yeast 5S DNA has been analyzed in detail using the N_3RdU photocrosslinking approach.[43] (Fig. 5.6) Despite the dissimilarity in promoter structure and the presence of an additional factor, the locations of the various TFIIIB and TFIIIC subunits relative to one another and relative to the initiation site are extremely similar on the 5S gene to those found on a tRNA gene. This is best illustrated by the fact that τ_{95}, which is situated around +30 on a tRNA promoter in association with the A-block, is also positioned in the vicinity of +30 on a 5S gene, even though in this case the A-block is located 30 bp further downstream.[43] The primary binding site for yeast TFIIIA is the C-block region, to give a DNase I footprint that extends from +69 to +97, but does not reach as far as the A-block.[32,46,139,140] However, TFIIIA is accessible to photoreactive nucleotides placed as far upstream as +48 and as far downstream as +127.[43] Its anisometric structure[140] therefore enables it to extend over a considerable distance and to be available for multiple associations with the various subunits of TFIIIC. The absence of a TFIIIA footprint over much of this region suggests that binding is weak and the tolerance of muta-

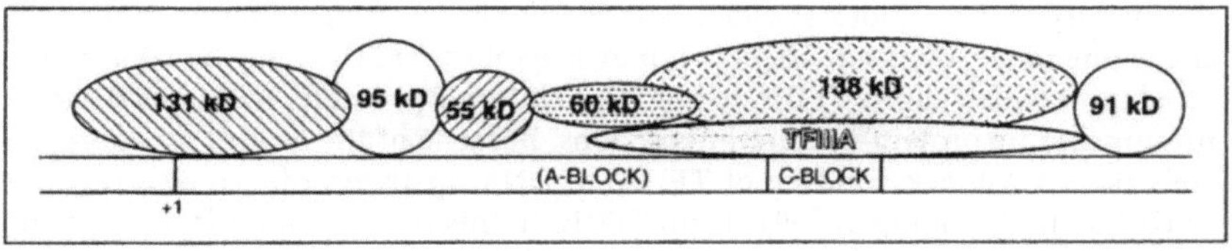

Fig. 5.6. Schematic illustration of the relative positions of TFIIIA and the various subunits of TFIIIC when bound to a 5S promoter in *S. cerevisiae.*

tions in the underlying sequence[138,139] implies that much of the interaction away from the C-block may not be sequence-specific. Like the frog protein, yeast TFIIIA induces a modest (~45°) bend in 5S DNA.[21]

Once bound, TFIIIA provides a scaffold across the 5S gene upon which the other factors can assemble. Binding of TFIIIB and TFIIIC to the yeast 5S gene is only observed in the presence of TFIIIA.[3,46] As in vertebrates, recruitment of TFIIIC stabilizes the TFIIIA/5S DNA complex.[3,139,141] The region of TFIIIA spanning fingers 1-3 is sufficient to bind 5S DNA and recruit TFIIIC.[142] The DNase I and DMS footprints produced by TFIIIA alone are extended both 5' and 3' by the addition of TFIIIC.[32,46,139] Nucleotides +62 to +124 of the TFIIIC/TFIIIA/5S gene complex are inaccessible to DNase I.[46] TFIIIC is not photocrosslinked to residues between +48 and +76, owing to the presence of TFIIIA. However, it straddles this region and extends far beyond it in both directions. Thus τ_{138} can be crosslinked to either +42 or +92.[43] 3' truncation of the 5S gene at +110 compromises TFIIIC binding although it has no major effect upon transcription levels.[21] τ_{91} associates with the extreme 3' end of the gene and can be crosslinked to nucleotides at +121 or +127.[43] τ_{95} interacts primarily with sequences from +20 to +35, far upstream of the A-block homology region, but in a similar position to that occupied on a tRNA gene.[43] As is also the case for a tRNA gene, τ_{131} is positioned furthest upstream and can be crosslinked from either 5' or 3' of the transcription start site.[43] However, a notable difference in the case of the 5S gene complex is that TFIIIB is absolutely required for any of the TFIIIC subunits to be accessible to photocrosslinking from upstream of +50.[43] The crosslinking of TFIIIA to nucleotides +48 and +59 is also enhanced by the recruitment of TFIIIB.[43] Addition of purified TFIIIB to a TFIIIC/TFIIIA/5S DNA complex also generates an upstream footprint extending over approximately 45 bp of 5' sequence.[32,46] The disposition of TFIIIB polypeptides on a 5S gene is very similar to that described for a tRNA gene, with yBRF preferentially photocrosslinked to residues near the start site and B" preferentially crosslinked from further upstream.[43] TFIIIB induces a strong bend of ~122° in 5S DNA, centered around -30.[21] This bend is almost identical to that produced upstream of the tRNA^{Glu} gene, despite an absence of sequence similarity between the two sites.[21] Finally, recruitment of pol III decreases the footprint from +18 to +34 and generates partial protection around the initiation site, from +13 to at least -7.[32] The fully assembled complex bends 5S DNA in the same net direction as does TFIIIB alone.[21]

Thus the polypeptides of yeast TFIIIB and TFIIIC have very similar locations in the TFIIIB/TFIIIC/TFIIIA/5S DNA complex to those occupied in a TFIIIB/TFIIIC/tDNA complex. Particularly striking is the positioning of τ_{95} around +30 in both complexes, despite the fact that the A-block is displaced 30 bp further downstream in the 5S promoter. None of the TFIIIC subunits are accessible to crosslinking from the vicinity of the A-block at +50 to +60.[43] A linker scanning mutation in the start site element, sse, does not affect complex assembly.[43] TFIIIA may be regarded as an adaptor which allows TFIIIC to bind to an internal promoter without a B-block, presumably by protein-protein interactions. Because of the lack of an appropriately placed A-block, binding of TFIIIC to DNA upstream of +50 also requires TFIIIB, so that binding of TFIIIB and TFIIIC in this region is interdependent and probably coordinate rather than sequential. This provides evidence in support of the idea that TFIIIB recruitment induces a structural transition in τ_{131}.

Transcription Complex Assembly on Type III Promoters

7SK and vertebrate U6 genes have distinct factor requirements from genes with type I or II promoters.[143-149] The lack of an ICR obviates the need for TFIIIA or TFIIIC2.[90,143-145,148,150] Instead, a PSE-binding factor is required.[144,148,150,151] However, TFIIIC1 is necessary for transcription directed by type III promoters.[90,148] Thus, TFIIIC1 is a truly general factor whereas TFIIIC2 is specific for genes with ICRs. The TFIIIB requirement of type III promoters is also distinct from that of types I and II.[146-149,152] This was first demonstrated by Lobo et al, who found that the TBP-containing 0.38M-TFIIIB is not needed for human U6 transcription.[152] U6 appears to employ a subcomplex of TFIIIB components.[146,149,152] This has been referred to as 0.48M-TFIIIB[152] or hTFIIIB-α,[146] although it is not clear whether the same species are described in each case. Conflicting results have been obtained concerning the need for hBRF by human U6 genes.[147,149,152,153] Recombinant TBP is required in addition to SNAPc in order to reconstitute U6 transcription in the presence of PC-B.[151] Unlike types I and II, type III promoters can recruit TBP without the assistance of other factors.[154-157] Free recombinant TBP can bind independently to the TATA box of U6 and 7SK genes.[154-157] However, little or no free TBP is found in extracts of mammalian cells.[158]

The very strict PSE-TATA box spacing requirement[154,159-161] strongly suggests that TATA-bound TBP or TFIIIB interacts with SNAPc/PTF. Indeed, subunits of SNAPc/PTF have been shown to bind directly to TBP.[153,162-165] Immunoprecipitation experiments suggest that the TBP that interacts with PTF is also associated with hBRF.[153] It may therefore be that SNAPc/PTF bound to the PSE interacts with a form of TFIIIB that has bound to the TATA box. Both the SNAP190/PTF_α and the SNAP50/PTF_β subunits can be photocrosslinked to PSE DNA in a sequence-specific manner.[148,163] SNAP50/PTF_β also interacts with SNAP43/PTF_δ.[163] TBP binds to SNAP45/PTF_γ and SNAP43/PTF_δ.[153,162-165] (Fig. 5.7)

The pathway for transcription complex assembly on type III promoters has yet to be firmly established. Both PSE- and TATA-binding are relatively slow[166-169] and complex formation takes longer for U6 genes than it does for tRNA or 5S genes.[168] The long lag phase seen in U6 transcription can be overcome by preincubating the template with PBP,[168] suggesting that PSE binding is the rate-limiting step. This can be potentiated by Oct-1 or Oct-2, which stimulate binding to the human U1, U2, U6 and 7SK PSEs.[169,170] Association of SNAPc with the PSE of the human U6 gene takes over an hour in the absence of Oct-1, but is complete in 15-30 minutes in its presence.[169] In the case of 7SK, Oct-1 produces a 10- to 20-fold increase in PSE occupancy and a similar increase in transcription.[170,171] The abilities of various Oct-1 mutants to stimulate PSE binding correlates with their potential to activate expression.[169] The Oct factors require both a functional PSE and its cognate factor in order to stimulate transcription.[170,171] However, the mouse U6 gene has a high

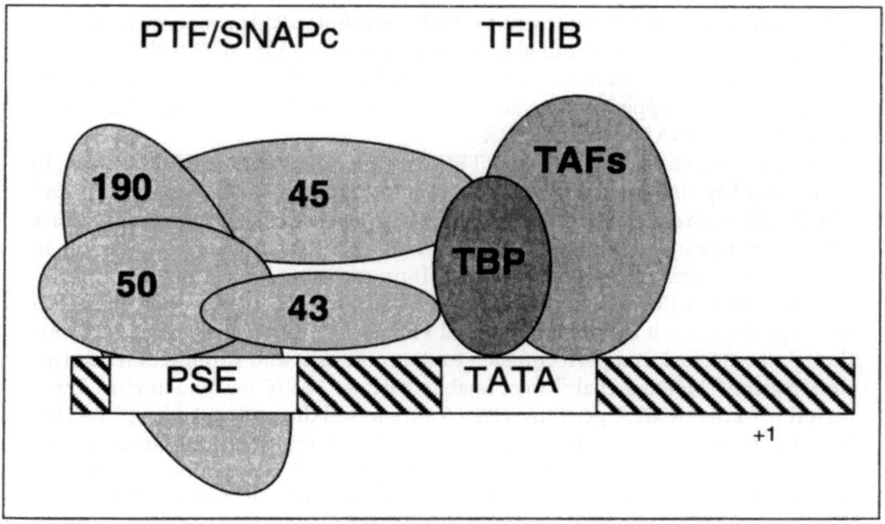

Fig. 5.7. Schematic diagram of the interactions between PTF/SNAPc and TFIIIB that may occur on a vertebrate U6 promoter.

affinity PSE which can efficiently recruit factors in the absence of Oct-1.[170] As a consequence, transcription of this gene, unlike the human U6 gene, is independent of the DSE in vitro.[144] Although Oct-1 binds readily in the absence of other factors, its interaction with DNA can be stabilized by the presence of PTF.[170] The evidence suggests a direct cooperative association between the Octs and PTF/SNAPc.[169,170] This may serve to accelerate the slow PSE binding step. Mutational analysis indicates that the POU-specific domain of Oct-1 is involved in interacting with SNAPc, regardless of the relative orientation of the octamer and PSE sites.[169] Although PTF_β/SNAP50 can contact DNA,[163] in the presence of Oct-1 Yoon et al[148] only detected crosslinking of PTF_α/SNAP190; the ability of Oct-1 to influence the interaction between PTF/SNAPc and PSE DNA may explain this discrepancy.

In ~70% of sequenced DSE elements, an octamer motif is located within 28 bp of a binding site for Staf.[172] These two elements can interact synergistically to stimulate transcription and the effect displays a marked dependence on their separation.[173] These observations suggest a functional cooperativity between Oct-1 and Staf.

Template competition experiments showed that the PSE is essential for stable complex formation on the human U6 promoter.[174] Mutation of the consensus octamer in the DSE reduced the stability of the U6 complex,[174] consistent with the results of Murphy et al.[170] Although the TATA box helped stabilize the complex, it was not essential for resistance to template competition.[174] Thus both kinetic and template commitment experiments suggest that the interaction with the PSE is of primary importance in assembling a transcription complex on a type III promoter. This step is rate-limiting[168] and is required to form a committed complex.[174] This can explain why it is the target for regulatory factors.[169,170] SNAPc/PTF may then help recruit TFIIIB to the TATA box. Mittal and Hernandez[157] demonstrated that

SNAPc enhances TBP recruitment to the U6 promoter. This cooperative interaction is not seen if the N-terminal 96 residues are deleted from TBP.[157] Full-length TBP also stimulates binding of SNAPc to the human PSE.[157] Thus, occupancy of the U6 promoter appears to be achieved by a series of cooperative interactions between Oct-1, SNAPc/PTF and TBP.

It remains to be established how TFIIIC1 is recruited to type III promoters. In all probability this involves protein-protein interactions between TFIIIC1 and TFIIIB, although contact with PTF/SNAPc is also a possibility. TFIIIC1 can increase PSE occupancy in a band shift assay.[90] TFIIIC1 recruitment may be a late step in transcription complex assembly on type III promoters.

The *Xenopus* tRNA^{Sec} promoter may be considered a hybrid between type II and type III, since it contains functional PSE, TATA and B-block elements.[175] Although the B-block is a perfect match to the consensus and stimulates transcription in vivo,[175] Meißner et al[176] were unable to detect TFIIIC binding in vitro. Transcription of the tRNA^{Sec} gene in a HeLa S100 showed the prolonged lag phase that is typical of type III promoters.[176] Oligonucleotide competition experiments demonstrated a clear requirement for a PSE-binding factor.[176] Expression was reported to increase following the addition of recombinant TBP and purified TFIIA.[176,177] The *Xenopus* tRNA^{Sec} promoter therefore behaves as a type III promoter when transcribed in a HeLa extract.[176] Similar results were obtained with the human Y3 gene.[176]

Complex Formation in the Absence of DNA

TFIIIB, TFIIIC, and pol III have been found associated in the absence of DNA both in *Drosophila* and human cell extracts.[96,178] Such complexes can be isolated from free factors and the bulk of cellular proteins by gel filtration or density gradient centrifugation, by virtue of their very large mass.[13,96,178] The inability to reassemble these complexes from separated and purified transcription factors in the absence of DNA makes them difficult to study. However, the stability of such complexes to 1 M KCl suggests that they are unlikely to be due to fortuitous aggregation.[178] Antibodies against TFIIICβ, but not preimmune sera, can quantitatively immunoprecipitate both TFIIIC2a and TFIIIC1 from unfractionated extracts.[179] Coprecipitation was still detected after washing with 1 M KCl.[179] This suggests that TFIIIC1 and TFIIIC2a associate stably in the absence of DNA. However, TFIIIC1 did not coprecipitate with TFIIIC2 when an anti-TFIIICα antibody was used to immunodeplete a PC-C fraction.[180] The interaction between TFIIIC2 and TFIIIC1 may be disrupted either by phosphocellulose chromatography or by the anti-TFIIICα antibody. Both TBP and hBRF can be immunoprecipitated from HeLa nuclear extract using an antibody against the C82 subunit of human pol III.[95] Conversely, antibodies against hBRF coprecipitate the C82, C53 and C20 pol III subunits.[95] These results provide further evidence that TFIIIB and pol III associate off the DNA.

Following addition of fractionated factors to a class III gene, a lag period occurs prior to the attainment of a linear transcription rate due to the time required for complex assembly.[2,5,168,181] In contrast, when gradient-purified protein complexes or unfractionated concentrated extracts are used, no lag is obtained.[13,178,181] This suggests that these large factor complexes are fully active and capable of carrying out transcription as soon as they have bound a gene. The exception to this observation is provided by 5S templates, which invariably display a time lag, consistent with the absence of TFIIIA from the gradient-purified pol III/TFIIIB/TFIIIC complexes.[13] However, Lagna et al[145] found that TFIIIA can be immunoprecipitated along with TFIIIC2 using antibodies against TFIIICα. This indicates an association between TFIIIA and TFIIIC2 in the absence of DNA. This interaction was stable to 0.5 M KCl.[145] Whether or not the putative TFIIIA/TFIIIC2 complex also contains TFIIIB and pol III remains to be determined. The existence of a preformed TFIIIA/TFIIIC2 complex might provide a means of bypassing the metastable TFIIIA/5S DNA stage during transcription complex assembly. Interactions between multifactor complexes and large DNA segments may provide the high degree of sequence specificity required for regulators to find their targets in the presence of the vast excess of nonspecific sites that occur in eukaryotic genomes. Protein-protein contacts among components of such complexes might in some situations partially compensate for the loss of normal protein-DNA contacts.

Stability of Complexes

Fully formed class III transcription complexes are extremely stable. The transition from metastability to a highly stable state generally accompanies a particular assembly step. For *Xenopus* 5S genes, TFIIIC confers stability upon the TFIIIA/5S DNA complex.[1,128] For VA_I and some tRNA genes, TFIIIC binding is sufficient to give a stable complex, but for most type II promoters the recruitment of TFIIIB is also required.[1,30,85,96,182] TFIIIC1 stabilizes binding of TFIIIC2 to $tDNA^{Arg}$ and VA_I.[5] Once formed, the stable preinitiation complexes support multiple rounds of transcription and remain refractory to template challenge.[1,85,122] 5S genes in chromatin isolated from frogs[183] or yeast[184] retain all the factors required for transcription except for pol III itself. Similarly, pol III is the only component missing from tRNA genes present on isolated recombinant SV40 minichromosomes.[185]

Stable complex formation on tRNA and 5S genes exhibits a strong temperature dependence.[26,186] The sharpness of this dependence suggests a cooperative structural transition and is reminiscent of the formation of open promoter complexes in prokaryotes, with polymerase directed helix melting. Single-stranded DNA interferes nonspecifically but selectively with the binding of TFIIIC to the 5' half of $tDNA^{Leu3}$, including the A-block, the initiation site, and further upstream.[26] Thus, helix unwinding may be involved in the formation of a stable preinitiation complex. This may contribute to the preference for the more easily melted A/T-rich sequences upstream of tRNA genes.

Dimethylsulphoxide stabilizes transcription complexes against dissociation by salt, and in general, the influence of ionic strength and dimethylsulphoxide on transcription complex stability correlates closely with the effects of these agents on transcription.[24] However, it is possible to find conditions under which class III genes are correctly transcribed despite being unable to form stable (preemptive) transcription complexes.[24,128,187]

Kassavetis et al[30] measured dissociation rates by preincubating a radiolabeled $tDNA^{Tyr}$ probe with yeast factors, and then adding excess unlabeled tDNA competitor, followed after various time intervals by DNase I. A TFIIIC/tDNA complex was found to dissociate with a half-life of 17 minutes, whereas a TFIIIB/TFIIIC/tDNA complex had a half-life of more than 10 hours. The efficiency of crosslinking of τ_{131} to upstream sequences increases 4- to 9-fold as TFIIIB is recruited into the complex, suggesting that it becomes locked into proximity to the DNA.[44] The B' fraction alone is sufficient to stabilize TFIIIC binding to DNA.[50] Indeed, upstream crosslinking of τ_{131} is stimulated to a greater extent by B' alone than it is by complete TFIIIB, and subsequent addition of B" reduces this crosslinking, perhaps by altering the topology of the B'/TFIIIC/tDNA complex. TFIIIC can also be crosslinked to the transcriptional start site in the presence of B' alone, but not if B" is also present.[50] Photocrosslinking of τ_{131} is enhanced slightly by yBRF alone, but TBP is also required to achieve the full effect.[7]

Once formed, class III transcription complexes display considerable stability towards high salt concentrations that would completely preclude their formation.[4,12,13,188] For example, ~45% of VA_I transcriptional activity remained after a 6.5- minute exposure of the preformed complex to 1 M KCl.[4] Preinitiation complexes on VA_I or U6 genes are stable to 0.05% sarkosyl, although pretreatment with 0.015% sarkosyl is sufficient to prevent their assembly.[174,189] Full transcriptional activity is retained by complexes formed on yeast tRNA and 5S genes after addition of 500 mM NaCl or 100 µg/ml heparin.[30,32]

In the yeast system, the interaction between TFIIIB and DNA is the most resistant to stripping by salt, whereas TFIIIC and TFIIIA are more readily dissociated from the complex.[30,32] Little or no exchange of stably bound TFIIIB to a second template occurs even at 1 M NaCl.[32] This is remarkable, because TFIIIB alone appears unable to recognize TATA-less class III genes.[12,27,30,32,85,190,191] Salt or heparin addition to a preformed yeast TFIIIB/TFIIIC/tDNA complex results in the disappearance of the A- and B-block protection, with little or no decrease in the strength of the upstream footprint.[30,32] A-block binding disappears at lower salt concentrations than B-block binding.[24] These conditions also abolish the gene-internal footprint of a preformed TFIIIB/TFIIIC/TFIIIA/5S DNA complex, but again, not the upstream footprint.[32] Therefore, although TFIIIB binding to the upstream region requires the presence of TFIIIC and, in the case of 5S genes, TFIIIA, once TFIIIB is recruited it is no longer dependent upon a continued association of TFIIIC or TFIIIA with A-, B-, or C-block sequences. yBRF, TBP and B" are all required for heparin-resistant binding, and each is retained in the heparin-resistant complex.[7]

Whereas direct interactions between TBP and various TATA-less class III genes are not readily detected,[6-8] the introduction of a TATA box upstream of the yeast SUP4 tRNA gene allows TBP to bind DNA directly.[7] The resultant complex is sufficiently stable to be observed in a gel retardation assay, but is not resistant to 250 µg/ml heparin.[7] Therefore, TBP recruited in association with TFIIIB is retained in a much more stable complex than is TBP that is recruited by recognition of a TATA box.

In a seminal study, Kassavetis et al[32] demonstrated that salt- or heparin-stripped complexes consisting only of TFIIIB bound to the upstream regions of tRNA or 5S genes is fully competent to recruit pol III for multiple rounds of accurately initiated transcription. Indeed, it can do so with the same efficiency as a fully assembled transcription complex containing TFIIIC and, for 5S genes, TFIIIA. The footprint produced at the start site by the polymerase is the same whether or not TFIIIC and TFIIIA are present.[32] Therefore, TFIIIB can be regarded as the only essential initiation factor proper in the yeast system, while TFIIIC and TFIIIA serve as assembly factors that are required to recruit and position TFIIIB on the promoter.

With frog or human factors the dissociation pathway is the reverse of that for association, with polymerase lost first and then TFIIIB, as the salt level is raised.[4,12,13,188,192] Thus, the exceptionally stable interaction between TFIIIB and upstream DNA that is seen in yeast appears not to occur in vertebrates. Few examples have been reported of upstream TFIIIB footprints on TATA-less promoters in vertebrate systems. One possible exception is provided by Wolffe and Morse,[136] who observed the time-dependent appearance of protection extending to -24 on a *Xenopus* somatic 5S gene. In vivo footprinting of a $tRNA^{Val}$ gene in human fibroblasts revealed weak protection at -7.[193] However, binding of lac repressor to sites as far as 47 bp upstream of a human $tRNA^{Ser}$ gene inhibits expression in human cells and extracts.[92,94] Furthermore, in *Xenopus* extracts GCN4 can repress transcription from tRNA or 5S genes that contain its binding site inserted 20-40 bp upstream of the initiation site.[93] These results suggest that vertebrate transcription complexes on internal promoters do indeed project upstream of the start site, most probably due to TFIIIB. However, if TFIIIB does directly contact upstream DNA in higher eukaryotes, the interaction is unlikely to be an avid one. In yeast, prebinding of GCN4 to 5' flanking sequences can redirect initiation to sites downstream of +1, and this downstream initiation persists even if GCN4 is subsequently removed.[47] In humans, upstream binding of lac repressor also results in aberrant initiation, but transcription reverts to the normal start site if lac is subsequently removed.[92] This difference in behavior provides another indication that TFIIIB binds less tightly to DNA in vertebrates than it does in *Saccharomyces*. The fact that this feature appears not to have been conserved suggests that the tight DNA binding displayed by yeast TFIIIB may not be essential for transcription.

References

1. Lassar AB, Martin PL, Roeder RG. Transcription of class III genes: formation of preinitiation complexes. Science 1983; 222:740-748.
2. Bieker JJ, Martin PL, Roeder RG. Formation of a rate-limiting intermediate in 5S RNA gene transcription. Cell 1985; 40:119-127.
3. Segall J. Assembly of a yeast 5S RNA gene transcription complex. J Biol Chem 1986; 261:11578-11584.
4. Carey MF, Gerrard SP, Cozzarelli NR. Analysis of RNA polymerase III transcription complexes by gel filtration. J Biol Chem 1986; 261:4309-4317.
5. Dean N, Berk AJ. Ordering promoter binding of class III transcription factors TFIIIC1 and TFIIIC2. Mol Cell Biol 1988; 8:3017-3025.
6. Huet J, Sentenac A. The TATA-binding protein participates in TFIIIB assembly on tRNA genes. Nucleic Acids Res 1993; 20:6451-6454.
7. Kassavetis GA, Joazeiro CAP, Pisano M et al. The role of the TATA-binding protein in the assembly and function of the multisubunit yeast RNA polymerase III transcription factor, TFIIIB. Cell 1992; 71:1055-1064.
8. White RJ, Jackson SP. Mechanism of TATA-binding protein recruitment to a TATA-less class III promoter. Cell 1992; 71:1041-1053.
9. Schultz MC, Reeder RH, Hahn S. Variants of the TATA-binding protein can distinguish subsets of RNA polymerase I, II, and III promoters. Cell 1992; 69:697-702.
10. Martinez E, Zhou Q, L'Etoile ND et al. Core promoter-specific function of a mutant transcription factor TFIID defective in TATA-box binding. Proc Natl Acad Sci USA 1995; 92:11864-11868.
11. Joazeiro CAP, Kassavetis GA, Geiduschek EP. Alternative outcomes in assembly of promoter complexes: the roles of TBP and a flexible linker in placing TFIIIB on tRNA genes. Genes Dev 1996; 10:725-739.
12. Setzer DR, Brown DD. Formation and stability of the 5S RNA transcription complex. J Biol Chem 1985; 260:2483-2492.
13. Jahn D, Wingender E, Seifart KH. Transcription complexes for various class III genes differ in parameters of formation and stability towards salt. J Mol Biol 1987; 193:303-313.

14. Simmen KA, Bernues J, Parry HD et al. TFIID is required for in vitro transcription of the human U6 gene by RNA polymerase III. EMBO J 1991; 10:1853-1862.
15. Marsolier M-C, Chaussivert N, Lefebvre O et al. Directing transcription of an RNA polymerase III gene via GAL4 sites. Proc Natl Acad Sci USA 1994; 91:11938-11942.
16. Stillman DJ, Better M, Geiduschek EP. Electron-microscopic examination of the binding of a large RNA polymerase III transcription factor to a tRNA gene. J Mol Biol 1985; 185:451-455.
17. Bazett-Jones DP, Brown ML. Electron microscopy reveals that transcription factor TFIIIA bends 5S DNA. Mol Cell Biol 1989; 9:336-341.
18. Schroth GP, Cook GR, Bradbury EM et al. Transcription factor IIIA induced bending of the *Xenopus* somatic 5S gene promoter. Nature 1989; 340:487-488.
19. Leveillard T, Kassavetis GA, Geiduschek EP. *Saccharomyces cerevisiae* transcription factors IIIB and IIIC bend the DNA of a $tRNA^{Gln}$ gene. J Biol Chem 1991; 266:5162-5168.
20. Schroth GP, Gottesfeld JM, Bradbury EM. TFIIIA induced DNA bending: effect of low ionic strength electrophoresis buffer conditions. Nucleic Acids Res 1991; 19:511-516.
21. Braun BR, Kassavetis GA, Geiduschek EP. Bending of the *Saccharomyces cerevisiae* 5S rRNA gene in transcription factor complexes. J Biol Chem 1992; 267:22562-22569.
22. Ruet A, Camier S, Smagowicz W et al. Isolation of a class C transcription factor which forms a stable complex with tRNA genes. EMBO J 1984; 3:343-350.
23. Stillman DJ, Geiduschek EP. Differential binding of a *S. cerevisiae* RNA polymerase III transcription factor to two promoter segments of a tRNA gene. EMBO J 1984; 3:847-853.
24. Stillman DJ, Sivertsen A, Zentner PG et al. Correlations between transcription of a yeast tRNA gene and transcription factor-DNA interactions. J Biol Chem 1984; 259:7955-7962.
25. Camier S, Gabrielsen O, Baker R et al. A split binding site for transcription factor τ on the $tRNA^{Glu3}$ gene. EMBO J 1985; 4:491-500.
26. Stillman DJ, Caspers P, Geiduschek EP. Effects of temperature and single-stranded DNA on the interaction of an RNA polymerase III transcription factor with a tRNA gene. Cell 1985; 40:311-317.
27. Huibregtse JM, Evans CF, Engelke DR. Comparison of tRNA gene transcription complexes formed in vitro and in nuclei. Mol Cell Biol 1987; 7:3212-3220.
28. Engelke DR. Interaction of tRNA transcription factors with satellite I DNA from *Xenopus laevis*. Gene 1988; 62:323-330.
29. Johnson DL, Nichols M, Bolger MB et al. Interaction of yeast transcription factor IIIC with dimeric *Schizosaccharomyces pombe* $tRNA^{Ser}$-$tRNA^{Met}$ genes. J Biol Chem 1989; 264:19221-19227.
30. Kassavetis GA, Riggs DL, Negri R et al. Transcription factor IIIB generates extended DNA interactions in RNA polymerase III transcription complexes on tRNA genes. Mol Cell Biol 1989; 9:2551-2566.
31. Camier S, Baker RE, Sentenac A. On the flexible interaction of yeast factor τ with the bipartite promoter of tRNA genes. Nucleic Acids Res 1990; 18:4571-4578.
32. Kassavetis GA, Braun BR, Nguyen LH et al. *S. cerevisiae* TFIIIB is the transcription initiation factor proper of RNA polymerase III, while TFIIIA and TFIIIC are assembly factors. Cell 1990; 60:235-245.
33. Baker RE, Camier S, Sentenac A et al. Gene size differentially affects the binding of yeast transcription factor τ to two intragenic regions. Proc Natl Acad Sci USA 1987; 84:8768-8772.

34. Fabrizio P, Coppo A, Fruscoloni P et al. Comparative mutational analysis of wild-type and stretched $tRNA^{Leu3}$ gene promoters. Proc Natl Acad Sci USA 1987; 84:8763-8767.
35. Schultz P, Marzouki N, Marck C et al. The two DNA-binding domains of yeast transcription factor τ as observed by scanning transmission electron microscopy. EMBO J 1989; 8:3815-3824.
36. Eschenlauer JB, Kaiser MW, Gerlach VL et al. Architecture of a yeast U6 RNA gene promoter. Mol Cell Biol 1993; 13:3015-3026.
37. Burnol A-F, Margottin F, Schultz P et al. Basal promoter and enhancer element of yeast U6 snRNA gene. J Mol Biol 1993; 233:644-658.
38. Reyes VM, Newman A, Abelson J. Mutational analysis of the coordinate expression of the yeast $tRNA^{Arg}$-$tRNA^{Asp}$ gene tandem. Mol Cell Biol 1986; 6:2436-2442.
39. Baker RE, Gabrielsen O, Hall BD. Effects of $tRNA^{Tyr}$ point mutations on the binding of yeast RNA polymerase III transcription factor C. J Biol Chem 1986; 261:5275-5282.
40. Lefebvre O, Carles C, Conesa C et al. *TFC3*: gene encoding the B-block binding subunit of the yeast transcription factor IIIC. Proc Natl Acad Sci USA 1992; 89:10512-10516.
41. Lilley DM. HMG has DNA wrapped up. Nature 1992; 357:282-283.
42. Bartholomew B, Kassavetis GA, Braun BR et al. The subunit structure of *Saccharomyces cerevisiae* transcription factor IIIC probed with a novel photocrosslinking reagent. EMBO J 1990; 9:2197-2205.
43. Braun BR, Bartholomew B, Kassavetis GA et al. Topography of transcription factor complexes on the *Saccharomyces cerevisiae* 5S RNA gene. J Mol Biol 1992; 228:1063-1077.
44. Bartholomew B, Kassavetis GA, Geiduschek EP. Two components of *Saccharomyces cerevisiae* transcription factor IIIB (TFIIIB) are stereospecifically located upstream of a tRNA gene and interact with the second-largest subunit of TFIIIC. Mol Cell Biol 1991; 11:5181-5189.
45. Bartholomew B, Braun BR, Kassavetis GA et al. Probing close DNA contacts of RNA polymerase III transcription complexes with the photoactive nucleoside 4-thiodeoxythymidine. J Biol Chem 1994; 269:18090-18095.
46. Braun BR, Riggs DL, Kassavetis GA et al. Multiple states of protein-DNA interaction in the assembly of transcription complexes on *Saccharomyces cerevisiae* 5S ribosomal RNA genes. Proc Natl Acad Sci USA 1989; 86:2530-2534.
47. Leveillard T, Kassavetis GA, Geiduschek EP. Repression and redirection of *Saccharomyces cerevisiae* tRNA synthesis from upstream of the transcriptional start site. J Biol Chem 1993; 268:3594-3603.
48. Dingermann T, Frank-Stoll U, Werner H et al. RNA polymerase III catalyzed transcription can be regulated in *Saccharomyces cerevisiae* by the bacterial tetracycline repressor-operator system. EMBO J 1992; 11:1487-1492.
49. Palida F, Hale C, Sprague KU. Transcription of a silkworm $tRNA^{AlaC}$ gene is directed by two AT-rich upstream sequence elements. Nucleic Acids Res 1993; 21:5875-5881.
50. Kassavetis GA, Bartholomew B, Blanco JA et al. Two essential components of the *Saccharomyces cerevisiae* transcription factor TFIIIB: transcription and DNA-binding properties. Proc Natl Acad Sci USA 1991; 88:7308-7312.
51. Khoo B, Brophy B, Jackson SP. Conserved functional domains of the RNA polymerase III general transcription factor BRF. Genes Dev 1994; 8:2879-2890.
52. Chaussivert N, Conesa C, Shaaban S et al. Complex interactions between yeast TFIIIB and TFIIIC. J Biol Chem 1995; 270:15353-15358.

53. Ruth J, Conesa C, Dieci G et al. A suppressor of mutations in the class III transcription system encodes a component of yeast TFIIIB. EMBO J 1996; 15:1941-1949.
54. Willis I, Oksman A, Lopez-De-Leon A. The PCF1-1 mutation increases the activity of the transcription factor (TF) IIIB fraction from *Saccharomyces cerevisiae.* Nucleic Acids Res 1992; 20:3725-3730.
55. Rameau G, Puglia K, Crowe A et al. A mutation in the second largest subunit of TFIIIC increases a rate-limiting step in transcription by RNA polymerase III. Mol Cell Biol 1994; 14:822-830.
56. Librizzi MD, Moir RD, Brenowitz M et al. Expression and purification of the RNA polymerase III transcription specificity factor $IIIB_{70}$ from *Saccharomyces cerevisiae* and its cooperative binding with TATA-binding protein. J Biol Chem 1996; 271:32695-32701.
57. Lopez-de-Leon A, Librizzi M, Tuglia K et al. PCF4 encodes an RNA polymerase III transcription factor with homology to TFIIB. Cell 1992; 71:211-220.
58. Sethy I, Moir RD, Librizzi M et al. In vitro evidence for growth regulation of tRNA gene transcription in yeast. J Biol Chem 1995; 270:28463-28470.
59. Huet J, Conesa C, Manaud N et al. Interactions between yeast TFIIIB components. Nucleic Acids Res 1994; 22:3433-3439.
60. Kumar A, Kassavetis GA, Geiduschek EP et al. Functional dissection of the B" component of RNA polymerase III transcription factor IIIB: a scaffolding protein with multiple roles in assembly and initiation of transcription. Mol Cell Biol 1997; 17:1868-1880.
61. Dombroski AJ, Walter WA, Record MTJ et al. Polypeptides containing highly conserved regions of transcription initiation factor σ^{70} exhibit specificity of binding to promoter DNA. Cell 1992; 70:501-512.
62. Hupp TR, Meek DW, Midgley CA et al. Regulation of the specific DNA binding function of p53. Cell 1992; 71:875-886.
63. Persinger J, Bartholomew B. Mapping the contacts of yeast TFIIIB and RNA polymerase III at various distances from the major groove of DNA by DNA photoaffinity labeling. J Biol Chem 1996; 271:33039-33046.
64. Whitehall SK, Kassavetis GA, Geiduschek EP. The symmetry of the yeast U6 RNA gene's TATA box and the orientation of the TATA-binding protein in yeast TFIIIB. Genes Dev 1995; 9:2974-2985.
65. Strubin M, Struhl K. Yeast and human TFIID with altered DNA-binding specificity for TATA elements. Cell 1992; 68:721-730.
66. Arndt KM, Wobbe CR, Ricupero-Hovasse S et al. Equivalent mutations in the two repeats of yeast TATA-binding protein confer distinct TATA recognition specificities. Mol Cell Biol 1994; 14:3719-3728.
67. Wang Y, Stumph WE. RNA polymerase II/III transcription specificity determined by TATA box orientation. Proc Natl Acad Sci USA 1995; 92:8606-8610.
68. Huang W, Wong JM, Bateman E. TATA elements direct bi-directional transcription by RNA polymerases II and III. Nucleic Acids Res 1996; 24:1158-1163.
69. Wang Y, Jensen RC, Stumph WE. Role of TATA box sequence and orientation in determining RNA polymerase II/III transcription specificity. Nucleic Acids Res 1996; 24:3100-3106.
70. Dieci G, Duimio L, Coda-Zabetta F et al. A novel RNA polymerase III transcription factor fraction that is not required for template commitment. J Biol Chem 1993; 268:11199-11207.
71. Greenblatt J. RNA polymerase-associated transcription factors. Trends Biochem Sci 1991; 16:408-411.
72. Roeder RG. The complexities of eukaryotic transcription initiation: regulation of preinitiation complex assembly. Trends Biochem Sci 1991; 16:402-408.

73. Werner M, Chaussivert N, Willis IM et al. Interaction between a complex of RNA polymerase III subunits and the 70-kDa component of transcription factor IIIB. J Biol Chem 1993; 268:20721-20724.
74. Bartholomew B, Durkovich D, Kassavetis GA et al. Orientation and topography of RNA polymerase III in transcription complexes. Mol Cell Biol 1993; 13:942-952.
75. Kassavetis GA, Blanco JA, Johnson TE et al. Formation of open and elongating transcription complexes by RNA polymerase III. J Mol Biol 1992; 226:47-58.
76. Brow DA, Guthrie C. Transcription of a yeast U6 snRNA gene requires a polymerase III promoter element in a novel position. Genes Dev 1990; 4:1345-1356.
77. Moenne A, Camier S, Anderson G et al. The U6 gene of *Saccharomyces cerevisiae* is transcribed by RNA polymerase C (III) in vivo and in vitro. EMBO J 1990; 9:271-277.
78. Burnol A-F, Margottin F, Huet J et al. TFIIIC relieves repression of U6 snRNA transcription by chromatin. Nature 1993; 362:475-477.
79. Joazeiro CAP, Kassavetis GA, Geiduschek EP. Identical components of yeast transcription factor IIIB are required and sufficient for transcription of TATA box-containing and TATA-less genes. Mol Cell Biol 1994; 14:2798-2808.
80. Ottonello S, Rivier DH, Doolittle GM et al. The properties of a new polymerase III transcription factor reveal that transcription complexes can assemble by more than one pathway. EMBO J 1987; 6:1921-1927.
81. Young LS, Rivier DH, Sprague KU. Sequences far downstream from the classical tRNA promoter elements bind RNA polymerase III transcription factors. Mol Cell Biol 1991; 11:1382-1392.
82. Young LS, Ahnert N, Sprague KU. Silkworm TFIIIB binds both constitutive and silk gland-specific $tRNA^{Ala}$ promoters but protects only the constitutive promoter from DNase I cleavage. Mol Cell Biol 1996; 16:1256-1266.
83. Young LS, Takahashi N, Sprague KU. Upstream sequences confer distinctive transcriptional properties on genes encoding silkgland-specific $tRNA^{Ala}$. Proc Natl Acad Sci USA 1986; 83:374-378.
84. Sullivan HS, Young LS, White CN et al. Silk gland-specific $tRNA^{Ala}$ genes interact more weakly than constitutive $tRNA^{Ala}$ genes with silkworm TFIIIB and polymerase III fractions. Mol Cell Biol 1994; 14:1806-1814.
85. Fuhrman SA, Engelke DR, Geiduschek EP. HeLa cell RNA polymerase III transcription factors. J Biol Chem 1984; 259:1934-1943.
86. Cromlish JA, Roeder RG. Human transcription factor IIIC (TFIIIC). J Biol Chem 1989; 264:18100-18109.
87. Schneider HR, Waldschmidt R, Seifart KH. Human transcription factor IIIC contains a polypeptide of 55 kDa specifically binding to Pol III genes. Nucleic Acids Res 1990; 18:4743-4750.
88. Wang Z, Roeder RG. TFIIIC1 acts through a downstream region to stabilize TFIIIC2 binding to RNA polymerase III promoters. Mol Cell Biol 1996; 16:6841-6850.
89. Yoshinaga SK, Boulanger PA, Berk AJ. Resolution of human transcription factor TFIIIC into two functional components. Proc Natl Acad Sci USA 1987; 84:3585-3589.
90. Oettel S, Hartel F, Kober I et al. Human transcription factors $IIIC_2$, $IIIC_1$ and a novel component $IIIC_0$ fulfill different aspects of DNA binding to various pol III genes. Nucleic Acids Res 1997; 25:2440-2447.
91. Boulanger PA, Yoshinaga SK, Berk AJ. DNA-binding properties and characterization of human transcription factor TFIIIC2. J Biol Chem 1987; 262:15098-15105.
92. Tapping RI, Syroid DE, Capone JP. Upstream interactions of functional mammalian tRNA gene transcription complexes probed using a heterologous DNA-binding protein. J Biol Chem 1994; 269:21812-21819.

93. McBryant SJ, Kassavetis GA, Gottesfeld JM. Repression of vertebrate RNA polymerase III transcription by DNA binding proteins located upstream from the transcription start site. J Mol Biol 1995; 250:315-326.
94. Syroid DE, Tapping RI, Capone JP. Regulated expression of a mammalian nonsense suppressor tRNA gene in vivo and in vitro using the *lac* operator/repressor system. Mol Cell Biol 1992; 12:4271-4278.
95. Wang Z, Roeder RG. Three human RNA polymerase III-specific subunits form a subcomplex with a selective function in specific transcription initiation. Genes Dev 1997; 11:1315-1326.
96. Burke DJ, Soll D. Functional analysis of fractionated Drosophila Kc cell tRNA gene transcription components. J Biol Chem 1985; 260:816-823.
97. Johnson DL, Fan RS, Treinies ML. Analysis of the molecular mechanisms for the species-specific transcription of *Drosophila* and human tRNA gene transcription components. J Biol Chem 1991; 266:16037-16043.
98. Bieker JJ, Roeder RG. Physical properties and DNA-binding stoichiometry of a 5S gene-specific transcription factor. J Biol Chem 1984; 259:6158-6164.
99. Smith DR, Jackson IJ, Brown DD. Domains of the positive transcription factor specific for the *Xenopus* 5S RNA gene. Cell 1984; 37:645-652.
100. Taylor W, Jackson IJ, Siegel N et al. The developmental expression of the gene for TFIIIA in *Xenopus laevis*. Nucleic Acids Res 1986; 14:6185-6195.
101. Vrana KE, Churchill MEA, Tullius TD et al. Mapping functional regions of transcription factor TFIIIA. Mol Cell Biol 1988; 8:1684-1696.
102. Engelke DR, Ng S-Y, Shastry BS et al. Specific interaction of a purified transcription factor with an internal control region of 5S RNA genes. Cell 1980; 19:717-728.
103. Sakonju S, Brown DD, Engelke D et al. The binding of a transcription factor to deletion mutants of a 5S ribosomal RNA gene. Cell 1981; 23:665-669.
104. Sakonju S, Brown DD. Contact points between a positive transcription factor and the *Xenopus* 5S RNA gene. Cell 1982; 31:395-405.
105. Fairall L, Rhodes D, Klug A. Mapping of the site of protection on a 5S RNA gene by the *Xenopus* transcription factor IIIA. J Mol Biol 1986; 192:577-591.
106. Fradkin LG, Yoshinaga SK, Berk AJ et al. Human transcription factor TFIIIC2 specifically interacts with a unique sequence in the *Xenopus laevis* 5S rRNA gene. Mol Cell Biol 1989; 9:4941-4950.
107. Xing YY, Worcel A. The C-terminal domain of transcription factor IIIA interacts differently with different 5S RNA genes. Mol Cell Biol 1989; 9:499-514.
108. Hayes JJ, Tullius TD. Structure of the TFIIIA-5S DNA complex. J Mol Biol 1992; 227:407-417.
109. Hayes J, Tullius TD, Wolffe AP. A protein-protein interaction is essential for stable complex formation on a 5S RNA gene. J Biol Chem 1989; 264:6009-6012.
110. McConkey GA, Bogenhagen DF. TFIIIA binds with equal affinity to somatic and major oocyte 5S RNA genes. Genes Dev 1988; 2:205-214.
111. Moorefield B, Roeder RG. Purification and characterization of human transcription factor IIIA. J Biol Chem 1994; 269:20857-20865.
112. Clemens KR, Liao X, Wolf V et al. Definition of the binding sites of individual zinc fingers in the transcription factor IIIA-5S RNA gene complex. Proc Natl Acad Sci USA 1992; 89:10822-10826.
113. Hayes JJ, Clemens KR. Locations of contacts between individual zinc fingers of *Xenopus laevis* transcription factor IIIA and the internal control region of a 5S RNA gene. Biochemistry 1992; 31:11600-11605.
114. Fairall L, Rhodes D. A new approach to the analysis of DNase I footprinting data and its application to the TFIIIA/5S DNA complex. Nucleic Acids Res 1992; 20:4727-4731.

115. Christensen JH, Hansen PK, Lillelund O et al. Sequence-specific binding of the N-terminal three-finger fragment of *Xenopus* transcription factor IIIA to the internal control region of a 5S RNA gene. FEBS Lett 1991; 281:181-184.
116. Darby MK, Joho KE. Differential binding of zinc fingers from *Xenopus* TFIIIA and p43 to 5S RNA and the 5S RNA gene. Mol Cell Biol 1992; 12:3155-3164.
117. Liao XB, Clemens KR, Tennant L et al. Specific interaction of the first three zinc fingers of TFIIIA with the internal control region of the *Xenopus* 5S RNA gene. J Mol Biol 1992; 223:857-871.
118. Theunissen O, Rudt F, Guddat U et al. RNA and DNA binding zinc fingers in *Xenopus* TFIIIA. Cell 1992; 71:679-690.
119. Gottesfeld JM, Neely L, Trauger JW et al. Regulation of gene expression by small molecules. Nature 1997; 387:202-205.
120. Pavletich NP, Pabo CO. Zinc finger-DNA recognition: crystal structure of a Zif268-DNA complex at 2.1 A. Science 1991; 252:809-817.
121. Berg JM. Zinc finger domains: hypotheses and current knowledge. Annu Rev Biophys Chem 1990; 19:405-421.
122. Bogenhagen DF, Wormington WM, Brown DD. Stable transcription complexes of *Xenopus* 5S RNA genes: a means to maintain the differentiated state. Cell 1982; 28:413-421.
123. Hanas JS, Bogenhagen DF, Wu C-W. Binding of *Xenopus* transcription factor A to 5S RNA and to single-stranded DNA. Nucleic Acids Res 1984; 12:2745-2758.
124. Keller HJ, Romaniuk PJ, Gottesfeld JM. Interaction of *Xenopus* TFIIIC with the TFIIIA.5S RNA gene complex. J Biol Chem 1992; 267:18190-18198.
125. Keller HJ, You QM, Romaniuk PJ et al. Additional intragenic promoter elements of the *Xenopus* 5S RNA genes upstream from the TFIIIA-binding site. Mol Cell Biol 1990; 10:5166-5176.
126. Majowski K, Mentzel H, Pieler T. A split binding site for TFIIIC on the *Xenopus* 5S gene. EMBO J 1987; 6:3057-3063.
127. Pieler T, Hamm J, Roeder RG. The 5S gene internal control region is composed of three distinct sequence elements, organized as two functional domains with variable spacing. Cell 1987; 48:91-100.
128. Wolffe AP. Transcription fraction TFIIIC can regulate differential *Xenopus* 5S RNA gene transcription in vitro. EMBO J 1988; 4:1071-1079.
129. Wolffe AP, Brown DD. DNA replication in vitro erases a *Xenopus* 5S RNA gene transcription complex. Cell 1986; 47:217-227.
130. Peck LJ, Bartilson M, DeRisi L. Bead-shift isolation of protein-DNA complexes on a 5S RNA gene. Nucleic Acids Res 1994; 22:443-449.
131. Sturges MR, Bartilson M, Peck LJ. Interaction of *Xenopus* TFIIIC with a 5S RNA gene. Nucleic Acids Res 1995; 23:1551-1556.
132. Schneider HR, Waldschmidt R, Jahn D et al. Purification of human transcription factor IIIC and its binding to the gene for ribosomal 5S RNA. Nucleic Acids Res 1989; 17:5003-5016.
133. Sands MS, Bogenhagen DF. TFIIIA binds to different domains of 5S RNA and the *Xenopus borealis* 5S RNA gene. Mol Cell Biol 1987; 7:3985-3993.
134. Taylor MJ, Segall J. Characterization of factors and DNA sequences required for accurate transcription of the *Saccharomyces cerevisiae* 5S RNA gene. J Biol Chem 1985; 260:4531-4540.
135. Sharp SJ, Garcia AD. Transcription of the *Drosophila melanogaster* 5S RNA gene requires an upstream promoter and four intragenic sequence elements. Mol Cell Biol 1988; 8:1266-1274.

136. Wolffe AP, Morse RH. The transcription complex of the *Xenopus* somatic 5 S RNA gene. A functional analysis of protein-DNA interactions outside of the internal control region. J Biol Chem 1990; 265:4592-4599.
137. Smith TPL, Young LS, Bender LB et al. Silkworm TFIIIA requires additional class III factors for committment to transcription complex assembly on a 5S RNA gene. Nucleic Acids Res 1995; 23:1244-1251.
138. Lee Y, Erkine AM, Van Ryk DI et al. In vivo analyses of the internal control region in the 5S rRNA gene from *Saccharomyces cerevisiae*. Nucleic Acids Res 1995; 23:634-640.
139. Challice JM, Segall J. Transcription of the 5S rRNA gene of *Saccharomyces cerevisiae* requires a promoter element at +1 and a 14 base-pair internal control region. J Biol Chem 1989; 264:20060-20067.
140. Wang CK, Weil PA. Purification and characterization of *Saccharomyces cerevisiae* transcription factor IIIA. J Biol Chem 1989; 264:1092-1099.
141. Struksnes K, Forus A, Gabrielsen OS et al. Yeast TFIIIA + TFIIIC/τ-factor, but not yeast TFIIIA alone, interacts with the *Xenopus* 5S rRNA gene. Nucleic Acids Res 1991; 19:565-571.
142. Milne CA, Segall J. Mapping regions of yeast transcription factor IIIA required for DNA binding, interaction with transcription factor IIIC, and transcription activity. J Biol Chem 1993; 268:11364-11371.
143. Reddy R. Transcription of a U6 small nuclear RNA gene in vitro. J Biol Chem 1988; 263:15980-15984.
144. Waldschmidt R, Wanandi I, Seifart KH. Identification of transcription factors required for the expression of mammalian U6 genes in vitro. EMBO J 1991; 10:2595-2603.
145. Lagna G, Kovelman R, Sukegawa J et al. Cloning and characterization of an evolutionarily divergent DNA-binding subunit of mammalian TFIIIC. Mol Cell Biol 1994; 14:3053-3064.
146. Teichmann M, Seifart KH. Physical separation of two different forms of human TFIIIB active in the transcription of the U6 or the VAI gene in vitro. EMBO J 1995; 14:5974-5983.
147. Wang Z, Roeder RG. Structure and function of a human transcription factor TFIIIB subunit that is evolutionarily conserved and contains both TFIIB- and high-mobility-group protein 2-related domains. Proc Natl Acad Sci USA 1995; 92:7026-7030.
148. Yoon J-B, Murphy S, Bai L et al. Proximal sequence element-binding transcription factor (PTF) is a multisubunit complex required for transcription of both RNA polymerase II- and RNA polymerase III-dependent small nuclear RNA genes. Mol Cell Biol 1995; 15:2019-2027.
149. Mital R, Kobayashi R, Hernandez N. RNA polymerase III transcription from the human U6 and adenovirus type 2 VAI promoters has different requirements for human BRF, a subunit of human TFIIIB. Mol Cell Biol 1996; 16:7031-7042.
150. Bernues J, Simmen KA, Lewis JD et al. Common and unique transcription factor requirements of human U1 and U6 snRNA genes. EMBO J 1993; 12:3573-3585.
151. Sadowski CL, Henry RW, Lobo SM et al. Targeting TBP to a non-TATA box *cis*-regulatory element: a TBP-containing complex activates transcription from snRNA promoters through the PSE. Genes Dev 1993; 7:1535-1548.
152. Lobo SM, Tanaka M, Sullivan ML et al. A TBP complex essential for transcription from TATA-less but not TATA-containing RNA polymerase III promoters is part of the TFIIIB fraction. Cell 1992; 71:1029-1040.
153. Bai L, Wang Z, Yoon J-B et al. Cloning and characterization of the β subunit of human proximal sequence element-binding transcription factor and its involvement in transcription of small nuclear RNA genes by RNA polymerases II and III. Mol Cell Biol 1996; 16:5419-5426.

154. Lobo SM, Lister J, Sullivan ML et al. The cloned RNA polymerase II transcription factor IID selects RNA polymerase III to transcribe the human U6 gene in vitro. Genes Dev 1991; 5:1477-1489.
155. Waldschmidt R, Seifart KH. TFIIA is required for in vitro transcription of mammalian U6 genes by RNA polymerase III. J Biol Chem 1992; 267:16359-16364.
156. Boyd DC, Turner PC, Watkins NJ et al. Functional redundancy of promoter elements ensures efficient transcription of the human 7SK gene in vivo. J Mol Biol 1995; 253:677-690.
157. Mittal V, Hernandez N. Role for the amino-terminal region of human TBP in U6 snRNA transcription. Science 1997; 275:1136-1140.
158. Timmers HT, Sharp PA. The mammalian TFIID protein is present in two functionally distinct complexes. Genes Dev 1991; 5:1946-1956.
159. Waibel F, Filipowicz W. RNA-polymerase specificity of transcription of Arabidopsis U snRNA genes determined by promoter element spacing. Nature 1990; 346:199-202.
160. Lescure A, Carbon P, Krol A. The different positioning of the proximal sequence element in the *Xenopus* RNA polymerase II and III snRNA promoters is a key determinant which confers RNA polymerase III specificity. Nucleic Acids Res 1991; 19:435-441.
161. Goomer RS, Kunkel GR. The transcriptional start site for a human U6 small nuclear RNA gene is dictated by a compound promoter element consisting of the PSE and the TATA box. Nucleic Acids Res 1992; 20:4903-4912.
162. Henry RW, Sadowski CL, Kobayashi R et al. A TBP-TAF complex required for transcription of human snRNA genes by RNA polymerases II and III. Nature 1995; 374:653-656.
163. Henry RW, Ma B, Sadowski CL et al. Cloning and characterization of SNAP50, a subunit of the snRNA-activating protein complex $SNAP_c$. EMBO J 1996; 15:7129-7136.
164. Sadowski CL, Henry RW, Kobayashi R et al. The SNAP45 subunit of the small nuclear RNA (snRNA) activating protein complex is required for RNA polymerase II and III snRNA gene transcription and interacts with the TATA box binding protein. Proc Natl Acad Sci USA 1996; 93:4289-4293.
165. Yoon J-B, Roeder RG. Cloning of two proximal sequence element-binding transcription factor subunits (γ and δ) that are required for transcription of small nuclear RNA genes by RNA polymerases II and III interact with the TATA-binding protein. Mol Cell Biol 1996; 16:1-9.
166. Schmidt MC, Zhou Q, Berk AJ. Sp1 activates transcription without enhancing DNA-binding activity of the TATA box factor. Mol Cell Biol 1989; 9:3299-3307.
167. Kuddus R, Schmidt MC. Effect of the non-conserved N-terminus on the DNA binding activity of the yeast TATA binding protein. Nucleic Acids Res 1993; 21:1789-1796.
168. Wanandi I, Waldschmidt R, Seifart KH. Mammalian transcription factor PBP. J Biol Chem 1993; 268:6629-6640.
169. Mittal V, Cleary MA, Herr W et al. The Oct-1 POU-specific domain can stimulate small nuclear RNA gene transcription by stabilizing the basal transcription complex $SNAP_c$. Mol Cell Biol 1996; 16:1955-1965.
170. Murphy S, Yoon JB, Gerster T et al. Oct-1 and Oct-2 potentiate functional interactions of a transcription factor with the proximal sequence element of small nuclear RNA genes. Mol Cell Biol 1992; 12:3247-3261.
171. Murphy S, Pierani A, Scheidereit C et al. Purified octamer binding transcription factors stimulate RNA polymerase III-mediated transcription of the 7SK RNA gene. Cell 1989; 59:1071-1080.

172. Schaub M, Myslinski E, Schuster C et al. Staf, a promiscuous activator for enhanced transcription by RNA polymerases II and III. EMBO J 1997; 16:173-181.
173. Myslinski E, Schuster C, Krol A et al. Promoter strength and structure dictate module composition in RNA polymerase III transcriptional activator elements. J Mol Biol 1993; 234:311-318.
174. Kunkel GR, Danzeiser DA. Formation of a template committed complex on the promoter of a gene for the U6 small nuclear RNA from the human requires multiple sequence elements, including the distal region. J Biol Chem 1992; 267:14250-14258.
175. Carbon P, Krol A. Transcription of the *Xenopus laevis* selenocysteine $tRNA^{(Ser)Sec}$ gene: a system that combines an internal B box and upstream elements also found in U6 snRNA genes. EMBO J 1991; 10:599-606.
176. Meißner W, Wanandi I, Carbon P et al. Transcription factors required for the expression of *Xenopus laevis* selenocysteine tRNA in vitro. Nucleic Acids Res 1994; 22:553-559.
177. Myslinski E, Schuster C, Huet J et al. Point mutations 5' to the tRNA selenocysteine TATA box alter RNA polymerase III transcription by affecting the binding of TBP. Nucleic Acids Res 1993; 21:5852-5858.
178. Wingender E, Jahn D, Seifart KH. Association of RNA polymerase III with transcription factors in the absence of DNA. J Biol Chem 1986; 261:1409-1413.
179. Sinn E, Wang Z, Kovelman R et al. Cloning and characterization of a TFIIIC2 subunit (TFIIICβ) whose presence correlates with activation of RNA polymerase III-mediated transcription by adenovirus E1A expression and serum factors. Genes Dev 1995; 9:675-685.
180. L'Etoile ND, Fahnestock ML, Shen Y et al. Human transcription factor IIIC box B binding subunit. Proc Natl Acad Sci USA 1994; 91:1652-1656.
181. Greenberg GR, St. Louis D, Duncan L et al. Partial purification of stable transcription complexes with cloned tRNA genes of *Drosophila melanogaster*. J Biol Chem 1985; 260:6248-6253.
182. Baker RE, Hall BD. Structural features of yeast tRNA genes which affect transcription factor binding. EMBO J 1984; 3:2793-2800.
183. Parker CS, Roeder RG. Selective and accurate transcription of the *Xenopus laevis* 5S RNA genes in isolated chromatin by purified RNA polymerase III. Proc Natl Acad Sci USA 1977; 74:44-48.
184. Tekamp PA, Garcea RL, Rutter WJ. Transcription and in vitro processing of yeast 5S rRNA. J Biol Chem 1980; 255:9501-9506.
185. Lassar AB, Hamer DH, Roeder RG. Stable transcription complex on a class III gene in a minichromosome. Mol Cell Biol 1985; 5:40-45.
186. Andrews DL, Millstein L, Hamkalo BA et al. Competition between *Xenopus* satellite I sequences and pol III genes for stable transcription complex formation. Nucleic Acids Res 1984; 12:7753-7769.
187. Wolffe AP, Brown DD. Differential 5S RNA gene expression in vitro. Cell 1987; 51:733-740.
188. Kasher MS, Pintel D, Ward DC. Rapid enrichment of HeLa transcription factors IIIB and IIIC by using affinity chromatography based on avidin-biotin interactions. Mol Cell Biol 1986; 6:3117-3127.
189. Kovelman R, Roeder RG. Sarkosyl defines three intermediate steps in transcription initiation by RNA polymerase III: application to stimulation of transcription by E1A. Genes Dev 1990; 4:646-658.

190. Klekamp MS, Weil PA. Partial purification and characterization of the *Saccharomyces cerevisiae* transcription factor TFIIIB. J Biol Chem 1986; 261:2819-2827.
191. Waldschmidt R, Jahn D, Seifart KH. Purification of transcription factor IIIB from HeLa cells. J Biol Chem 1988; 263:13350-13356.
192. Wu G-R. Formation of large, sedimentable transcription complexes with VARNA genes and other related genes. J Biol Chem 1988; 264:9753-9761.
193. Dammann R, Pfeifer GP. Lack of gene- and strand-specific DNA repair in RNA polymerase III-transcribed human tRNA genes. Mol Cell Biol 1997; 17:219-229.

CHAPTER 6

Transcription

Initiation

Once a preinitiation complex has formed on a yeast tRNA gene, RNA chain initiation requires a further 5 min at 22°C (half-life ~2 min).[1] During this period, three successive steps occur: recruitment of pol III, melting of the DNA helix, and initiation of RNA synthesis. Recruitment and melting are the slow steps, whereas the subsequent synthesis of a seventeen base transcript is completed within 10 seconds.[1]

On its assembly into a preinitiation complex, yeast pol III generates a footprint from -3 to +21 on the SUP4 $tRNA^{Tyr}$ gene and from -10 to +13 on the 5S RNA gene.[2] A weak sequence homology between various class III genes occurs at the initiation site and mutations in this region can reduce transcription strongly, in some instances.[3-8] Since the homology lies within the footprint of pol III, it may be recognized directly by the polymerase itself. Six different subunits of yeast pol III can be crosslinked specifically to DNA within the assembled preinitiation complex.[9,10] The C160 and C128 subunits are sufficiently extended along the DNA to be accessible to photoreactive residues situated anywhere between -17 and +20, while C34 extends from -21 to +6.[9] Its projection from the trailing edge of the polymerase allows C34 to interact with TFIIIB. In contrast, ABC27 is situated at the leading edge of the polymerase before initiation begins.[9] C82, C53, C31, and AC40 or C37 (these polypeptides could not be clearly resolved) can all be crosslinked to DNA at more restricted and centrally placed positions within the preinitiation complex.[9]

Pol III "melts" the DNA at the start site in a process that can be monitored by footprinting with potassium permanganate.[2,6] Permanganate oxidizes thymine in single-stranded DNA but has little effect upon thymine in duplex DNA.[11] Permanganate probing experiments indicate that prior to initiation pol III induces a reversible change in the conformation of the DNA due to opening of the helix.[2,6] The melted region at the SUP4 tRNA promoter extends from -11 to +11.[6] The A/T-rich sequences that tend to flank transcription start sites may facilitate promoter opening due to the lower thermodynamic stability of A:T base pairs. Although pol III can bind to the preinitiation complex at 0°C, the process of template melting is highly temperature-dependent and increases progressively from 10°C to 40°C, with a sharp transition between 10°C and 15°C.[6] The extent of DNA opening and its temperature dependence are reminiscent of the melting process that occurs at bacterial promoters.[12] However, whereas promoter bound *E. coli* RNA polymerase is

RNA Polymerase III Transcription, Second Edition, by Robert J. White.

in rapid equilibrium with free enzyme as long as the template remains closed, yeast pol III associates stably with the preinitiation complex even prior to promoter melting.[6]

Elongation

The footprints generated by transcription complexes assembled on yeast tRNA and 5S genes change as transcription initiates.[2] Pol III subunits move out of the range of photoactive nucleotides placed at -9 or further upstream.[9] If initiation occurs in the absence of a full set of nucleotides so as to limit the nascent RNA chain to fewer than six bases, then the transcription bubble is not displaced.[6] Unlike an open preinitiation complex, such complexes maintain the bubble when cooled to 0°C.[6] Upon extending the RNA chain to 17 nucleotides, the size of the transcription bubble diminishes from 22 to between 13 and 17 bases.[6] The start site becomes more accessible to DNase I, presumably due to displacement of RNA polymerase, and protection increases between about +20 and +30.[2] Pol III subunits move out of the range of photoreactive residues in the 5' flanking region.[13]

Although it is readily stripped from a preinitiation complex by addition of heparin, salt or sarkosyl, pol III becomes more resistant to such treatment once transcription has commenced.[2,8,14-18] Formation of the first phosphodiester bond is not sufficient for stabilization, but heparin-resistance is achieved after transcripts of five nucleotides have been synthesized.[8] However, pol III is less able to undergo the transition to a stable elongating state if the start of the RNA is very pyrimidine-rich.[8] In such cases, transcript slippage can occur and the RNA is more readily released as an abortive initiation product.[8] The fact that purine-rich transcripts are more avidly bound by pol III may explain the preponderance of purines at the start of class III genes.[8]

The bubble of unwound DNA moves downstream with the nascent transcript during the elongation phase of transcription.[2] The extent of the RNA-DNA hybrid in an arrested class III transcription bubble may be as little as 7 bp, which is much less than that observed in *E. coli*.[2] A shorter RNA-DNA duplex in pol III transcription would correlate with the relatively simple chain termination process that occurs in this system.

At least five and possibly seven subunits of yeast pol III can be crosslinked to an N_3RdU residue situated in the middle of the transcription bubble of an arrested elongation complex.[9] C160, C128 and C34 are labeled strongly, whereas C82, ABC27, and AC40 or C37 are labeled weakly from the center of the bubble.[9] In contrast, C128, C31 and especially C34 are labeled by a photoactive nucleotide placed five base pairs upstream of the arrested elongation complex.[9] Only C160 and C128 are accessible to crosslinking from the leading edge of the elongation bubble, with C128 being the more strongly labeled.[9] Crosslinking studies in which the photoaffinity analog is incorporated into the nascent RNA rather than the DNA template also label the two largest subunits of pol III.[19]

At 20°C, yeast pol III elongates RNA at an average rate of 20-22 nt/second.[20,21] This is similar to the chain elongation rate of pol II in vivo.[20] However, elongation does not proceed at a uniform rate.[20] For example, at 20°C and with 100 μM nucleotides it takes pol III 3.0 seconds to traverse from +17 to +46 of the SUP4 gene and 4.1 seconds to travel the next 9 nt.[20] This uneven progress results from pausing at internal sites, and the rate of extension at individual nucleotides can vary by 31-

fold.[20,21] Adding UMP or AMP to U-terminated transcripts tends to be slow, whereas GMP or CMP additions generally proceed more rapidly.[20] At pause sites, it is thought that pol III enters an "inchworming mode", in which the distance between the active site and the downstream border of the enzyme shrinks.[21] Passage past the pause site may then involve the 3' boundary of the polymerase leaping forward several bases.[21] Mutations in domain f of C160 can impair this transition.[21]

A large transcription complex within the transcribed region of a gene might be expected to either occlude progression of polymerase or to be displaced by it. However, assembled transcription complexes are not removed from class III genes by multiple passages of pol III.[22,23] Furthermore, removing yeast TFIIIC from the SUP4 $tRNA^{Tyr}$ gene made no significant difference to the rate of RNA chain elongation.[20,24] During transcription in the normal direction, the presence of TFIIIC delays pol III for just 0.2 seconds at a single site upstream of the B-block.[20] Since the time required for promoter clearance limits initiation rates to below 0.5/second, a downstream delay of 0.2 seconds will make no difference to the overall level of transcription.[24] However, if pol III is made to transcribe in the antisense direction and encounters TFIIIC from downstream, it pauses for around 9 seconds before continuing through the B-block.[24] In contrast, TFIIIB prevents the passage of pol III approaching from downstream for over an hour.[24] A full *Xenopus* 5S transcription complex has also been found not to impede transcription of either DNA strand by SP6 RNA polymerase, and the complex remains stably bound following multiple polymerase transits.[25] However, a TFIIIA/5S gene complex in the absence of TFIIIB and TFIIIC is dissociated by passage of either SP6 RNA polymerase or pol III.[25,26] Similarly, TFIIIC alone is rapidly displaced from DNA by pol III.[24] These results suggest that the multiple contacts made by a complete transcription complex may be essential for continued integrity. The interaction between τ_{95} and DNA diminishes as pol III advances into the vicinity of the A-block, as revealed by photolabeling.[9] However, crosslinking studies using a photoreactive ribonucleotide analog found that two large polypeptides which may be part of TFIIIC come into contact with the nascent RNA chain as elongation reaches the A-block region.[19] At the same time, labeling of the polymerase appears to diminish.[19] RNA polymerase may transiently displace a given factor from its binding site as it transcribes through the gene, but the factor may remain stably associated with the complex by protein-protein contacts with other factors bound to DNA sites not in the process of being transcribed. The association of TFIIIB with DNA upstream of the transcription start site may be particularly important in preventing TFIIIC and TFIIIA from being displaced from the template as polymerase transcribes through their DNA binding sites.

It remains to be determined how pol III is able to displace TFIIIC and TFIIIA during chain elongation. One possibility would involve specific protein-protein interactions inducing conformational changes that trigger release of the factors. However, the ability of phage polymerases to dislodge TFIIIA[25,26] argues against a specific interaction between pol III and the factors in its path. It may be that the energy consumed in translocation generates sufficient force to displace obstructive proteins. Alternatively, the process of transcription may indirectly weaken the binding of TFIIIA and TFIIIC due to DNA strand separation or the generation of positive supercoils ahead of pol III.

Termination

Whereas pol III requires a variety of accessory factors in order to initiate transcription specifically, it can recognize termination signals accurately and efficiently in the apparent absence of other factors. This was first demonstrated using highly purified pol III from *Xenopus* ovaries[27] and was later shown to be the case also for the yeast[28] and mammalian enzymes.[29] Mutations in the C128 subunit of yeast pol III alter the efficiency of termination in vivo.[30-32]

Bogenhagen and Brown[33] deduced a simple consensus sequence for termination of 5S gene transcription, viz 4 or more T residues embedded in a G/C-rich environment. Oligo(dT) clusters also serve as terminator signals for tRNA genes[34-36] and, indeed, for most class III templates.[37] As such, short runs of U residues form the 3' ends of most pol III transcripts. Termination can occur at runs of Ts even when they are placed at ectopic positions far downstream of a gene's coding region.[38] Human pol III may terminate to some extent at stretches shorter than T_4.[39] However, termination signals seem to be more stringent in yeast, since shortening T_7 to T_5 downstream of the SUP4 $tRNA^{Tyr}$ gene results in substantial readthrough, and truncation to T_4 abolishes termination completely.[35] Creation of a T_4 sequence within the SUP4 coding region does not result in premature termination.[40] All known yeast tRNA genes are followed by a cluster of at least 6 consecutive T residues,[35] in contrast to *Xenopus* class III genes, where T_4 termination signals are common and transcription can terminate at T_3.[33,41] The *Drosophila* transcription apparatus also appears to require a stronger termination signal than does the vertebrate machinery.[42] Several examples of termination sites that do not conform with the oligo(dT) consensus have also been reported.[36] For instance, human pol III has been reported to terminate at clusters of A or alternating A/T residues.[43-45] In one case, the sequence of the termination site was shown to affect the subsequent processing of the transcript.[46] Long-range effects have also been reported, such that mutations around the start site or ICR can affect termination site utilization.[47]

Heterogeneous termination within a T cluster is common.[39,48,49] Termination of yeast pol III at the T_7GT_6 sequence of the SUP4 $tRNA^{Tyr}$ gene is progressive.[20] Transcripts are formed with five, six or seven terminal U residues.[20] There is also some readthrough of the T_7 stretch, but this only occurs after extensive pausing.[20] After the slow passage past the T_7 sequence, the following GU_3 is added very rapidly, which suggests that the single G is sufficient to restore the elongation rate.[20]

Cambell and Setzer[50] showed that termination signal recognition by purified *Xenopus* pol III can be uncoupled experimentally from the process of polymerase release. Polymerase release is dependent upon the displacement of the nascent RNA strand from the DNA template, whereas termination signal recognition is not.[50] Pol III pauses at a strong termination site even if strand displacement has not occurred, but in this case elongation then continues and aberrant readthrough transcripts result.[50] A strong barrier to Bal-31 nuclease digestion from downstream of a tRNA gene occurs at the termination site.[51] This may be due to the presence of paused polymerase molecules.

It has been proposed that the La autoantigen is involved in termination by pol III.[49,52,53] Newly synthesized pol III products are bound by La.[52,54-60] In the absence of La, very few pol III transcripts are synthesized and these are truncated at their 3' termini by 1 to 5 U residues.[49,52] Addition of purified La to immobilized DNA templates increases the release of RNA and the overall level of transcription.[53] These

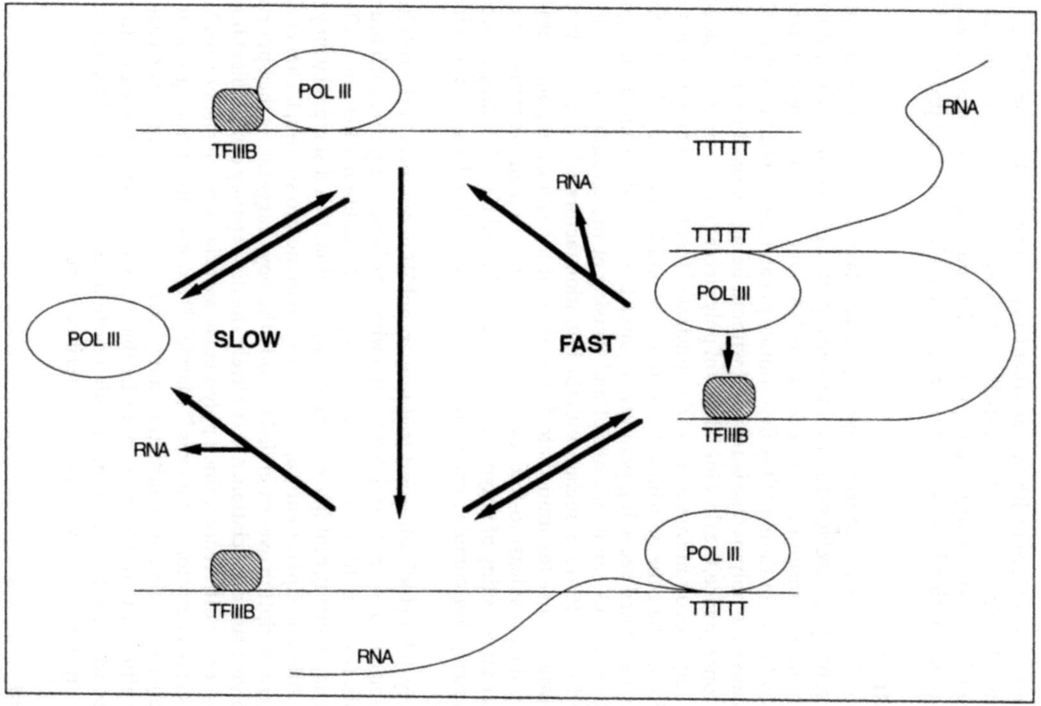

Fig. 6.1. Schematic illustration of the alternative pathways employed by pol III for transcription reinitiation. TTTTT symbolizes the termination site.

observations are consistent with the idea that La is involved in the production and release of full-length pol III transcripts.[49,52,53] However, the gene encoding La is not required for growth of yeast[60] and purified *Xenopus* pol III has been shown to terminate transcription accurately in the apparent absence of other proteins.[27,50]

Reinitiation

After the initial round of transcription, a yeast class III preinitiation complex can direct subsequent cycles 5- to 10-fold more rapidly than the first.[1] Thus, during multiple round transcription, synthesis of a tRNA molecule takes approximately 35 seconds, whereas initiation of the first transcript can take ~5 minutes (at 22°C).[1] This is because the polymerase is recycled without being released from the template; as a consequence, the slow initial step of polymerase recruitment is avoided.[1] (Fig. 6.1) Template commitment assays demonstrated that the polymerase remains associated with the gene on which it first initiated and is not released into a free pool.[1] This facilitated recycling pathway requires the natural termination signal and is not seen with runoff transcription.[1] Pausing at the termination site may allow time for pol III to reassociate with the initiation factors, whereas runoff termination would cause the immediate dissociation of the elongating polymerase. Recycling pol III is at least 10-fold less sensitive to heparin than the initiating enzyme.[1] Internal recycling of polymerase may be facilitated if the bending induced by the transcription complex brings the two ends of a class III gene into close proximity.

Similarly, Jahn et al[23] observed that human pol III is retained in the original transcription complex on VA and tRNA genes without dissociating after each round of synthesis. However, Bieker et al[61] found that pol III equilibrates between different 5S templates during multiple rounds of transcription. This discrepancy may be due to the different experimental approaches used in the two studies; Bieker et al[61] measured transcription over extended periods, thereby exaggerating the opportunity for polymerase dissociation. Alternatively, the discrepancy could reflect bona fide gene-specific differences, since polymerase associates more stably with VA_I than with 5S transcription complexes.[62] Maraia[53,63] reported that limiting amounts of pol III can be recycled to preinitiation complexes on the VA_I promoter at least 5-fold more efficiently in the presence of La than in its absence. It remains to be determined whether La allows mammalian systems to utilize a facilitated recycling pathway that is similar to the one described in yeast.

References

1. Dieci G, Sentenac A. Facilitated recycling pathway for RNA polymerase III. Cell 1996; 84:245-252.
2. Kassavetis GA, Braun BR, Nguyen LH et al. *S. cerevisiae* TFIIIB is the transcription initiation factor proper of RNA polymerase III, while TFIIIA and TFIIIC are assembly factors. Cell 1990; 60:235-245.
3. Raymond GJ, Johnson JD. The role of non-coding DNA sequences in transcription and processing of yeast tRNA. Nucleic Acids Res 1983; 11:5969-5988.
4. Johnson JD, Raymond GJ. Three regions of a yeast $tRNA_3^{Leu}$ gene promote RNA polymerase III transcription. J Biol Chem 1984; 259:5990-5994.
5. Raymond KC, Raymond GJ, Johnson JD. In vivo modulation of yeast tRNA gene expression by 5'-flanking sequences. EMBO J 1985; 4:2649-2656.
6. Kassavetis GA, Blanco JA, Johnson TE et al. Formation of open and elongating transcription complexes by RNA polymerase III. J Mol Biol 1992; 226:47-58.

7. Fruscoloni P, Zamboni M, Panetta G et al. Mutational analysis of the transcription start site of the yeast $tRNA^{Leu3}$ gene. Nucleic Acids Res 1995; 23:2914-2918.
8. Zecherle GN, Whelen S, Hall BD. Purines are required at the 5' ends of newly initiated RNAs for optimal RNA polymerase III gene expression. Mol Cell Biol 1996; 16:5801-5810.
9. Bartholomew B, Durkovich D, Kassavetis GA et al. Orientation and topography of RNA polymerase III in transcription complexes. Mol Cell Biol 1993; 13:942-952.
10. Persinger J, Bartholomew B. Mapping the contacts of yeast TFIIIB and RNA polymerase III at various distances from the major groove of DNA by DNA photoaffinity labeling. J Biol Chem 1996; 271:33039-33046.
11. Hayatsu H, Ukita T. The selective degradation of pyrimidines in nucleic acids by permanganate oxidation. Biochem Biophys Res Commun 1967; 29:556-561.
12. von Hippel PH, Bear DG, Morgan WD et al. Protein-nucleic acid interactions in transcription: a molecular analysis. Annu Rev Biochem 1984; 53:389-446.
13. Bartholomew B, Braun BR, Kassavetis GA et al. Probing close DNA contacts of RNA polymerase III transcription complexes with the photoactive nucleoside 4-thiodeoxythymidine. J Biol Chem 1994; 269:18090-18095.
14. Klekamp MS, Weil PA. Yeast class III gene transcription factors and homologous RNA polymerase III form ternary transcription complexes stable to disruption by N-lauroyl-sarcosine (Sarcosyl). Archives Biochem Biophys 1986; 246:783-800.
15. Braun BR, Riggs DL, Kassavetis GA et al. Multiple states of protein-DNA interaction in the assembly of transcription complexes on *Saccharomyces cerevisiae* 5S ribosomal RNA genes. Proc Natl Acad Sci USA 1989; 86:2530-2534.
16. Kassavetis GA, Riggs DL, Negri R et al. Transcription factor IIIB generates extended DNA interactions in RNA polymerase III transcription complexes on tRNA genes. Mol Cell Biol 1989; 9:2551-2566.
17. Kovelman R, Roeder RG. Sarkosyl defines three intermediate steps in transcription initiation by RNA polymerase III: application to stimulation of transcription by E1A. Genes Dev 1990; 4:646-658.
18. Kunkel GR, Danzeiser DA. Formation of a template committed complex on the promoter of a gene for the U6 small nuclear RNA from the human requires multiple sequence elements, including the distal region. J Biol Chem 1992; 267:14250-14258.
19. Bartholomew B, Meares CF, Dahmus ME. Photoaffinity labeling of RNA polymerase III transcription complexes by nascent RNA. J Biol Chem 1990; 265:3731-3737.
20. Matsuzaki H, Kassavetis GA, Geiduschek EP. Analysis of RNA chain elongation and termination by *Saccharomyces cerevisiae* RNA polymerase III. J Mol Biol 1994; 235:1173-1192.
21. Thuillier V, Brun I, Sentenac A et al. Mutations in the α-amanitin conserved domain of the largest subunit of yeast RNA polymerase III affect pausing, RNA cleavage and transcriptional transitions. EMBO J 1996; 15:618-629.
22. Bogenhagen DF, Wormington WM, Brown DD. Stable transcription complexes of *Xenopus* 5S RNA genes: a means to maintain the differentiated state. Cell 1982; 28:413-421.
23. Jahn D, Wingender E, Seifart KH. Transcription complexes for various class III genes differ in parameters of formation and stability towards salt. J Mol Biol 1987; 193:303-313.
24. Bardeleben C, Kassavetis GA, Geiduschek EP. Encounters of *Saccharomyces cerevisiae* RNA polymerase III with its transcription factors during RNA chain elongation. J Mol Biol 1994; 235:1193-1205.

25. Wolffe AP, Jordan E, Brown DD. A bacteriophage RNA polymerase transcribes through a *Xenopus* 5S RNA gene transcription complex without disrupting it. Cell 1986; 44:381-389.
26. Campbell FE, Setzer DR. Displacement of *Xenopus* transcription factor IIIA from a 5S rRNA gene by a transcribing RNA polymerase. Mol Cell Biol 1991; 11:3978-3986.
27. Cozzarelli NR, Gerrard SP, Schlissel M et al. Purified RNA polymerase III accurately and efficiently terminates transcription of 5S RNA genes. Cell 1983; 34:829-835.
28. Hammond CI, Holland MJ. Purification of yeast RNA polymerases using heparin agarose affinity chromatography. Transcriptional properties of the purified enzymes on defined templates. J Biol Chem 1983; 258:3230-3241.
29. Watson JB, Chandler DW, Gralla JD. Specific termination of in vitro transcription by calf thymus RNA polymerase III. Nucleic Acids Res 1984; 12:5369-5384.
30. James P, Hall BD. *ret1-1*, a yeast mutant affecting transcription termination by RNA polymerase III. Genetics 1990; 125:293-303.
31. James P, Whelen S, Hall BD. The RET1 gene of yeast encodes the second-largest subunit of RNA polymerase III. Structural analysis of the wild-type and ret1-1 mutant alleles. J Biol Chem 1991; 266:5616-5624.
32. Shaaban SA, Krupp BM, Hall BD. Termination-altering mutations in the second-largest subunit of yeast RNA polymerase III. Mol Cell Biol 1995; 15:1467-1478.
33. Bogenhagen DF, Brown DD. Nucleotide sequence in *Xenopus* 5S DNA required for transcription termination. Cell 1981; 24:261-270.
34. Hipskind RA, Clarkson SG. 5'-flanking sequences that inhibit in vitro transcription of a *Xenopus laevis* tRNA gene. Cell 1983; 34:881-890.
35. Allison DS, Hall BD. Effects of alterations in the 3' flanking sequence on in vivo and in vitro expression of the yeast SUP4-o $tRNA^{Tyr}$ gene. EMBO J 1985; 4:2657-2664.
36. Mazabraud A, Scherly D, Muller F et al. Structure and transcription termination of a lysine tRNA gene from *Xenopus laevis*. J Mol Biol 1987; 195:835-845.
37. Geiduschek EP, Tocchini-Valentini GP. Transcription by RNA polymerase III. Annu Rev Biochem 1988; 57:873-914.
38. Ciliberto G, Castagnoli L, Melton DA et al. Promoter of a eukaryotic $tRNA^{Pro}$ gene is composed of three noncontiguous regions. Proc Natl Acad Sci USA 1982; 79:1195-1199.
39. Haynes SR, Jelinek WR. Low molecular weight RNAs transcribed in vitro by RNA polymerase III from Alu-type dispersed repeats in Chinese hamster DNA are also found in vivo. Proc Natl Acad Sci USA 1981; 78:6130-6134.
40. Koski RA, Allison DS, Worthington M et al. An in vitro RNA polymerase III system from *S. cerevisiae*: effects of deletions and point mutations upon SUP4 gene transcription. Nucleic Acids Res 1982; 10:8127-8143.
41. Galli G, Hofstetter H, Birnstiel ML. Two conserved sequence blocks within eukaryotic tRNA genes are major promoter elements. Nature 1981; 294:626-631.
42. Schaack J, Sharp S, Dingermann T et al. Transcription of eukaryotic tRNA genes in vitro. II, Formation of stable complexes. J Biol Chem 1983; 258:2447-2453.
43. Emerson BM, Roeder RG. DNA sequences and transcription factor interactions of active and inactive forms of mammalian 5S RNA genes. J Biol Chem 1984; 259:7926-7935.
44. Hess J, Perez-Stable C, Wu GJ et al. End-to-end transcription of an Alu family repeat. A new type of polymerase-III-dependent terminator and its evolutionary implication. J Mol Biol 1985; 184:7-21.
45. Matsumoto K, Takii T, Okada N. Characterization of a new termination signal for RNA polymerase III responsible for generation of a discrete-sized RNA tran-

scribed from salmon total genomic DNA in a HeLa cell extract. J Biol Chem 1989; 264:1124-1131.
46. Maraia RJ, Chang DY, Wolffe AP et al. The RNA polymerase III terminator used by a B1-Alu element can modulate 3' processing of the intermediate RNA product. Mol Cell Biol 1992; 12:1500-1506.
47. Railey JF, Wu G-J. Organization of multiple regulatory elements in the control region of the adenovirus type 2-specific VARNA1 gene: fine mapping with linker-scanning mutants. Mol Cell Biol 1988; 8:1147-1159.
48. Zasloff M, Santos T, Romeo P et al. Transcription and precursor processing of normal and mutant human $tRNA_i^{Met}$ genes in a homologous cell-free system. J Biol Chem 1982; 257:7857-7863.
49. Gottlieb E, Steitz JA. Function of the mammalian La protein: evidence for its action in transcription termination by RNA polymerase III. EMBO J 1989; 8:851-861.
50. Campbell FE, Setzer DR. Transcription termination by RNA polymerase III: uncoupling of polymerase release from termination signal recognition. Mol Cell Biol 1992; 12:2260-2272.
51. Scanlon SR, Folk WR. Nuclease *Bal-31* mapping of proteins bound to a $tRNA^{tyr}$ gene in SV40 minichromosomes. Nucleic Acids Res 1991; 19:7185-7192.
52. Gottlieb E, Steitz JA. The RNA binding protein La influences both the accuracy and the efficiency of RNA polymerase III transcription in vitro. EMBO J 1989; 8:841-850.
53. Maraia RJ, Kenan DJ, Keene JD. Eukaryotic transcription termination factor La mediates transcript release and facilitates reinitiation by RNA polymerase III. Mol Cell Biol 1994; 14:2147-2158.
54. Rinke J, Steitz JA. Precursor molecules of both human 5S ribosomal RNA and transfer RNAs are bound by a cellular protein reactive with Anti-La Lupus antibodies. Cell 1982; 29:149-159.
55. Gottesfeld JM, Andrews DL, Hoch SA. Association of an RNA polymerase III transcription factor with a ribonucleoprotein complex recognized by autoimmune sera. Nucleic Acids Res 1984; 12:3185-3200.
56. Stefano JE. Purified lupus antigen La recognizes an oligouridylate stretch common to the 3' termini of RNA polymerase III transcripts. Cell 1984; 36:145-154.
57. Guddat U, Bakken AH, Pieler T. Protein-mediated nuclear export of RNA: 5S rRNA containing small RNPs in *Xenopus* oocytes. Cell 1990; 60:619-628.
58. Kramerov DA, Tillib SV, Shumyatsky GP et al. The most abundant nascent $poly(A)^+$ RNAs are transcribed by RNA polymerase III in murine tumor cells. Nucleic Acids Res 1990; 18:4499-4506.
59. Schwemmle M, Clemens MJ, Hilse K et al. Localization of Epstein-Barr virus-encoded RNAs EBER-1 and EBER-2 in interphase and mitotic Burkitt lymphoma cells. Proc Natl Acad Sci USA 1992; 89:10292-10296.
60. Yoo CJ, Wolin SL. La proteins from *Drosophila melanogaster* and *Saccharomyces cerevisiae*: a yeast homolog of the La autoantigen is dispensable for growth. Mol Cell Biol 1994; 14:5412-5424.
61. Bieker JJ, Martin PL, Roeder RG. Formation of a rate-limiting intermediate in 5S RNA gene transcription. Cell 1985; 40:119-127.
62. Carey MF, Gerrard SP, Cozzarelli NR. Analysis of RNA polymerase III transcription complexes by gel filtration. J Biol Chem 1986; 261:4309-4317.
63. Maraia RJ. Transcription termination factor La is also an initiation factor for RNA polymerase III. Proc Natl Acad Sci USA 1996; 93:3383-3387.

CHAPTER 7

Chromatin Structure of Class III Genes

The chromatin structure of a gene can be a major determinant of its transcriptional activity (reviewed in refs. 1-8). In chromatin, 146 bp of DNA is wrapped approximately twice around a nucleosome core comprising two molecules each of histones H2A, H2B, H3 and H4, arranged as two H2A/H2B dimers associated with a central $(H3/H4)_2$ tetramer.[9-12] Each of the core histones has a very similar C-terminal domain structure, containing three or four α-helices arranged in a "histone fold".[12,13] Within this structure, a long central helix forms a dimerization interface and is flanked on each side by a loop and a shorter helix.[12,13] Dimerization creates the DNA-binding surfaces, although the specific interfaces between the histone heterodimers are not extensive and have the potential for conformational flexibility.[11-13] The C-terminal domains of the core histones make substantial protein-DNA contacts.[11,12,14] Indeed, continuous contact of the histones with DNA is required for stable binding.[11] The N-terminal tails of the core histones protrude outside the nucleosome.[12,13]

The flanking DNA is preferentially bound by a single molecule of a linker histone, most usually H1.[11,15] Linker histones contain a DNA-binding domain called the "winged helix", which consists of a bundle of three α-helices attached to a three-stranded anti-parallel β-sheet.[13] Linker histones also have basic N- and C-terminal domains that influence the path of the linker DNA between nucleosomes.[13] H2A/H2B heterodimers need to be present to allow the stable association of linker histones with nucleosomal DNA.[14,16] The linker histone interacts predominantly with one end of the nucleosomal DNA close to the surface of the octamer core and protects an additional 20 bp of DNA from micrococcal nuclease digestion, ~5 bp to one side of the core and ~15 bp to the other side.[11,16,17] It may be able to adopt a variety of positions.[11] Traditionally, the linker histone was thought to be located just outside the nucleosome, clamping the DNA as it enters and exits.[18] However, the globular domain of H5 has been shown to interact in a more intimate manner with a nucleosome positioned on the *X. borealis* somatic 5S gene, making direct contacts with H2A and H2B, and binding DNA at a single location inside the superhelical gyres that wrap around the core particle.[14,19] Association of the linker histone may cause allosteric changes in the histone octamer that could stabilize core histone contacts at the periphery of the nucleosome.[11,14]

The structure of DNA can be severely distorted by assembly into nucleosomes, due to both bending and changes in the helical periodicity.[5,9,11,12,20-23] One face of the helix is occluded by the histone core and the adjacent helix further prevents

RNA Polymerase III Transcription, Second Edition, by Robert J. White.

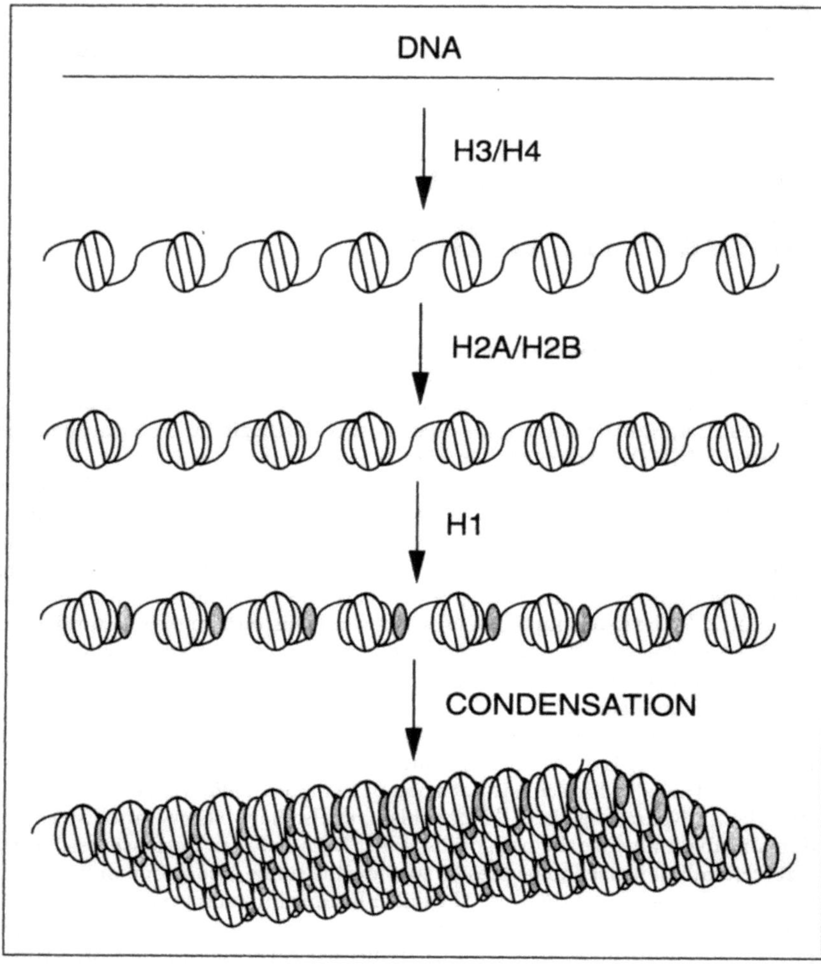

Fig. 7.1. Schematic representation of the sequential stages of DNA assembly into chromatin. $(H3/H4)_2$ tetramers (represented as lightly shaded ellipsoids) bind first to DNA. This is followed by the recruitment of two H2A/H2B dimers (represented as medium shaded ellipsoids) to each $(H3/H4)_2$ tetramer to form the nucleosome core. A single molecule of histone H1 (represented as a heavily shaded ellipsoid) then binds to the linker DNA between each nucleosome core. Finally, the arrays of nucleosomes fold together to give the heavily compacted 30 nm fiber structures.

access.[5,11,24] Arrays of nucleosomes fold together into compacted structures called 30 nm fibers in which the accessibility of genes to transcription factors is likely to be severely limited[1] (Fig. 7.1).

A variety of studies have found that pol III transcription can be inhibited by the presence of histones.[25-44] For example, Morse[30] recruited nucleosomes onto *Xenopus* 5S genes using purified chicken core histones and selected for plasmids bearing positioned nucleosomes by treating the reconstituted molecules with restriction enzymes. A nucleosome on or near the 5S ICR was found to inhibit transcription initiation in an oocyte nuclear extract.[30] Furthermore, downstream nucleosomes were shown to block transcription elongation.[30] The efficiency of both initiation and elongation decreases progressively as the density of nucleosomes and the level of compaction increases.[32,35,36,39] Class III genes that have been assembled into highly condensed solenoid-like structures are incapable of supporting significant levels of transcription.[36] 95% or more of transcriptionally competent B2 genes are subject to histone H1-dependent repression in murine chromatin.[45] Fewer than 1% of potentially active Alu genes escape from chromatin-mediated silencing.[46] However, deleting the H1 gene in *Tetrahymena thermophila* makes no difference to the steady-state levels of tRNA and 5S rRNA, despite producing a general decondensation of chromatin.[47] *Tetrahymena* may be exceptional in this regard, since H1 in this ciliate lacks the central globular domain that is found in the linker histones of multicellular eukaryotes.[47]

In the absence of replication, transcription complexes can be extremely stable and resistant to displacement by histones, although pol III itself is only loosely bound. 5S genes in chromatin isolated from frog or yeast cells are associated with all the factors required for transcription, except for pol III.[25,28,48-51] This is also true of tRNA genes in isolated recombinant SV40 minichromosomes.[52] Even in nucleated *Xenopus* erythrocytes, which are transcriptionally inactive and devoid of polymerase, transcription complexes remain associated with class III genes for days, and perhaps weeks; this was shown by the ability of nuclei isolated from these cells to synthesize tRNA and 5S RNA when supplemented with purified pol III.[53] The transcription pattern of isolated nuclei or chromatin reflects that of the cells of origin. Thus, cultured *Xenopus* kidney cells express somatic but not oocyte 5S genes, and this remains true following isolation of the nuclei or chromatin.[25,28,54] The chromatin structure of active 5S genes is clearly different from that of silent 5S genes in somatic cells.[55,56] However, repression is not irreversible and the inactive oocyte 5S genes can be reactivated if somatic nuclei are injected into oocytes.[57]

Although preformed stable transcription complexes may not be disrupted by histones, they are displaced by the passage of a replication fork. Wolffe and Brown[58] mixed a HeLa cell replication extract with a *Xenopus* oocyte transcription extract to achieve compromise conditions which allowed both replication and weak transcription of 5S genes. They found that a transcription complex provided no obstacle to the passage of a replication fork.[58] Furthermore, no factors remained associated with the replicated template, as assayed either by transcription or by footprinting.[58] Preformed nucleosomes are also disrupted during replication.[59] These results have important implications for gene control, since an absence of epigenetic memory would mean that the transcriptional phenotype of a cell is liable to change each time that it replicates its DNA. They also suggest that whether or not a given gene is transcribed following replication will be decided by whether an active transcription complex or a repressive chromatin structure forms first. Once established, either may be stable and could remain until disrupted by the next round of DNA replication. Almouzni et al[31,34] used *Xenopus* egg extracts to demonstrate a competition between the formation of transcription complexes and of nucleosomes on replicating 5S DNA. The rapid and efficient assembly of replicating templates into chromatin can inhibit 5S transcription due to a failure to assemble competent transcription complexes.[31] Small reductions in the efficiency of chromatin assembly may be sufficient to allow the expression of newly replicated 5S genes.[31] Thus, at each cell cycle there may be competition between histones and transcription factors for binding to a promoter. This provides an opportunity for either reestablishing or altering the set of genes that are actively expressed.[27]

5S Genes

Chromatin formation begins with the rapid binding of $(H3/H4)_2$ tetramers to DNA, followed by the recruitment of H2A and H2B to generate ordered arrays of nucleosomes.[34,60,61] When positioned on the somatic 5S gene of *X. laevis*, $(H3/H4)_2$ tetramers are sufficient to obscure the entire TFIIIA binding site and thereby block the formation of a transcription complex.[23,62,63] In contrast, histones adopt a different position on the somatic 5S gene of *X. borealis* (Fig. 7.2). In this case, an $(H3/H4)_2$ tetramer generates a footprint that only includes sequences upstream of +65.[64] Since the C-block remains accessible in such a complex, TFIIIA is able to bind under these conditions.[63,64] As a consequence, transcription of the *X. borealis* somatic 5S gene is possible in the presence of histone tetramers, albeit at reduced levels.[33-35,39] However, histone octamers containing H2A and H2B produce footprints on this gene that extend as far as +90.[63,64] In such complexes the C-block is obscured, TFIIIA binding is inhibited and transcription is repressed.[35,39,44,63,64] Thus, the core histones are together necessary and sufficient in vitro to establish a repressed state which is refractory to the subsequent addition of transcription factors. In these respects the response of *X. borealis* somatic 5S transcription to the presence of histones is similar to that of many class II genes.[65-67]

Variations in the sequence of a 5S gene have been shown to affect the positioning of nucleosomes in *Lytechinus variegatus*.[68] This presumably accounts for the different locations of nucleosomes positioned on somatic 5S genes from different frog species. As a result, $(H3/H4)_2$ tetramers can prevent TFIIIA from binding in *X. laevis*, but not in *X. borealis*.[63] Since the expression and regulation of somatic 5S genes appears to be the same in *X. laevis* as it is in *X. borealis*, the precise positioning of nucleosomes is unlikely to be a major determinant of 5S gene activity. Indeed, a nucleosome does not need to be precisely positioned in order to repress 5S transcription.[31,34,42,69] Transcriptional repression can be achieved not only by the occlusion of promoter sequences through direct contact with histones but also through the compaction of nucleosomal arrays.[36,39] Interactions between the nucleosome cores are sufficient to drive compaction, with histone octamers allowing more extensive compaction than H3/H4 tetramers alone.[36,39] Fully repressed higher order chromatin structures are stabilized by the presence of linker histones.[36]

The association of linker histones H1 or H5 with the nucleosome core further extends the region of 5S DNA that is occluded.[14,15] A histone octamer plus one molecule of linker histone protects *X. borealis* somatic 5S sequences between -75 and +94 from micrococcal nuclease digestion.[14,15] The central globular domain of H5, in the absence of its positively charged N- and C-terminal tails, is sufficient to extend the protected region of DNA to either side of the nucleosome core.[14] Under circumstances in which nucleosomes are only loosely packed, the binding of H1 may be required to establish complete repression.[28,69] H1 can influence the interactions between core histones and DNA, as well as inhibiting nucleosome sliding.[38,42,43] Movement of histone octamers may allow transcription factors to gain access to promoters and thereby facilitate transcription; this process is discouraged by the linker histones H1, B4 or H5.[42-44] Thus, linker histones can inhibit transcription in

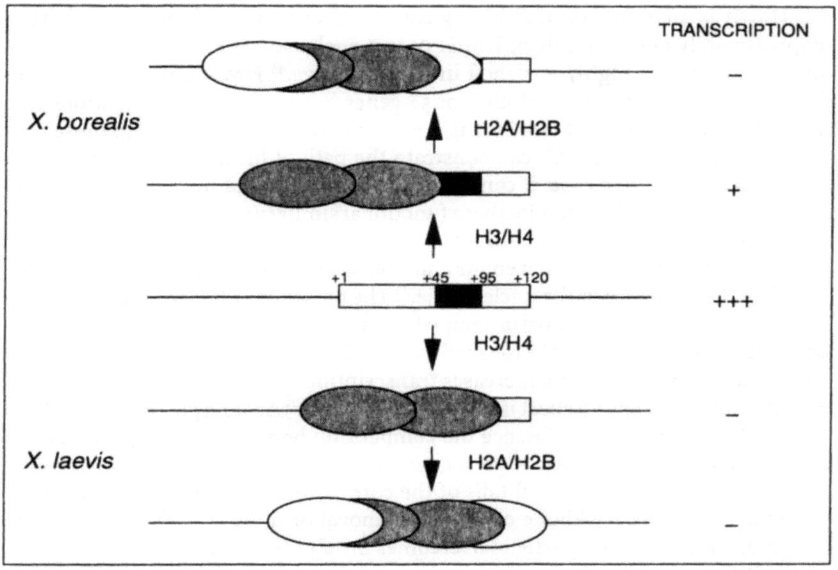

Fig. 7.2. Positioning of core histones on somatic 5S genes of *X. borealis* and *X. laevis*. The coding region of a 5S gene is drawn as a box and the ICR is filled in. H3/H4 pairs are represented as heavily shaded ellipsoids and H2A/H2B pairs are represented as lightly shaded ellipsoids. The 5S gene promoter becomes progressively obscured as it is assembled into a positioned nucleosome and this results in a block to transcription. A positioned $(H3/H4)_2$ tetramer is sufficient to inhibit transcription of *X. laevis* somatic 5S genes. However, because an $(H3/H4)_2$ tetramer is positioned further upstream of an *X. borealis* somatic 5S gene, some transcription is possible in this case and full inhibition requires a positioned octamer of core histones.

several different ways: by directly occluding regions of DNA; by reducing the mobility of histone octamers; and by encouraging chromatin condensation. Nevertheless, only certain types of genes are sensitive to these effects in vivo.[38,40,47]

Pruss et al[19] used photocrosslinking to map the location of the linker histone on an *X. borealis* somatic 5S gene reconstituted into a nucleosome. The globular domain of H5 was found to interact with the major groove of the DNA 60-68 bp from the dyad axis.[19] This places it just within the confines of the core particle. The globular domain makes close contact with H2A and H3.[19] The basic C-terminal tail of H5 points along the DNA at the very edge of the nucleosome core.[19] This allows it to interact with the linker DNA, thereby facilitating chromatin compaction. Hayes[70] mapped H1° to the same position on this gene using a different approach. He used disulfide exchange to attach to H1° an Fe(II)-EDTA complex; this catalyzes strand cleavage of nearby DNA by generating hydroxyl radicals.[70] Both mapping approaches place the linker histone in an asymmetrical position on the somatic 5S nucleosome, within the gyres of DNA that also wrap around the core particle.[19,70] Such a location is likely to restrict the mobility of the histone octamer with respect to DNA. The presence of linker histones will restrict the movement of nucleosomes and lock them in position over essential promoter elements.[42]

HMG1 is an abundant and highly conserved chromosomal protein. There is approximately one molecule of HMG1 per 20 nucleosomes in an average somatic cell, although it is highly enriched in *Xenopus* eggs.[43] HMG1 can bind to linker DNA and repress the transcription of 5S genes.[43] The structural transitions that result from association of HMG1 with a nucleosome are very similar to those produced by histone H1: HMG1 can constrain the path of linker DNA and protect it from micrococcal nuclease; it can direct the positioning of nucleosomes and restrict their mobility.[17,43] Despite these functional similarities, HMG1 is unrelated to the linker histones in sequence or structure.

HMG-17 can bind DNA as it enters and exits the histone core particle and thereby stabilize the structure of the nucleosome.[71] The presence of HMG-17 can stimulate transcription from a chromatin template (but not naked DNA) by 5.5-fold for a somatic 5S gene and by 4-fold for satellite I.[71] HMG-17 only stabilizes the structure of the nucleosomal core and increases transcription if it is deposited during chromatin assembly; it has no effect if added subsequently.[71] An interplay between HMG-17 and the histones may influence the competition between transcription factors and the nucleosome in favor of gene expression.

The lysine-rich N-terminal tails of the core histones interact extensively with the phosphodiester backbone of DNA.[72] Removal of these tails by tryptic digestion of nucleosome core particles overcomes the ability to exclude TFIIIA from 5S genes.[63] A similar effect can be achieved by acetylation of the histone tails.[63] TFIIIA binds to acetylated 5S gene nucleosomes with a speed and affinity comparable to that observed with naked 5S DNA.[63] The physiological levels of acetylation that are sufficient to produce this effect do not cause any major change in the conformation of the nucleosomes or alter their position or mobility with respect to the DNA.[44,63] H1 recruitment also occurs with equal efficiency whether or not the histone octamer is acetylated.[44] However, the shape of the nucleosome changes sufficiently upon acetylation to decrease the number of times that DNA winds around its central axis.[73] This may reflect changes in the stability with which the N-terminal tails of the core histones contact DNA. Transcription of a dinucleosomal 5S gene fragment containing just the core histones can be stimulated 8-fold by high levels of acetylation.[44] However, H1 restricts nucleosomal mobility and inhibits transcription regardless of the acetylation state of the core histones.[44]

Acetylation of histones is associated with nascent nucleosomes deposited on newly replicated DNA.[3,74,75] As chromatin matures the histones are progressively deacetylated.[3] Genes in new chromatin may remain accessible to transcription factors during the time required for the deacetylation events associated with maturation.[63] For *X. laevis* oocyte 5S genes and human 5S genes, binding of TFIIIA is sufficient to preempt inhibition during subsequent chromatin assembly.[26,41] However, the *X. laevis* somatic 5S and yeast 5S genes need to assemble a complete transcription complex before mature chromatin forms in order to avoid repression.[32,76]

Under some circumstances chromatin can have a dominant effect that can repress 5S genes even after complete transcription complexes have assembled.[31,32,34,69,77] For example, active complexes preformed in vitro on newly replicated *Xenopus* 5S DNA can be disrupted by the reorganization of disordered chromatin into physiologically spaced nucleosomal repeats.[34] This finding correlates with the observation that in chromatin isolated from *Xenopus* erythrocytes the repressed oocyte 5S genes are assembled into positioned arrays of nucleosomes, whereas the potentially active somatic 5S genes are not.[56] Several studies have observed nucleosomes positioned at the promoters of silent 5S genes.[28,41,56,78,79] Genomic footprinting failed to detect a uniquely positioned nucleosome associated with the oocyte 5S genes in cultured *Xenopus* kidney cells, although multiple nucleosome positions were compatible with the data.[55] DNase I footprinting of isolated chromatin suggests that the promoters of silent oocyte 5S genes are incorporated into weakly positioned nucleosomal arrays, whereas transcription factors are bound to the 5'-flanking region and ICR of active somatic 5S genes.[56] Active 5S genes remain accessible to nuclease digestion in chromatin, presumably due to the presence of stably bound transcription factors. This has been shown to be true not only in *Xenopus*,[55,56] but also in *Drosophila*,[80,81] and *Saccharomyces*.[32] However, a high-resolution analysis of *S. cerevisiae* chromatin found that the 5S repeat region is completely covered by nucleosomes in vivo.[82] The individual core particles occupy alternative positions that are characterized by a unique helical phase.[82] This suggests that each nucleosome can choose among several positions of similar energy value. Maintenance of nucleosome positioning in vivo may require the continuous presence of sequence-specific DNA binding proteins.[83]

Histone H1 plays a major role in the specific repression of the oocyte 5S genes in somatic cell chromatin.[28,38,40,56,77] Disruption of this chromatin by the selective removal of histone H1 allows oocyte 5S genes to form active transcription complexes following reconstitution with a complete set of factors.[28,56] Readdition of H1 silences these genes again.[28,56] These results have been corroborated in vivo.[38,40] Thus, transcription of oocyte 5S genes in gastrulating embryos was decreased when H1 was overexpressed from microinjected mRNA.[38] In contrast, transcription of these genes increased when H1 was depleted from the embryos using a hammerhead ribozyme.[38,40] Sliding of nucleosomes following H1 removal might allow transcription factors access to the promoter.

H1-mediated repression may not result in the complete displacement of factors from the 5S genes, since addition of pol III to chromatin after the H1 has been removed can result in a low level of transcription in some cases.[56] Stronger expression is obtained if the complete set of factors is used.[56] These observations suggest that the repression induced by histone H1 may be achieved partly by displacing factors and partly by the compaction of chromatin templates that retain at least some transcription complexes but are inaccessible to polymerase. However, Gurdon et al[84] showed that higher order chromatin structure is not required to maintain repression. They treated somatic cell nuclei with micrococcal nuclease in order to fragment chromatin into pieces shorter than 600 bp.[84] On injecting these nuclei into oocytes, somatic 5S genes were expressed whereas oocyte 5S genes were not.[84] If the DNA was deproteinized prior to injection, most 5S RNA expression was of the oocyte type.[84] Therefore 5S gene repression can occur on DNA fragments that are too small to form compacted higher order structures.

The dominant repression of transcription by chromatin can have differential effects upon class III genes. For example, transcription of newly replicated *X. borealis* somatic 5S DNA can be repressed 10- to 20-fold by chromatin assembly in a *Xenopus* egg extract under conditions in which satellite I (OAX) transcription is decreased by no more than 3-fold.[31,34] This effect is produced by the recruitment of histones H2A and H2B, but does not require precisely positioned nucleosomes.[31,34] Although some TFIIIA and TFIIIC may be displaced from 5S genes in this reaction, repression is primarily due to the loss of TFIIIB.[34] The 5S and the satellite I genes have similar competitive strengths and form complexes of equivalent stability to template challenge in an egg extract.[31] Nevertheless, the somatic 5S complex would appear to be more readily displaced by core histones. However, precisely the opposite order of stability is observed in response to somatic histone H1. Wolffe[77] found that addition of purified somatic H1 to embryonic or sperm chromatin containing preformed transcription complexes can result in the selective inactivation of satellite I and oocyte 5S genes while somatic 5S and tRNA genes are relatively unaffected. In developing embryos, raising the level of histone H1 represses expression of oocyte 5S genes but does not affect the activity of somatic 5S, tRNA, U1 or U2 genes.[38] In contrast, complexes formed on oocyte 5S genes in chromatin from tissue culture cells are resistant to H1-mediated repression.[28] The dominant repression of satellite I and oocyte 5S genes in embryonic chromatin by histone H1 can occur in the presence of excess transcription factors and requires some feature of chromatin structure since it does not occur with naked genomic DNA.[77] *Xenopus* egg and oocyte extracts are deficient in the normal somatic form of histone H1 and do not appear to assemble endogenous H1 into chromatin.[3,31] This may account for the stability of satellite I genes in the egg extract.[31] Taken together, these findings show that different types of genes can respond differentially to the various aspects of chromatin formation.

tRNA Genes

tRNA transcription appears to be relatively resistant to repression by histones. Manipulating the level of histone H1 in frog embryos has little effect on the synthesis of tRNA.[38] Furthermore, removal of H1 from murine chromatin makes little or no difference to the number of tRNA genes that are accessible to transcription factors.[45] TFIIIB and TFIIIC were found to remain associated with a *Xenopus* $tRNA^{Met1}$ gene in an H1-containing recombinant SV40 minichromosome in which a somatic 5S gene was repressed.[52] A nucleosome deficiency in yeast that activates several class II genes does not affect tRNA expression.[85] The SUP4 $tRNA^{Tyr}$ gene could even remain active in yeast cells when fused to nucleosome positioning signals that are capable of suppressing both pol II transcription and the initiation of DNA replication.[86] These signals were able to incorporate the start site and A-block of a mutated tRNA gene within the center of a nucleosome.[86] However, the wild-type SUP4 gene could override the positioning signals and remain free of nucleosomes.[86] A yeast $tRNA^{Glu}$ gene also remained active after prolonged incubation in a *Xenopus* egg extract that assembled nucleosomes with a regular and physiological spacing.[37]

Although tRNA genes are able to override certain nucleosome positioning signals, they can be incorporated into inactive chromatin in some situations. For example, a yeast tRNA gene is repressed following its insertion into the silent HMR mating-type locus.[87] The structural sequence of several tRNA genes was found to begin approximately 20 bp inside of the nucleosome core in chick embryos.[88] In *Xenopus,* the regular spacing of nucleosomes on tRNA genes was shown to be restricted to transcriptionally inactive erythrocytes and not to apply to tissues that express tRNA.[89]

DeLotto and Schedl[90] used digestion with several different nucleases to investigate the organization of tRNA genes in *Drosophila* chromatin. They examined 8 tRNA genes in a 15 kb cluster on chromosome 3, as well as some additional tRNA genes on chromosome 2, and found the ICR sequences of each to be preferentially exposed to micrococcal nuclease and DNase I in chromatin.[90] Assuming these results to be representative, it would appear that most, and perhaps all, functional tRNA genes can be assembled into active complexes in vivo that are distinct from the structure of bulk chromatin.[90] An additional weaker cleavage site was observed ~40 bp upstream of each gene.[90] Sequences within the interblock region of the tRNA gene were preferentially cleaved by the single strand-specific S1 nuclease, consistent with a distortion or bending of the DNA.[90] Bal-31 nuclease mapping of proteins bound to $tRNA^{Tyr}$ genes embedded in SV40 minichromosomes revealed binding to the upstream flanking region, the A- and B-blocks, the interblock region and the termination site.[91] A detailed analysis of tDNA structure in vivo has been obtained by genomic footprinting in yeast.[92-94] Chromosomal $tRNA^{Tyr}$, $tRNA^{Leu}$ and $tRNA^{Ser}$ genes showed patterns of protection and enhanced cleavages closely resembling those observed in vitro, thereby validating the results obtained using cell-free extracts. Protection was observed over and around the A- and B-blocks, with hypersensitivities located in the interblock regions.[92,94] An additional footprint was detected 5' to the $tRNA^{Leu}$ gene, extending from -40 to +15.[92,93] The relative amounts of upstream and internal protection varied considerably according to the conditions of the footprinting.[92,93] Although one cannot exclude the possibility that transcription complex structure is disrupted during cell lysis and chromatin isolation, the absence of footprints at tRNA genes bearing ICR mutations strongly suggests that the protection patterns observed with wild-type genes in these studies[92,93] reflect bona fide transcription complexes rather than fortuitous protein binding. Genomic footprinting of a $tRNA^{Val}$ gene in human fibroblasts revealed weak protection in the vicinity of the A- and B-blocks and also at -7.[95] Protection was also detected near the TATA box of the $tRNA^{Sec}$ gene.[95]

U6 Genes

The chromatin structure of the yeast U6 gene has been examined in vivo using micrococcal nuclease and DNase I footprinting.[96,97] A region of strong protection was seen around the TATA box, and the pattern of this footprint correlates closely with the pattern obtained in vitro using purified TFIIIB.[96,97] A clear footprint was seen over the B-block by Gerlach et al,[96] but not by Marsolier et al.[97] Neither group observed protection of the A-block.[96,97] These findings are consistent with electron microscopic studies, which revealed a strong interaction between TFIIIC and the B-block but no association with the A-block.[98] The data suggest that the association of TFIIIC with the yeast U6 gene is less stable and/or more transitory than the binding of TFIIIB.

Deleting 42 bp between the terminator and the B-block produced a substantial disruption of the footprint at the TATA box, over 150 bp upstream.[96] This can be explained if the downstream deletion disrupts TFIIIC binding, which in turn reduces the recruitment of TFIIIB. Given the extreme separation of the A- and B-blocks in the yeast U6 gene, one might have expected that deleting part of the intervening region would promote TFIIIC binding. The fact that removing some of this sequence proved detrimental has led to the suggestion that the interblock chromatin is folded in such a way as to facilitate TFIIIC recruitment.[96,99] Indeed, ~100 bp of interblock DNA is strongly protected against micrococcal nuclease.[96] This protection remains if the B-block is inactivated, and is therefore unlikely to be due to TFIIIC.[96] The size of this protected region is insufficient to accommodate a positioned nucleosome, but nonhistone chromosomal proteins could be involved in folding this extended transcription unit.

The yeast U6 gene is flanked on both sides by series of positioned nucleosomes.[97] This structure was lost upon inactivation of the B-block by a 2 bp deletion.[97] These results provided the first evidence that a class III gene can organize the chromatin around it. It remains to be established whether TFIIIB, TFIIIC or the process of transcription is responsible for this positioning effect. The ability of the U6 gene to organize surrounding chromatin was lost when it was transferred from its chromosomal location to a centromeric vector.[97]

The *S. cerevisiae* U6 gene was repressed by a *Xenopus* egg extract that assembled nucleosomes with a regular and physiological spacing.[37,98] However, U6 transcription could be restored by the subsequent addition of affinity-purified TFIIIC.[37,98] TBP or TFIIIB did not produce this effect.[37] The restoration of U6 transcription by TFIIIC required the presence of an intact B-block[37] or an A-block that had been strengthened by substitution towards the consensus.[98] In contrast, transcription of the *S. cerevisiae* U6 gene using highly purified components in the absence of histones is independent of both the B-block and TFIIIC.[37,100,101] These results suggest that TFIIIC performs an additional role besides serving as an assembly factor for recruiting TFIIIB: namely, to prevent the association of nucleosomes with the transcribed region of at least some class III genes.

The ability of pol III factors to compete successfully with nucleosomes in vivo is strongly indicated by the fact that inactivation of the histone H4 gene makes no difference to the level of U6 expression.[97] However, expression of U6 mutants bearing a crippled TATA box, A-block or B-block was significantly enhanced by H4 deletion.[97] Thus, whereas the wild-type U6 gene competes efficiently with nucleosomes, mutant forms with weakened promoters are subject to histone-dependent repression. Inactivation was not associated with incorporation of the gene into positioned nucleosomes.[97] The repressive effect may therefore result from variably placed nucleosomes. The ability of TFIIIC to compete with histones can be seen in vitro even after a gene has been incorporated into chromatin.[37] It may therefore serve to relieve some of the competitive pressure that follows DNA replication. The greater susceptibility of 5S genes to repression by core histones may be due to their lack of a B-block. Satellite I genes, which have a B-block, are relatively resistant to repression by the core histones under conditions in which 5S transcription is silenced.[31,34]

N-terminal deletions of residues 4-30 of histone H3 or residues 4-28 of histone H4 had no effect upon the expression of wild-type or crippled U6 genes in yeast cells.[97] Such mutations produced a substantial reduction in the transcription of several, but not all, class II genes.[97] Substituting acetylation sites in histone H3 increased expression of U6 genes with mutated B-blocks.[97] These results contrast with those obtained using *Xenopus* 5S genes, where acetylation or digestion of the N-terminal tails of histones activates transcription.[63]

SINEs

The genomes of higher organisms contain a huge number of SINEs. For example, the 500,000 Alu elements in the haploid human genome represent 5% of the total chromosomal DNA. This number of templates could potentially provide an enormous sink for transcription factors which might be highly detrimental due to competition with essential class III genes. Whereas satellite sequences are generally masked in heterochromatic domains, SINEs are concentrated in the chromosomal regions in which active genes are located, often close to pol II transcription units.[102] It may therefore be of considerable importance for a cell to suppress the majority of SINE copies. Although many SINEs may lack the necessary flanking sequences for efficient transcription,[103] a large number of potentially active templates are nevertheless present in mammalian genomes. Specific mechanisms appear to repress the expression of these repetitive elements in vivo. Thus, although Alu genes can be strongly transcribed in vitro, their expression in somatic cells is extremely low.[104] Genomic DNA contains sufficient functional Alu templates to sustain a high rate of transcription, but most of these are repressed in chromatin from HeLa cells.[46] Indeed, chromatin may silence ~99% of potentially active Alu genes.[46] In contrast, the majority of tRNA and 5S genes are accessible to transcription factors in HeLa cell chromatin.[46] Reconstitution experiments using purified nucleosomes have shown that Alu elements direct translational and rotational positioning of histone octamers and tetramers.[105] This ability to position nucleosomes is intrinsic to the Alu sequence and independent of the flanking DNA.[105,106] One positioned nucleosome is centered over the start site and occludes the region from -80 to +69, which includes the A-block but not the B-block of the Alu promoter.[105] A second nucleosome is located over the right Alu monomer.[105] The dimeric structure of Alu repeats allows a pair of nucleosomes to be positioned with a ~200 bp spacing, which is optimal for stable binding.[105] (Fig. 7.3) Furthermore, A-rich regions between the two monomers and downstream of the right monomer are predicted to form barriers to nucleosomal movement.[105] Intranuclear footprinting has shown that a substantial proportion of Alu elements position nucleosomes in this way in native chromatin.[106] The functional consequence of these interactions is that Alu transcription is substantially repressed by histone octamers and partially repressed by tetramers.[105] Silencing of Alus by nucleosomal core particles is so efficient that inclusion of H1 has little additional effect.[46] Indeed, removal of H1 from HeLa cell chromatin raises Alu expression by only ~2-fold.[46] In contrast, transcription of B2 genes increases by ~20-fold when H1 is depleted from the chromatin of 3T3 cells.[45,46] Thus, although chromatin-mediated repression may be common to SINEs, the molecular mechanism can vary between different families of repetitive element.

Methylation

In vertebrate chromatin, transcriptional repression and nuclease resistance is often accompanied by methylation of CpG dinucleotides.[107,108] The Alu consensus sequence contains an unusually high CpG density, 9-fold above the average for the human genome.[105] Indeed, Alu CpGs account for about one-third of the potential methylation sites in human DNA. The maintenance of this high CpG content is consistent with the idea that evolution exerts pressure to silence the expression of Alu elements. Most Alu genes are highly methylated in DNA from a wide range of somatic human cell types.[109,110] Alu repeats are also heavily methylated in DNA from oocytes, although they are hypomethylated in sperm DNA.[110,111] This pattern is highly unusual, since most DNA is more methylated in sperm than in oocytes.[111] Sperm nuclei contain a 60 kD protein called SABP, that binds specifically to Alu genes just downstream of the A-block.[112] SABP protects Alu elements from CpG methyltransferases under conditions in which the flanking DNA becomes fully methylated.[112] This factor may therefore be responsible for Alus being unmethylated in the male germ line. Such parent-specific differential modification could be associated with genomic imprinting.

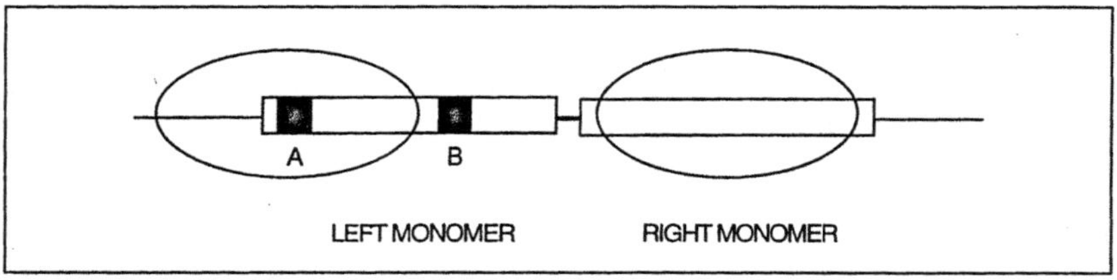

Fig. 7.3. Schematic illustration of the positions of nucleosomes relative to an Alu gene. The locations of the A- and B-blocks and the two monomeric halves of the Alu gene are indicated.

Treatment of HeLa cells with 5-azacytidine results in hypomethylation of several consensus Alu CpGs and a concomitant 5- to 8-fold increase in the abundance of Alu transcripts.[113] Template methylation has been shown to repress pol III transcription of Alu, tRNA and VA genes in vitro and of VA_I in transfected cells.[110,114-116] It can also inhibit the expression of tRNA genes in injected *Xenopus* oocytes, although a 5S gene was unaffected in the same assay.[117] When *Xenopus* oocyte 5S genes were purified from blood cells, in which they are not expressed, every CpG in the repeat unit was found to be heavily methylated.[118,119] When this heavily methylated DNA was injected into *Xenopus* oocytes, it was transcribed as efficiently as cloned 5S genes, which lack CpG methylation.[120] Thus, methylation appears to affect only specific categories of class III gene. It may contribute to the suppression of inappropriate pol III transcription in higher organisms, such as that of SINEs.

It is not yet clear how methylation is able to influence pol III activity. Juttermann et al[114] suggested that methylation of the ICR can block the access of class III factors. However, repression of methylated tRNA or Alu genes in vitro can be specifically relieved by the presence of methylated competitor DNA.[115,116] This suggests that the repression is due to one or more factors that bind to methylated DNA and can be removed by the competitor. Factors called MeCP1 and MeCP2 have been described that can bind to methylated DNA in a sequence-independent fashion.[107,108,121] Both these factors have been shown to repress transcription of methylated class II genes.[122,123] MeCP1 and MeCP2 are deficient in EC and ES cells[107] and this deficiency correlates with an unusually high level of pol III transcription in these cell types.[124] Therefore MeCP1 and MeCP2 may be involved in methylation-dependent repression of class III genes.

In addition, some workers have found that DNA methylation can increase the binding of linker histones H1 and H5.[116,125] If so, then this in itself may be sufficient to account for the correlation between methylation and transcriptional repression in chromatin. Johnson et al[116] found that methylation increases the sensitivity of a tRNA gene to repression by H1 in vitro. However, this study was carried out by adding linker histones to naked DNA in the absence of nucleosomes, a rather artificial situation.[116] Other workers have found little or no effect of methylation on H1 binding to naked DNA.[126,127] More importantly, it was shown that methylation makes no difference to H1 recruitment in the presence of nucleosomes.[126,127] Complete methylation of an *X. borealis* 5S gene had no qualitative or quantitative influence upon its interaction with core or linker histones.[127] The affinity of histone octamers for an Alu gene was also unaffected by methylation.[105] However, CpG methylation increased the efficiency with which a histone tetramer could obstruct access to an Alu promoter and thereby repress transcription.[105] Thus, histone tetramers reduce the expression of an unmethylated Alu gene by 2.5-fold but inhibit the same template by over 50-fold when it is methylated.[105] When this gene was not reconstituted with histones, methylation decreased its transcription in a nuclear extract by approximately 2-fold.[105] Therefore, at least two different mechanisms may contribute to the ability of methylation to repress pol III transcription; alterations in histone binding and/or direct effects of methyl-CpG-binding factors. These two alternatives are not mutually exclusive.

References

1. Wolffe AP. New approaches to chromatin function. New Biol 1990; 2:211-218.
2. Kornberg RD, Lorch Y. Irresistible force meets immovable object: transcription and the nucleosome. Cell 1991; 67:833-836.
3. Wolffe AP. Developmental regulation of chromatin structure and function. Trends Cell Biol 1991; 1:61-66.
4. Felsenfeld G. Chromatin as an essential part of the transcriptional mechanism. Nature 1992; 355:219-223.
5. Hayes JJ, Wolffe AP. The interaction of transcription factors with nucleosomal DNA. BioEssays 1992; 14:1-7.
6. Wolffe AP. Transcription: in tune with the histones. Cell 1994; 77:13-16.
7. Wolffe AP. The transcription of chromatin templates. Current Opinion in Genetics and Development 1994; 4:245-254.
8. Wolffe A. Chromatin Structure and Function. Academic Press Ltd, London 1995.
9. Richmond TJ, Finch JT, Rushton B et al. Structure of the nucleosome core particle at 7A resolution. Nature 1984; 311:532-537.
10. Burlingame RW, Love WE, Wang B-C et al. Crystallographic structure of the octameric histone core of the nucleosome at a resolution of 3.3A. Science 1985; 228:546-553.
11. Pruss D, Hayes JJ, Wolffe AP. Nucleosomal anatomy—where are the histones? Bioessays 1995; 17:161-170.
12. Luger K, Mader AW, Richmond RK et al. Crystal structure of the nucleosome core particle at 2.8A resolution. Nature 1997; 389:251-260.
13. Wolffe AP, Pruss D. Deviant nucleosomes: the functional specialization of chromatin. Trends Genet 1996; 12:58-62.
14. Hayes JJ, Pruss D, Wolffe AP. Contacts of the globular domain of histone H5 and core histones with DNA in a "chromatosome". Proc Natl Acad Sci USA 1994; 91:7817-7821.
15. Hayes JJ, Wolffe AP. Preferential and asymmetric interaction of linker histones with 5S DNA in the nucleosome. Proc Natl Acad Sci USA 1993; 90:6415-6419.
16. Hayes JJ, Pruss D, Wolffe AP. Histone domains required to assemble a chromatosome including the *Xenopus borealis* somatic 5S rRNA gene. Proc Natl Acad Sci USA 1994; 91:7817-7821.
17. Nightingale K, Dimitrov S, Reeves R et al. Evidence for a shared structural role for HMG1 and linker histones B4 and H1 in organizing chromatin. EMBO J 1996; 15:548-561.
18. Crane-Robinson C. Where is the globular domain of linker histone located on the nucleosome? Trends Biochem Sci 1997; 22:75-77.
19. Pruss D, Bartholomew B, Persinger J et al. An asymmetric model for the nucleosome: a binding site for linker histones inside the DNA gyres. Science 1996; 274:614-617.

20. Drew HR, Travers AA. DNA bending and its relation to nucleosome positioning. J Mol Biol 1985; 186:773-790.
21. Satchwell SC, Drew HR, Travers AA. Sequence periodicities in chicken nucleosome core DNA. J Mol Biol 1986; 191:659-675.
22. Hayes JJ, Tullius TD, Wolffe AP. The structure of DNA in a nucleosome. Proc Natl Acad Sci USA 1990; 87:7405-7409.
23. Hayes JJ, Clark DJ, Wolffe AP. Histone contributions to the structure of DNA in the nucleosome. Proc Natl Acad Sci USA 1991; 88:6829-6833.
24. Klug A, Lutter LC. The helical periodicity of DNA on the nucleosome. Nucleic Acids Res 1981; 9:4267-4283.
25. Bogenhagen DF, Wormington WM, Brown DD. Stable transcription complexes of *Xenopus* 5S RNA genes: a means to maintain the differentiated state. Cell 1982; 28:413-421.
26. Gottesfeld JM, Bloomer LS. Assembly of transcriptionally active 5S RNA gene chromatin in vitro. Cell 1982; 28:781-791.
27. Brown DD. The role of stable complexes that repress and activate eukaryotic genes. Cell 1984; 37:359-365.
28. Schlissel MS, Brown DD. The transcriptional regulation of *Xenopus* 5S RNA genes in chromatin: the roles of active stable transcription complexes and histone H1. Cell 1984; 37:903-913.
29. Shimamura A, Tremethick D, Worcel A. Characterization of the repressed 5S DNA minichromosomes assembled in vitro with a high-speed supernatant of *Xenopus laevis* oocytes. Mol Cell Biol 1988; 8:4257-4269.
30. Morse RH. Nucleosomes inhibit both transcriptional initiation and elongation by RNA polymerase III in vitro. EMBO J 1989; 8:2343-2351.
31. Almouzni G, Mechali M, Wolffe AP. Competition between transcription complex assembly and chromatin assembly on replicating DNA. EMBO J 1990; 9:573-582.
32. Felts SJ, Weil PA, Chalkley R. Transcription factor requirements for in vitro formation of transcriptionally competent 5S rRNA gene chromatin. Mol Cell Biol 1990; 10:2390-2401.
33. Tremethick D, Zucker K, Worcel A. The transcription complex of the 5 S RNA gene, but not transcription factor IIIA alone, prevents nucleosomal repression of transcription. J Biol Chem 1990; 265:5014-5023.
34. Almouzni G, Mechali M, Wolffe AP. Transcription complex disruption caused by a transition in chromatin structure. Mol Cell Biol 1991; 11:655-665.
35. Clark DJ, Wolffe AP. Superhelical stress and nucleosome-mediated repression of 5S RNA gene transcription in vitro. EMBO J 1991; 10:3419-3428.
36. Hansen JC, Wolffe AP. Influence of chromatin folding on transcription initiation and elongation by RNA polymerase III. Biochemistry 1992; 31:7977-7988.
37. Burnol A-F, Margottin F, Huet J et al. TFIIIC relieves repression of U6 snRNA transcription by chromatin. Nature 1993; 362:475-477.
38. Bouvet P, Dimitrov S, Wolffe AP. Specific regulation of *Xenopus* chromosomal 5S rRNA gene transcription in vivo by histone H1. Genes Dev 1994; 8:1147-1159.
39. Hansen JC, Wolffe AP. A role for histones H2A/H2B in chromatin folding and transcriptional repression. Proc Natl Acad Sci USA 1994; 91:2339-2343.
40. Kandolf H. The H1A histone variant is an in vivo repressor of oocyte-type 5S gene transcription in *Xenopus laevis* embryos. Proc Natl Acad Sci USA 1994; 91:7257-7261.
41. Stunkel W, Kober I, Kauer M et al. Human TFIIIA alone is sufficient to prevent nucleosomal repression of a homologous 5S gene. Nucleic Acids Res 1995; 23:109-116.

42. Ura K, Hayes JJ, Wolffe AP. A positive role for nucleosome mobility in the transcriptional activity of chromatin templates: restriction by linker histones. EMBO J 1995; 14:3752-3765.
43. Ura K, Nightingale K, Wolffe AP. Differential association of HMG1 and linker histones B4 and H1 with dinucleosomal DNA: structural transitions and transcriptional repression. EMBO J 1996; 15:4959-4969.
44. Ura K, Kurumizaka H, Dimitrov S et al. Histone acetylation: influence on transcription, nucleosome mobility and positioning, and linker histone-dependent transcriptional repression. EMBO J 1997; 16:2096-2107.
45. Carey MF, Singh K. Enhanced B2 transcription in simian virus 40-transformed cells is mediated through the formation of RNA polymerase III transcription complexes on previously inactive genes. Proc Natl Acad Sci USA 1988; 85:7059-7063.
46. Russanova VR, Driscoll CT, Howard BH. Adenovirus type 2 preferentially stimulates polymerase III transcription of *Alu* elements by relieving repression: a potential role for chromatin. Mol Cell Biol 1995; 15:4282-4290.
47. Shen X, Gorovsky MA. Linker histone H1 regulates specific gene expression but not global transcription in vivo. Cell 1996; 86:475-483.
48. Parker CS, Roeder RG. Selective and accurate transcription of the *Xenopus laevis* 5S RNA genes in isolated chromatin by purified RNA polymerase III. Proc Natl Acad Sci USA 1977; 74:44-48.
49. Tekamp PA, Garcea RL, Rutter WJ. Transcription and in vitro processing of yeast 5S rRNA. J Biol Chem 1980; 255:9501-9506.
50. Wormington WM, Brown DD. Onset of 5S RNA gene regulation during *Xenopus* embryogenesis. Dev Biol 1983; 99:248-257.
51. Darby MK, Andrews TM, Brown DD. Transcription complexes that program *Xenopus* 5S RNA genes are stable in vivo. Proc Natl Acad Sci USA 1988; 85:5516-5520.
52. Lassar AB, Hamer DH, Roeder RG. Stable transcription complex on a class III gene in a minichromosome. Mol Cell Biol 1985; 5:40-45.
53. Cozzarelli NR, Gerrard SP, Schlissel M et al. Purified RNA polymerase III accurately and efficiently terminates transcription of 5S RNA genes. Cell 1983; 34:829-835.
54. Shastry BS, Honda BM, Roeder RG. Altered levels of a 5S gene-specific transcription factor (TFIIIA) during oogenesis and embryonic development of *Xenopus laevis*. J Biol Chem 1984; 259:11373-11382.
55. Engelke DR, Gottesfeld JM. Chromosomal footprinting of transcriptionally active and inactive oocyte-type 5S RNA genes of *Xenopus laevis*. Nucleic Acids Res 1990; 18:6031-6037.
56. Chipev CC, Wolffe AP. Chromosomal organization of *Xenopus laevis* oocyte and somatic 5S rRNA genes in vivo. Mol Cell Biol 1992; 12:45-55.
57. Korn LJ, Gurdon JB. The reactivation of developmentally inert 5S genes in somatic nuclei injected into *Xenopus* oocytes. Nature 1981; 289:461-465.
58. Wolffe AP, Brown DD. DNA replication in vitro erases a *Xenopus* 5S RNA gene transcription complex. Cell 1986; 47:217-227.
59. Jackson V, Chalkley R. Histone synthesis and deposition in the G1 and S phases of hepatoma tissue culture cells. Biochemistry 1985; 24:6921-6930.
60. Camerini-Otero RD, Sollner-Webb B, Felsenfeld G. The organization of histones and DNA in chromatin: evidence for an arginine-rich histone kernel. Cell 1976; 8:333-347.
61. Worcel A, Han S, Wong ML. Assembly of newly replicated chromatin. Cell 1978; 15:969-977.
62. Gottesfeld JM. DNA sequence-directed nucleosome reconstitution on 5S RNA genes of *Xenopus laevis*. Mol Cell Biol 1987; 7:1612-1622.

63. Lee DY, Hayes JJ, Pruss D et al. A positive role for histone acetylation in transcription factor access to nucleosomal DNA. Cell 1993; 72:73-84.
64. Hayes JJ, Wolffe AP. Histones H2A/H2B inhibit the interaction of transcription factor IIIA with the *Xenopus borealis* somatic 5S RNA gene in a nucleosome. Proc Natl Acad Sci USA 1992; 89:1229-1233.
65. Knezetic JA, Luse DS. The presence of nucleosomes on a DNA template prevents initiation by RNA polymerase II in vitro. Cell 1986; 45:95-104.
66. Workman JL, Roeder RG. Binding of transcription factor TFIID to the major late promoter during in vitro nucleosome assembly potentiates subsequent initiation by RNA polymerase II. Cell 1987; 51:613-622.
67. Workman JL, Taylor IC, Kingston RE. Activation domains of stably bound GAL4 derivatives alleviate repression of promoters by nucleosomes. Cell 1991; 64:533-544.
68. FitzGerald PC, Simpson RT. Effects of sequence alterations in a DNA segment containing the 5S rRNA gene from *Lytechinus variegatus* on positioning of a nucleosome core particle in vitro. J Biol Chem 1985; 260:15318-15324.
69. Shimamura A, Sapp M, Rodriquez-Campos A et al. Histone H1 represses transcription from minichromosomes assembled in vitro. Mol Cell Biol 1989; 9:5573-5584.
70. Hayes JJ. Site-directed cleavage of DNA by a linker histone-Fe(II) EDTA conjugate: localization of a globular domain binding site within a nucleosome. Biochemistry 1996; 35:11931-11937.
71. Crippa MP, Trieschmann L, Alfonso PJ et al. Deposition of chromosomal protein HMG-17 during replication affects the nucleosomal ladder and transcriptional potential of nascent chromatin. EMBO J 1993; 12:3855-3864.
72. Walker IO. Differential dissociation of histone tails from core chromatin. Biochemistry 1984; 23:5622-5628.
73. Bauer WR, Hayes JJ, White JH et al. Nucleosome structural changes due to acetylation. J Mol Biol 1994; 236:685-690.
74. Waterborg JH, Matthews HR. Patterns of histone acetylation in *Physarum polycephalum*. Eur J Biochem 1984; 142:329-333.
75. Perry CA, Annunziato AT. Influence of histone acetylation on the solubility, H1 content and DNAase I sensitivity of newly replicated chromatin. Nucleic Acids Res 1989; 17:4275-4291.
76. Tremethick D, Zucker K, Worcel A. The transcription complex of the 5 S RNA gene, but not transcription factor IIIA alone, prevents nucleosomal repression of transcription. J Biol Chem 1990; 265:5014-5023.
77. Wolffe AP. Dominant and specific repression of *Xenopus* oocyte 5S RNA genes and satellite I DNA by histone H1. EMBO J 1989; 8:527-537.
78. Gottesfeld JM, Bloomer LS. Nonrandom alignment of nucleosomes on 5S RNA genes of *X. laevis*. Cell 1980; 21:751-760.
79. Young D, Carroll D. Regular arrangement of nucleosomes on 5S rRNA genes in *Xenopus laevis*. Mol Cell Biol 1983; 3:720-730.
80. Louis C, Schedl P, Samal B et al. Chromatin structure of the 5S RNA genes of *D. melanogaster*. Cell 1980; 22:387-392.
81. Cartwright IL, Elgin SCR. Chemical footprinting of 5S RNA chromatin in embryos of *Drosophila melanogaster*. EMBO J 1984; 3:3101-3108.
82. Buttinelli M, Di Mauro E, Negri R. Multiple nucleosome positioning with unique rotational setting for the *Saccharomyces cerevisiae* 5S rRNA gene in vitro and in vivo. Proc Natl Acad Sci USA 1993; 90:9315-9319.
83. Pazin MJ, Bhargava P, Geiduschek EP et al. Nucleosome mobility and the maintenance of nucleosome positioning. Science 1997; 276:809-811.

84. Gurdon JB, Dingwall C, Laskey RA et al. Developmental inactivity of 5S RNA genes persists when chromosomes are cut between genes. Nature 1982; 299:652-653.
85. Han M, Grunstein M. Nucleosome loss activates yeast downstream promoters in vivo. Cell 1988; 55:1137-1145.
86. Morse RH, Roth SY, Simpson RT. A transcriptionally active tRNA gene interferes with nucleosome positioning in vivo. Mol Cell Biol 1992; 12:4015-4025.
87. Schnell R, Rine J. A position effect on the expression of a tRNA gene mediated by the SIR genes in *Saccharomyces cerevisiae.* Mol Cell Biol 1986; 6:494-501.
88. Wittig B, Wittig S. A phase relationship associates tRNA structural gene sequences with nucleosome cores. Cell 1979; 18:1173-1183.
89. Bryan PN, Hofstetter H, Birnstiel ML. Nucleosome arrangement on tRNA genes of *Xenopus laevis.* Cell 1981; 27:459-466.
90. DeLotto R, Schedl P. Internal promoter elements of transfer RNA genes are preferentially exposed in chromatin. J Mol Biol 1984; 179:607-628.
91. Scanlon SR, Folk WR. Nuclease *Bal-31* mapping of proteins bound to a $tRNA^{tyr}$ gene in SV40 minichromosomes. Nucleic Acids Res 1991; 19:7185-7192.
92. Huibregtse JM, Evans CF, Engelke DR. Comparison of tRNA gene transcription complexes formed in vitro and in nuclei. Mol Cell Biol 1987; 7:3212-3220.
93. Huibregtse JM, Engelke DR. Genomic footprinting of a yeast tRNA gene reveals stable complexes over the 5'-flanking region. Mol Cell Biol 1989; 9:3244-3252.
94. Hull MW, Erickson J, Johnston M et al. tRNA genes as transcriptional repressor elements. Mol Cell Biol 1994; 14:1266-1277.
95. Dammann R, Pfeifer GP. Lack of gene- and strand-specific DNA repair in RNA polymerase III-transcribed human tRNA genes. Mol Cell Biol 1997; 17:219-229.
96. Gerlach VL, Whitehall SK, Geiduschek EP et al. TFIIIB placement on a yeast U6 RNA gene in vivo is directed primarily by TFIIIC rather than by sequence-specific DNA contacts. Mol Cell Biol 1995; 15:1455-1466.
97. Marsolier M-C, Tanaka S, Livingstone-Zatchej M et al. Reciprocal interferences between nucleosomal organization and transcriptional activity of the yeast *SNR6* gene. Genes Dev 1995; 9:410-422.
98. Burnol A-F, Margottin F, Schultz P et al. Basal promoter and enhancer element of yeast U6 snRNA gene. J Mol Biol 1993; 233:644-658.
99. Eschenlauer JB, Kaiser MW, Gerlach VL et al. Architecture of a yeast U6 RNA gene promoter. Mol Cell Biol 1993; 13:3015-3026.
100. Moenne A, Camier S, Anderson G et al. The U6 gene of *Saccharomyces cerevisiae* is transcribed by RNA polymerase C (III) in vivo and in vitro. EMBO J 1990; 9:271-277.
101. Margottin F, Dujardin G, Gerard M et al. Participation of the TATA factor in transcription of the yeast U6 gene by RNA polymerase C. Science 1991; 251:424-426.
102. Korenberg JR, Thermann E, Denniston C. Hotspots and functional organization of human chromosomes. Hum Genet 1978; 43:13-22.
103. Ullu E, Weiner AM. Upstream sequences modulate the internal promoter of the human 7SL RNA gene. Nature 1985; 318:371-374.
104. Matera G, Hellman U, Schmid CW. A transpositionally and transcriptionally competent *Alu* subfamily. Mol Cell Biol 1990; 10:5424-5432.
105. Englander EW, Wolffe AP, Howard BH. Nucleosome interactions with a human Alu element. J Biol Chem 1993; 268:19565-19573.
106. Englander EW, Howard BH. Nucleosome positioning by human Alu elements in chromatin. J Biol Chem 1995; 270:10091-10096.
107. Meehan R, Lewis J, Cross S et al. Transcriptional repression by methylation of CpG. J Cell Sci (Supp) 1992; 16:9-14.

108. Tate PH, Bird AP. Effects of DNA methylation on DNA-binding proteins and gene expression. Curr Opin Genet Dev 1993; 3:226-231.
109. Schmid CW. Human Alu subfamilies and their methylation revealed by blot hybridization. Nucleic Acids Res 1991; 19:5613-5617.
110. Kochanek S, Renz D, Doerfler W. DNA methylation in the *Alu* sequences of diploid and haploid primary human cells. EMBO J 1993; 12:1141-1151.
111. Rubin CM, VandeVoort CA, Teplitz RL et al. *Alu* repeated DNAs are differentially methylated in primate germ cells. Nucleic Acids Res 1994; 22:5121-5127.
112. Chesnokov IN, Schmid CW. Specific Alu binding protein from human sperm chromatin prevents DNA methylation. J Biol Chem 1995; 270:18539-18542.
113. Liu W-M, Maraia RJ, Rubin CM et al. Alu transcripts: cytoplasmic localization and regulation by DNA methylation. Nucleic Acids Res 1994; 22:1087-1095.
114. Juttermann R, Hosokawa K, Kochanek S et al. Adenovirus type 2 VAI RNA transcription by polymerase III is blocked by sequence-specific methylation. J Virol 1991; 65:1735-1742.
115. Liu W-M, Schmid CW. Proposed roles for DNA methylation in *Alu* transcriptional repression and mutational inactivation. Nucleic Acids Res 1993; 21:1351-1359.
116. Johnson CA, Goddard JP, Adams RLP. The effect of histone H1 and DNA methylation on transcription. Biochem J 1995; 305:791-798.
117. Besser D, Gotz F, Schulze FK et al. DNA methylation inhibits transcription by RNA polymerase III of a tRNA gene, but not of a 5S rRNA gene. FEBS Lett 1990; 269:358-362.
118. Fedoroff NV, Brown DD. The nucleotide sequence of oocyte 5S DNA in *Xenopus laevis*. I. The AT-rich spacer. Cell 1978; 13:701-716.
119. Miller JR, Cartwright EM, Brownlee GG et al. The nucleotide sequence of oocyte 5S DNA in *Xenopus laevis*. II. The GC-rich region. Cell 1978; 13:717-725.
120. Brown DD, Gurdon J. High-fidelity transcription of 5S DNA injected into *Xenopus* oocytes. Proc Natl Acad Sci USA 1977; 74:2064-2068.
121. Bird A. The essentials of DNA methylation. Cell 1992; 70:5-8.
122. Boyes J, Bird A. DNA methylation inhibits transcription indirectly via a methyl-CpG binding protein. Cell 1991; 64:1123-1134.
123. Nan X, Campoy FJ, Bird A. MeCP2 is a transcriptional repressor with abundant binding sites in genomic chromatin. Cell 1997; 88:471-481.
124. White RJ, Stott D, Rigby PWJ. Regulation of RNA polymerase III transcription in response to F9 embryonal carcinoma stem cell differentiation. Cell 1989; 59:1081-1092.
125. McArthur M, Thomas JO. A preference of histone H1 for methylated DNA. EMBO J 1996; 15:1705-1714.
126. Campoy FJ, Meehan RR, McKay S et al. Binding of histone H1 is indifferent to methylation at CpG sequences. J Biol Chem 1995; 270:26473-26481.
127. Nightingale K, Wolffe AP. Methylation at CpG sequences does not influence histone H1 binding to a nucleosome including a *Xenopus borealis* 5S rRNA gene. J Biol Chem 1995; 270:4197-4200.

CHAPTER 8

Proteins that Modulate the Rate of RNA Polymerase III Transcription

So far I have described the basal pol III transcription apparatus and how this functions to allow expression of class III genes. The level of transcription can be modulated in either a positive or a negative fashion. This chapter aims to describe the transcription factors, kinases and phosphatases that have been shown to do this. In chapter 9, some of these proteins will be put into a more physiological context in order to explain how pol III is regulated under particular cellular and environmental conditions.

For many class III genes, basal transcription is very efficient and so the major form of regulation appears to be repression. I shall therefore begin by describing several key factors that have been shown to play a role in suppressing the level of pol III transcription.

Activities that Reduce Pol III Transcription

Dr1

Dr1 is a 19 kD nuclear phosphoprotein that was isolated from HeLa cells because of its ability to repress pol II transcription.[1] The importance of this factor is indicated by a high degree of evolutionary conservation. For example, human Dr1 is 44% identical to a homologue in *Arabidopsis thaliana*[2] and 37% identical to a homologue in *Saccharomyces cerevisiae*.[3] In yeast, Dr1 is essential for viability.[3] A 28 kD corepressor called DRAP1 has been shown to bind Dr1 and stimulate its ability to inhibit transcription.[4] *Saccharomyces cerevisiae* has a protein that is 37% identical to human DRAP1 and is also required for viability.[3] DRAP1 and Dr1 have also been named NC2α and NC2β, respectively.[5] These two proteins associate through regions in their N-termini that contain histone-fold motifs[4,5] (Fig. 8.1). DRAP1 is inactive on its own, whereas Dr1 contains a repression domain that maps to a glutamine- and alanine-rich α-helical region towards its C-terminus.[4] A proportion of cellular Dr1 is thought to function independently of DRAP1.[3,4] This DRAP1-independent repression by Dr1 does not require the histone-fold motif.[6] Overexpression of Dr1 in yeast inhibits cell growth, whereas growth is maintained following overexpression of DRAP1.[3]

Dr1 has little affinity for DNA and functions by binding TBP.[1,3,6] This interaction is necessary for repression, since deletion of its TBP-binding domain inactivates Dr1.[6] Furthermore, overexpression of TBP can reverse Dr1-mediated regulation both in vitro and in vivo.[1,3,6] The growth-inhibitory effect of Dr1 can also be

RNA Polymerase III Transcription, Second Edition, by Robert J. White.

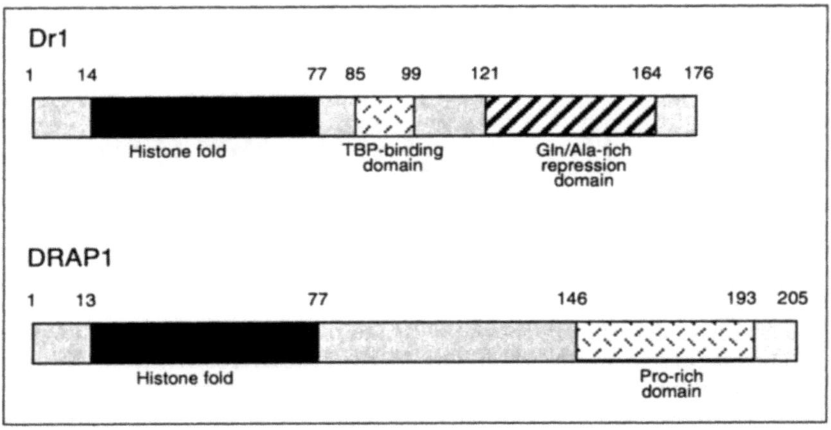

Fig. 8.1. Sequence features of human Dr1 and DRAP1. A putative histone fold domain is present in a similar location in both proteins. Dr1 contains a central TBP-binding domain and a C-terminal glutamine and alanine-rich repression domain. DRAP1 contains a C-terminal proline-rich domain.

overcome by overexpressing TBP.[3] Dr1 does not prevent TBP from binding DNA.[1] Instead, it represses pol II transcription by interfering with the interaction between TFIIB and TBP.[1,3] Dr1 does not bind TFIIB[6] and contacts a surface of TBP that is spatially removed from the TFIIB-docking site.[7] The Dr1-mediated exclusion of TFIIB is thought to result from conformational changes in the TBP/DNA complex.[4]

Highly purified native or recombinant Dr1 can repress the expression of class III genes when added to crude extracts or fractionated factors.[8] This is a specific effect, since pol I transcription is unaffected by Dr1.[8] These results were validated in vivo when it was shown that overexpression of Dr1 in yeast represses tRNA transcription without inhibiting pol I activity.[3] As is the case for pol II, Dr1-mediated repression of pol III can be relieved by raising the level of TBP.[8] TFIIIB fractions can also overcome repression, whereas TFIIIC has no effect.[8] Using recombinant proteins, it was shown that Dr1 can prevent the binding of TBP to BRF.[8] Thus, Dr1 appears to regulate pols II and III by very similar mechanisms, involving the displacement of TFIIB or its homologue BRF from essential interactions with TBP. Unlike TFIIB, however, BRF binds to TBP through two distinct sites.[9-11] By analogy to its effects on TFIIB, Dr1 may disrupt binding to the direct repeats of BRF through conformational changes in TBP. In contrast, the C-terminal domain of BRF appears to make a pol III-specific high-affinity interaction with the basic repeat region of TBP.[9] This interaction is prevented by point mutations in residues 133, 138 or 145 of TBP.[9,12] The binding of Dr1 to TBP is disrupted by mutation of residues 133, 145 or 151.[7] This suggests strongly that Dr1 and BRF compete directly for overlapping binding sites in the basic repeat region on the convex surface of TBP. Such competition can readily explain the ability of Dr1 to disrupt the interaction between TBP and BRF (Fig. 8.2).

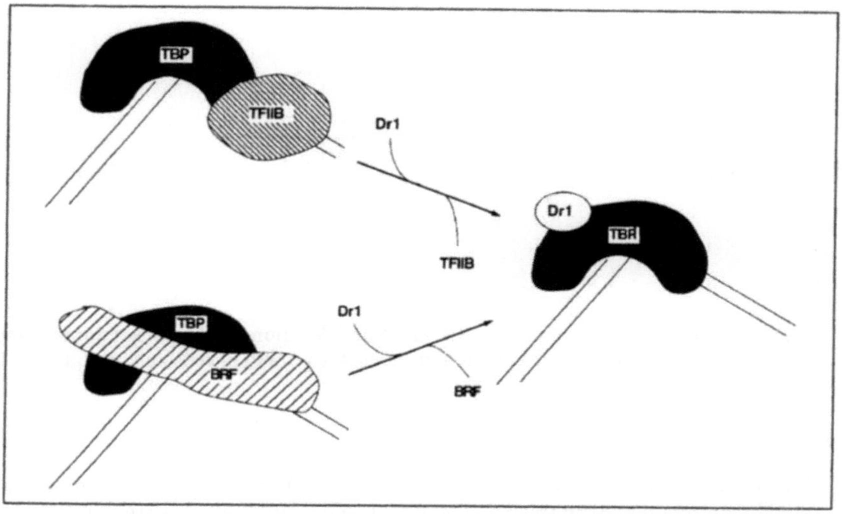

Fig. 8.2. Schematic diagram of the interactions made by TFIIB, BRF and Dr1 with TATA-bound TBP. X-ray crystallography has shown that TFIIB contacts the C-terminal stirrup of TBP. The N-terminal domain of BRF, that is homologous to TFIIB, is likely to make similar contacts with TBP. The C-terminal domain of BRF has high affinity for the basic repeat region on the convex surface of TBP, as shown by mutational analyses. Point mutagenesis has demonstrated that Dr1 binds to an overlapping area in the basic repeat region of TBP. Binding of Dr1 has been shown to displace TFIIB and BRF from their interactions with TBP.

RB

The retinoblastoma susceptibility gene Rb encodes a 105 kD nuclear phosphoprotein, RB, that is expressed almost ubiquitously in normal human and mouse cells.[13,14] The Rb gene was isolated because of its association with an inherited predisposition to retinoblastoma, a rare pediatric tumor of the retina.[15] Inactivating mutations in Rb have also been found in many other types of human tumor.[13,14] This genetic evidence suggested that Rb is a tumor suppressor and that loss of its function can contribute to oncogenesis. Strong support for this belief came from experiments in which the wild-type gene was introduced into cells that lacked its function; expression of exogenous Rb was found to suppress growth and proliferation, soft agar colony formation and tumorigenicity in nude mice.[16-18] Definitive proof of the importance of Rb in preventing tumor formation was provided by mutagenesis of this gene. Homozygous inactivation of Rb causes mouse embryos to die between days 14 and 15 of gestation, with neuronal and erythropoietic defects.[19-21] However, heterozygotes survive and display a clear predisposition to cancer.[20-25] These facts together provide a compelling case for regarding Rb as a bona fide tumor suppressor.

The ability of cells to arrest growth and proliferation is severely compromised when RB is inactivated.[13,14] The growth-suppressive function of RB is dependent on a region called the pocket, which is located between residues 379 and 792.[13,14]

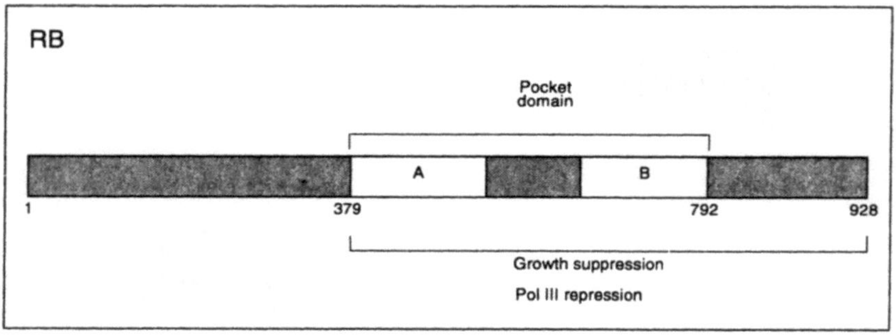

Fig. 8.3. Sequence features of human RB. The A and B regions of the pocket domain are shown. Residues 379 to 928 are sufficient to arrest the growth of cells and inhibit pol III transcription.

The pocket can be further divided into two subdomains, A and B, that are separated by an inessential spacer[13,14] (Fig. 8.3). The pocket domain is frequently mutated in cancers.[13,14] When growth factors are limiting, RB is instrumental in preventing passage through the restriction (R) point in late G1 phase.[13,14,26] If conditions are favorable, RB becomes hyperphosphorylated by cyclin D- and E-dependent kinases; this inactivates it and allows cells to progress into S phase[13,14,26] (Fig. 8.4). When RB function is lost, cells become less sensitive to normal regulatory signals.[13,14,26] This is a major step towards uncontrolled proliferation.

As has been reviewed recently,[27,28] White et al[29] demonstrated that RB can function as a potent repressor of pol III transcription. Thus, in transient transfection experiments, overexpressing RB was shown to inhibit transcription of VA_I without affecting a cotransfected HIV-CAT construct.[29] Furthermore, recombinant RB repressed expression of VA_I, B2, U6, tRNA, 5S and EBER2 genes in a system reconstituted with partially purified factors.[29] Residues 379 to 928 of RB are sufficient to inhibit pol III transcription.[29] This is also the minimal region of RB that is necessary to regulate cell growth and proliferation.[13,14] Deletion analysis suggested that the A and B domains of the RB pocket are both necessary for regulation of pol III.[29] A substitution and two small deletions within the B domain were shown to prevent RB from repressing pol III in vitro or in transfected cells.[29]

These experiments demonstrated that RB can inhibit pol III when it is overexpressed. A more important question is whether or not it performs this function when present at physiological concentrations within a cell. To address this issue, White et al[29] compared the pol III activity of two human osteosarcoma cell lines: SAOS2, which expresses only a nonfunctional truncated form of RB, and U2OS, which contains wild-type RB. SAOS2 cells were found to express a transfected VA_I gene 5-fold more actively than U2OS, despite having a substantially slower growth rate.[29] In vitro transcription using extracted proteins confirmed the higher activity of the pol III factors from the RB-negative SAOS2 cells.[29] A more rigorous test of the function of endogenous RB made use of mice in which the Rb gene had been inactivated by site-directed mutagenesis. Primary embryonic fibroblasts from these RB-knockout mice were shown by nuclear run-on assays to synthesize tRNA and 5S rRNA at 5-fold higher rates than equivalent cells from wild-type mice.[29] In con-

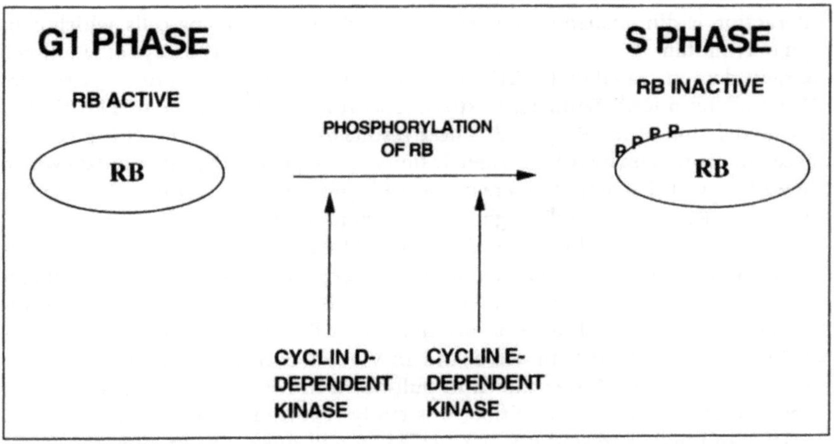

Fig. 8.4. Control of RB at the G1/S transition. During the early part of G1 phase RB is hypophosphorylated and active in regulating transcription. As cells progress into S phase, RB becomes phosphorylated at multiple sites through the sequential action of cyclin D- and cyclin E-dependent kinases. The hyperphosphorylated RB present during S, G2 and M phases is compromised in its ability to repress gene expression.

trast, the overall level of pol II transcription was not enhanced by deletion of the Rb gene.[29] In vitro assays using extracted factors again established that the increased expression in the RB-negative cells is due to a more active pol III transcription apparatus.[29] Since the only genetic difference between the $Rb^{+/+}$ and the $Rb^{-/-}$ fibroblasts is the presence of the Rb gene, these results established that endogenous RB plays a major role in suppressing the level of pol III transcription in vivo.

Larminie et al[30] demonstrated that TFIIIB is a specific target for repression by RB. The inhibitory effect of recombinant RB in a reconstituted system was overcome by the addition of partially purified TFIIIB, but not using pol III or TFIIIC fractions.[30] In the absence of added RB, TFIIIC was the limiting factor in this reconstituted system and TFIIIB was in relative excess.[30] This demonstrates that the addition of RB results in a specific decrease in the availability of active TFIIIB such that this factor becomes limiting.[30] Repression can also be partially relieved using affinity-purified TFIIIB TAFs, but not by recombinant TBP.[30] This suggests that the TAF component of TFIIIB may be targeted specifically by RB.[30] In pull-down assays, GST-RB was shown to interact with TBP and hBRF and to deplete an extract of TFIIIB activity.[30] More importantly, immunoprecipitation experiments demonstrated that endogenous RB associates with TFIIIB when these factors are present at physiological ratios.[30] This interaction is stable to 200 mM KCl and 0.1% NP-40.[30] Both TBP and hBRF were coprecipitated using an antibody that is specific for the hypophosphorylated (active) form of RB.[30] Immunoprecipitation could be blocked with the appropriate epitope and was not seen using control antibodies.[30] A population of endogenous RB molecules was also found to cofractionate with TFIIIB over a variety of ion-exchange columns and on glycerol gradients,[30] providing further evidence for a stable interaction between these factors. This interaction is diminished or abolished in SAOS2 osteosarcoma cells, which contain only mutant RB.[30] In addition to these approaches, direct assays of whole cell extracts demonstrated that TFIIIB activity is elevated in primary fibroblasts from RB-knockout mice.[30] Band shift experiments revealed no change in TFIIIC2 following deletion of the Rb gene.[30] These results establish that TFIIIB is a specific target for repression by RB. This conclusion is consistent with previous data which showed that TFIIIB activity increases as cells progress from G1 into S phase, the time when RB is silenced through hyperphosphorylation.[31]

A subsequent study by Chu et al[32] provided further evidence that RB represses TFIIIB. These workers demonstrated that overexpressing RB inhibits Alu expression in transfected cells and that recombinant RB inhibits Alu and VA_I transcription in vitro.[32] Clustered substitutions in the A or B domain of the RB pocket prevented repression both in vitro and in vivo.[32] Consistent with the earlier investigations, GST-RB was shown to pull-down hBRF and an anti-RB antibody coprecipitated hBRF.[32] In addition, this study reported an interaction between TFIIIC2 and overexpressed RB, as revealed by pull-down and immunoprecipitation assays.[32] A model was proposed in which RB binds either TFIIIB through the A domain of the pocket or TFIIIC2 through the B domain.[32] However, there was little correlation between TFIIIC2 binding and the ability of RB mutants to repress pol III transcription.[32] Indeed, U6 and 7SK genes are efficiently repressed by RB despite not utilizing TFIIIC2.[30,32] Chu et al[32] concluded that TFIIIB is the principal target for RB-mediated repression of pol III, but that a subsidiary interaction with TFIIIC2 can also contribute to the effect.

It remains to be determined how RB is able to repress class III gene expression. It is possible that RB disrupts TFIIIB in some way. Precedent for this is provided by Dr1, which blocks the interaction between TBP and BRF.[8] In contrast to the situation with Dr1, both TBP and hBRF remain associated with RB,[30] but another TFIIIB subunit may be displaced in this case. Alternatively, RB might disrupt the interactions between TFIIIB and TFIIIC, pol III or promoter DNA. RB has been reported to block DNA binding by the pol I factor UBF.[33] However, order of addition experiments showed that the pol III factors remain susceptible to RB even after they have been assembled into a stable preinitiation complex on VA_I.[30] This argues against a mechanism in which RB interferes with promoter recognition. Larminie et al[30] discovered that residues 96-326 of hBRF display 15% identity and 33% similarity to the B domain and C-terminal region (residues 665-910) of RB. These homologies raised the possibility of a structural similarity between RB and hBRF. Such a similarity might allow RB to mimic an interaction that is normally made by hBRF, thereby displacing hBRF and disrupting function. However, as the authors pointed out,[30] the level of homology is so low that one cannot be confident that it is significant. Previous workers have noticed a 21% identity between the A domain of RB and the C-terminal domain of TBP.[34] Crystallographic analysis subsequently demonstrated that the structure of the RB A domain bears no resemblance to that of TBP.[35] Thus, low levels of homology must be interpreted with extreme caution.

p53

The p53 gene is highly conserved among vertebrate species and has been shown to encode an important tumor suppressor.[36-39] p53 is lost or mutated in more than half of all human tumors and its inactivation is considered to be an important step in carcinogenesis.[40] Wild-type p53 can arrest cell growth.[41-44] However, p53 is not

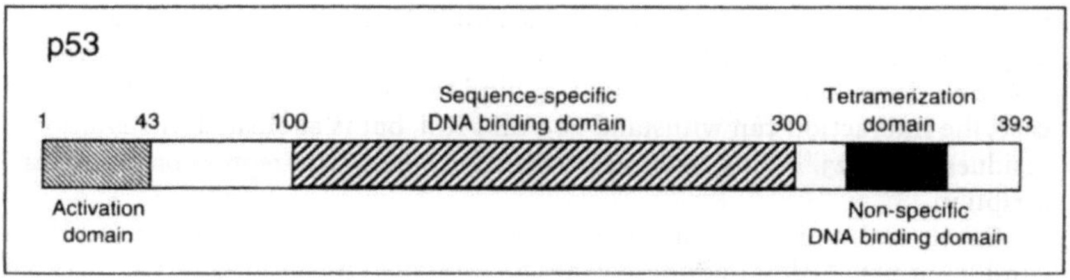

Fig. 8.5. Sequence features of human p53. A transcriptional activation domain that is rich in acidic residues is located within the N-terminal 43 residues. A sequence-specific DNA-binding domain is located within residues 100-300. The C-terminal 93 residues contain an autonomous domain that can bind nonspecifically to DNA. Residues 320-360 allow tetramerization.

an essential cell cycle regulator, since mice that are homozygously null for the p53 gene develop normally.[45] The $p53^{-/-}$ mice display a strong predisposition to cancer, such that 74% develop tumors by the age of six months.[45] This observation provides a clear indication that p53 contributes an important checkpoint against aberrant growth and neoplastic transformation.

The p53 protein can be divided into at least three domains.[39] The N-terminal 43 residues constitute an acidic domain that is important for transcriptional activation and repression.[39] A central core between residues 100 and 300 is responsible for sequence-specific DNA binding.[39] The C-terminal domain from amino acid 300 to 393 contains several activities, including tetramerization and nonspecific DNA-binding[39] (Fig. 8.5). p53 displays a variety of biochemical activities, including the ability to regulate transcription.[36-39] It can bind to DNA in a sequence-specific fashion and stimulate expression of proximal class II genes.[46-49] Transcriptional activation is mediated through an acidic domain at the N-terminus that binds directly to TBP[50-54] and several TAFs in TFIID.[55-57] This interaction can result in the cooperative binding of p53 and TFIID to promoters containing p53 recognition sequences.[58] In addition to activating genes with p53-binding sites, p53 can also repress promoters that lack its response element.[44,50,54,59-65] p53 has been shown to inhibit specifically the synthesis of c-fos,[66] PCNA[67] and cyclin A.[68] Since these gene products are involved in promoting cell cycle progression, it has been suggested that the transcriptional repression function of p53 may contribute to its ability to suppress proliferation and/or tumor formation.[37,39] This possibility is supported by the fact that many tumor-derived p53 mutants have lost the capacity to inhibit transcription.[44,50,59-64] Furthermore, two oncoproteins have been shown to block p53-mediated repression without affecting activation.[44]

Chesnokov et al[65] demonstrated that p53 can also regulate pol III transcription. When overexpressed in transfected COS or 293 cells, p53 repressed U6 and Alu genes.[65] U6 and Alu were also inhibited in vitro by recombinant p53.[65] However, the same amounts of p53 did not affect 7SL, 5S, VA_I and tRNA genes.[65] The authors suggested that U6 and Alu show a preferential response because these both have weak promoters. In support of this, upstream activating sequences from 7SL or an Alu source gene were found to confer p53-resistance to an Alu template with just an internal promoter.[65] A GST fusion protein containing the N-terminal 73 residues of p53 was sufficient to repress Alu but not U6 transcription in vitro.[65] This region of p53 contains a TBP-binding domain.[51,53] Substitution of residues 22 and 23 prevented p53 from repressing Alu expression both in vitro and in vivo.[65] In GST pull-down assays, p53 residues 1-73 bind to TBP, TFIIIB and TFIID.[65] In each case, the interaction can withstand 200 mM KCl, but is abolished by mutation of residues 22 and 23.[65] Thus, binding to TFIIIB correlates with repression of Alu transcription.

Cairns and White have also found that U6 and Alu genes are repressed by recombinant p53.[69] Although these genes are inhibited preferentially, VA, 5S, tRNA, B2 and EBER2 expression was also found to decrease in response to very active p53 preparations.[69] Deletion of either the N-terminal 123 residues or the C-terminal 158 residues abolished regulation of U6 and 5S genes.[69] In this respect pol III inhibition resembles the pol II situation, where both the N- and C-terminal domains of p53 are required for transcriptional repression.[54] In contrast to the case with RB,[30] preassembly of a pol III transcription complex confers resistance to p53.[69] Repression can be relieved specifically by the addition of excess TFIIIB, an effect that is dependent on TBP.[69] Furthermore, cofractionation and coimmunoprecipitation experiments demonstrated that p53 associates with endogenous TFIIIB at physiological ratios.[69] A significant role for cellular p53 as a general repressor of pol III transcription was suggested by the use of $p53^{-/-}$ knockout fibroblasts.[69] Nuclear run-on assays showed that transcription of tRNA and 5S genes is specifically elevated in $p53^{-/-}$ relative to $p53^{+/+}$ cells.[69] Extracts of the knockout fibroblasts transcribed VA, Alu, tRNA and 5S genes 3- to 10-fold more actively than extracts prepared in parallel from the equivalent wild-type cells.[69] This correlated with an increase in TFIIIB activity, whereas TFIIIC appeared to be unaffected by deletion of p53.[69] Add-back experiments confirmed that TFIIIB is limiting in the fibroblast cell extracts.[69] Together, these results suggest that p53 associates with TFIIIB and represses it. This can have a general effect upon the level of pol III transcription, although U6 and Alu genes appear to be most sensitive, probably because they have relatively weak promoters.

Calcium-Dependent Proteases

Treatment with 1.5 mM Ca^{2+} causes a rapid inactivation of the pol III transcription apparatus in extracts from whole *Xenopus* oocytes.[70] This effect can be prevented by leupeptin or E-64, protease inhibitors that block the action of the Ca^{2+}-dependent proteolytic enzymes calpains.[70] A synthetic calpain inhibitor peptide also protects against inactivation.[70] Both TFIIIC and, to a lesser extent, TFIIIB are vulnerable to Ca^{2+}, whereas pol III is relatively resistant.[70] These factors become much less susceptible to Ca^{2+}-dependent proteolysis once they have assembled into transcription complexes.[70]

Activities that Stimulate Pol III Transcription

TAP1

Di Segni et al[71] developed a genetic screen to look for mutations that affect pol III transcription in yeast. A B-block mutation that impaired expression but not tRNA function was introduced into a suppressor $tRNA^{Tyr}$ gene.[71] This gene was then used as the target to select for compensatory trans-acting mutations.[71] Two mutations in a gene named TAP1 resulted in increased levels of the suppressor

$tRNA^{Tyr}$ transcript.[71] The TAP1 gene was cloned by complementation of a temperature-sensitive mutation (tap1-1) and shown to be essential for cell viability.[71] Synthesis of tRNA is strongly inhibited 2 hours after shifting tap1-1 cells to the nonpermissive temperature; this is soon followed by a decrease in 5S rRNA, and then of the pol I rRNA products.[71]

Immunoblotting with anti-TAP1 antibodies showed that TAP1 does not copurify with any of the known class III factors.[72] TAP1 is identical to a gene called RAT1,[72] which had been cloned in a screen for defects in the export of poly(A)$^+$ RNA from the nucleus.[73] The TAP1 gene product also bears homology to a yeast DNA strand transfer protein with riboexonuclease activity.[71] On the basis of this homology and the pleiotropic effects of tap1 mutations, Aldrich et al[72] proposed that TAP1 acts on DNA in chromatin to facilitate both transcription and transcript release.

URP2

URP2 is an *S. cerevisiae* gene that was cloned as a multicopy suppressor of several temperature-sensitive mutations in pol III and in TFIIIC.[74] It is an essential single copy gene that is expressed at high levels.[74] Its 14 kD product, Urp2p, is 55% identical to the S20 ribosomal protein of rat, and is therefore presumed to be the yeast equivalent of this polypeptide.[74] Both Urp2p and rat S20 are significantly related to the S10 ribosomal subunit of *E. coli*, which has been shown to function as a transcriptional elongation factor.[74] Hermann-Le Denmat et al[74] speculated that Urp2p might perform an antitermination function for pol III, similar to the role of bacterial S10, and that this might account for its ability to suppress pol III-specific mutations. However, an identical suppression pattern is provided by FHL1, which encodes a polypeptide with a putative fork head DNA-binding domain.[75] FHL1 is required for pre-rRNA processing and some of the pol III mutations that are suppressed by FHL1 and URP2 are also defective in pre-rRNA processing.[74,75] It is therefore possible that URP2 can function as a dosage-dependent suppressor of mutations in pol III by facilitating ribosome production in some way.[74]

TFIIIR

Transcription of tRNA, 5S and Bmx genes in extracts of *Bombyx* silk glands requires an additional activity called TFIIIR.[76] The resistance of TFIIIR activity to heat, detergent, phenol, protease and DNase, and its sensitivity to alkali or RNase lead to the surprising conclusion that TFIIIR is made of RNA.[76] Extensive purification revealed that TFIIIR activity resides in a specific isoleucine tRNA ($tRNA^{IleIAU}$).[77] A $tRNA^{IleIAU}$ synthesized in vitro has TFIIIR activity.[77] Furthermore, natural RNA selected by hybridization to antisense $tRNA^{IleIAU}$ has a specific activity in complementation assays that is comparable to that of native TFIIIR.[77] Silkworm $tRNA^{AlaSG}$ is 600-fold less active in the same assay.[77] Even the other silkworm isoleucine isoacceptor tRNA is ~80-fold less active than $tRNA^{IleIAU}$.[77] However, a yeast $tRNA^{IleIAU}$ has comparable activity to its silkworm equivalent.[77] Thus, TFIIIR activity is highly specific to this isoacceptor species. The precise mode of action of TFIIIR is unclear. However, it appears to perform an indirect protective role by preventing a low-frequency cleavage of DNA that is inhibitory to transcription in vitro.[78] An intriguing speculation is that the level of $tRNA^{IleIAU}$ acts as a regulatory signal in silk glands to control DNA breakdown prior to the programmed degradation that accompanies the larval-to-pupal transition.[78]

In contrast to these observations in silkworm, transcription by highly purified TFIIIB, TFIIIC and pol III from yeast was not affected by treatment with a large excess of RNases.[79] TFIIIR may be a peculiarity of *Bombyx* or of the highly specialized silk glands from which it was isolated. Since its role in transcription is indirect or protective, it may not be considered a true transcription factor.[80]

TFIIA

Seifart's group reported that the class II transcription factor TFIIA can also stimulate pol III transcription of VA, tRNA, 5S and U6 genes.[81-83] This idea seemed plausible, because TFIIA can bind directly to TBP, a property that has been exploited for its affinity purification.[84-86] Meißner et al[82] reported that highly purified human TFIIA and recombinant yeast TFIIA could activate pol III transcription in a human system that had been chromatographically depleted of endogenous TFIIA. In the case of U6, TFIIA served to stimulate binding of TBP to the TATA box.[81] For TATA-less class III genes, Meißner et al[82] suggested that it may serve as an anti-inhibitor to prevent repressors from binding to TBP.

Kang et al[87] prepared extracts from yeast strains carrying temperature-sensitive mutations in TFIIA. Although pol II transcription was compromised, these TFIIA-depleted extracts showed no defect in the expression of tRNA, 5S or U6 genes.[87] Deletion of TFIIA genes was also shown to have no effect upon $tRNA^{Ile}$ synthesis in vivo.[87] These results establish unequivocally that TFIIA is not required for pol III transcription in yeast. The results of point mutational analyses suggest that yBRF and TFIIA bind to the same linker region between the direct repeats of TBP.[12] This overlap in the binding sites makes it unlikely that TFIIA and yBRF can bind simultaneously to TBP. Indeed, Roberts et al[88] showed that prebinding of yBRF and B" to TBP prevents association of TFIIA. Similarly, prebinding of TFIIA to TBP prevents recruitment of the pol III TAFs.[88]

The results in yeast cast doubt upon the earlier in vitro studies suggesting an involvement of TFIIA in mammalian pol III transcription.[81-83] It is possible that a novel role for TFIIA has evolved in higher organisms, but this seems unlikely since TFIIA, TBP and BRF are all strongly conserved. In my own laboratory we observe no activation of VA or U6 transcription when highly purified human TFIIA is added to HeLa extracts. Efficient transcription is also maintained following the chromatographic depletion of TFIIA.

STAF

Schuster et al[89] screened a *Xenopus* oocyte expression library with the distal activator element (-209 to -195) of the *X. laevis* $tRNA^{Sec}$ gene in order to isolate cognate binding factors. They cloned cDNAs encoding a novel factor, which they named Staf.[89] Staf is a 65 kD polypeptide of 600 amino acid residues.[89] Between residues 267 and 468 lie seven tandemly repeated zinc fingers of the C2-H2 class and this region alone can bind specifically to the activator element[89] (Fig. 8.6). Whereas the central finger domain is highly basic (pI 10), the remainder of the protein is very acidic (pI ~4).[89] When injected into oocytes, recombinant Staf stimulated transcription of the $tRNA^{Sec}$ gene via its activator element.[89] The acidic regions of Staf are sufficient to increase expression of $tRNA^{Sec}$, U6 and U1 genes when directed to the promoter by a heterologous DNA-binding domain.[89,90] Staf also stimulated pol II transcription of a tkCAT reporter gene that was linked to the

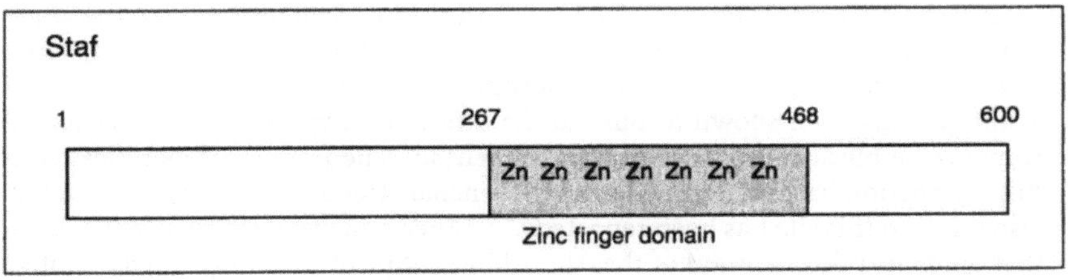

Fig. 8.6. Sequence features of Staf from *X. laevis*. Seven putative zinc fingers are located within residues 267-468.

tRNA^{Sec} activator element.[89] Thus, like Oct-1, Staf can activate both pol II and pol III. In addition to tRNA^{Sec} promoters, Staf was found to bind to the DSE elements of human U6, U4C, 7SK and Y4, mouse U6, and *Xenopus* MRP, U1b1, U2 and U5 genes.[90] In each of these cases, substitution of the Staf-binding motif reduced expression in vivo.[90] A consensus recognition sequence of YY(A/T)CCC(A/G)N(A/C)AT(G/C)C(A/C)YYRCR was deduced by binding site selection with recombinant Staf.[90] This highly extended binding motif will tolerate considerable substitution, and substantial degeneracy is observed in the naturally-occurring sites.[90] Whereas Staf sites occur alone upstream of the *Xenopus* tRNA^{Sec} and MRP genes, ~70% of all DSE elements contain an octamer motif within 28 bp of a Staf site.[90] Oct-1 may function cooperatively with Staf, since their respective binding motifs can activate synergistically if appropriately spaced.[90,91]

Additional Factors

A number of relatively uncharacterized factors have also been implicated in pol III regulation. For example, Oei and Pieler[92] reported that a protein present in both *Xenopus* and HeLa extracts can bind specifically to the 5'-flanking sequence (-34 to -9) of the *Xenopus* somatic 5S gene. Cross-competition experiments showed that this factor also binds to the upstream region (-30 to -5) of a *Xenopus* tRNA^{Met} gene, although this region shows little homology to the 5S binding site.[92] Oligonucleotides corresponding to these binding sequences compete specifically for 5S and tRNA transcription, presumably due to titration of a positively acting cognate factor.[92] A fraction containing this factor stimulates tRNA^{Met} transcription by 15-fold.[92] Further fractionation resolved this factor into DNA-binding and nonbinding components, both of which are required for transcriptional activation.[92] An uncharacterized cellular protein has also been reported to bind to the upstream regions of certain yeast tRNA genes.[93]

Giardina and Wu[94] described two antagonistic activities in *Xenopus* oocyte extracts that can modulate pol III transcription. TFIIIB and TFIIIC became inactivated during prolonged incubation with an activity termed fraction I.[94] Inactivation did not occur if a 5S or tRNA gene was included in the incubation.[94] TFIIIC that had been inactivated in this way could no longer bind DNA.[94] Inactivation required the presence of ATP or its nonhydrolyzable analog AMP-PNP.[94] The action of fraction I could be rapidly reversed by a second activity, termed fraction A2, that required a hydrolyzable form of ATP but did not require the presence of a gene.[94] Fraction A2 does not stimulate transcription in the absence of fraction I.[94]

A HeLa cell protein termed VBP binds to a site just 8 bp upstream of the VA_I B-block.[95] Although VBP does not seem to have a direct role in transcription, it may have an indirect negative one by sterically hindering TFIIIC binding.[95] SV40 large T antigen has been shown to bind to the interblock region of an Alu gene.[96] A comparable binding-site is present in B1 genes.[97] The functional significance of this interaction has not been assessed. An uncharacterized HeLa cell protein that also binds to this site has been reported.[98-100] Sperm nuclei contain a 60 kD factor that binds just downstream of the Alu A-block, at positions 25-33.[101] This protein migrates more rapidly in native gels than the HeLa cell factor that binds to the interblock region of Alu genes.[101] Chesnokov and Schmid[101] purified the 60 kD Alu-binding protein from human sperm to near homogeneity and named it SABP (for sperm Alu binding protein). Peptide sequencing revealed no homology to other proteins.[101] SABP can selectively protect Alu genes from methylation in vitro.[101]

In addition to Oct-1 and Oct-2, several other pol II factors have been implicated in pol III transcription on the basis of sequence motifs associated with class III genes. For example, GC boxes that resemble Sp1-binding sites are found in the promoters of the EBER2, mammalian 5S and *Xenopus* U6 genes.[102-104] Although purified Sp1 has been shown to bind to EBER2 and U6,[102,104] it has not yet been shown to regulate pol III. However, two tandem copies of an Sp1 activation domain have been shown to activate pol III transcription when fused to a heterologous DNA-binding domain.[105] Sp1 is one of a family of ubiquitous factors with homologous DNA-binding domains and similar binding specificities.[106] It therefore remains to be determined which members of this family can regulate the expression of class III genes with GC box promoter sequences. A similar example is provided by the ATF sites that contribute to the expression of EBER2 and 7SL genes.[102,107] Such sites are recognized by a large family of ATF and CREB factors[108] and it is not clear which of these can regulate pol III transcription. A subset of these factors form part of the cAMP-dependent signal transduction system.[108] Bredow et al[107] treated HeLa cells with forskolin in order to elevate cAMP levels. Extracts prepared from such cells displayed activated transcription of 7SL, but not of 7SK which lacks an ATF motif.[107] This suggests that at least one of the cAMP-responsive factors, such as ATF-1 or CREB itself, may interact with the pol III transcription machinery. CREB has been shown to bind to a site upstream of the VA_I gene,[109] although a functional significance for this interaction has yet to be reported. USF has been found to bind to an E-box that is required for expression of a sea urchin U6 gene.[110] An AP1 site has been shown to stimulate transcription of an Alu gene in vitro.[111] Recombinant YY1 was reported to bind to the A-block of a $tRNA^{Gln}$ gene, although the functional relevance of this interaction was not tested.[112]

Not all regulatory factors that can influence pol II transcription can do the same for pol III. For example, the upstream activating sequence from the yeast class II gene CYC1 has no effect on pol III transcription when placed upstream of the U6 gene.[113] The $\alpha 2$ protein of *S. cerevisiae* can repress transcription by pol I and pol II, but not by pol III.[114] In contrast, octamer sites that can activate pol II and pol III transcription do not stimulate pol I.[115] The polyomavirus enhancer, which is bound by many different class II factors, does not increase VA_I expression when linked in cis.[116] The class II factor NFI binds to two sites at the 3' end of VA_I, but does not affect its transcription.[95] The pol II activator GCN4 was not found to stimulate $tRNA^{Tyr}$ transcription when positioned at a variety of upstream sites.[117]

Thus, the activity of many regulatory factors is restricted to particular polymerases. Such differences are likely to prove informative concerning the mechanisms of transcriptional regulation.

Das et al[105] tested the response of U6, U2 and c-fos promoters to several well-defined activation domains. Short reiterated segments from the activation domains were fused to the DNA-binding domain of Gal4, and four DNA binding sites for Gal4 were inserted upstream of these promoters.[105] A fusion protein containing four tandem copies of a nineteen amino acid segment from the glutamine-rich activation domain of Oct-1 stimulated expression from the U6 and U2 promoters but not the c-fos promoter.[105] All three promoters were stimulated using four tandem copies of eighteen residues from the glutamine-rich activation domain of Oct-2.[105] Substitution of the glutamines for alanines in this fusion protein increased its ability to stimulate U6 transcription, whereas the same substitutions prevented activation of c-fos and had no effect upon U2.[105] Thus, although the glutamine-rich activation domains from Oct-1 and Oct-2 can regulate transcription by both pol II and pol III, different residues from within these domains appear to be important in each case. The Gal4 DNA-binding domain alone produced some stimulation of U6 transcription, although it had no effect upon c-fos or U2.[105] In contrast, a proline-rich segment from Oct-2 and an acidic segment from VP16 had no effect upon U6 or U2 but caused substantial activation of c-fos.[105,113] Clearly, the response to upstream activators is extremely dependent on the target gene. U6 expression could also be activated by short segments from the activation domains of Sp1 (glutamine-rich), CTF (proline-rich) and p53 (rich in prolines and acidic residues) when fused to Gal4.[105]

In the pol II system, the response to upstream activators often involves specific TAFs.[118-123] For example, activation by VP16 involves direct contact with $TAF_{II}40$.[124] The absence of this TAF from the class III transcription apparatus may therefore explain the inability of VP16 to regulate pol III. Similarly, stimulation of pol II transcription by Sp1 was reported to require $TAF_{II}110$.[125,126] However, U6 can respond to the Sp1 activation domain, even though the pol III system does not contain $TAF_{II}110$.[105] This raises the question as to whether TFIIIB contains a functional homologue of $TAF_{II}110$, or whether Sp1 can activate U6 transcription by a different mechanism. TAF-independent activation pathways have been identified.[123,127] It will be of considerable interest to learn which component(s) of the pol III transcription apparatus interact with the activation domains of factors such as Sp1 and Oct-1.

Kinases and Phosphatases

CDC2/cyclin B1

Cdc2, in complex with B-type cyclins, is the major protein kinase activity when cells enter mitosis.[128-130] Gottesfeld et al[131] used a GST-cyclin B1 fusion protein to affinity purify cdc2 from mitotic *Xenopus* extracts. This kinase was able to inactivate TFIIIB specifically when incubated with fractionated factors.[131] Repression could be blocked by the kinase inhibitor 6-dimethylaminopurine (DMAP) and reversed by alkaline phosphatase.[131] 5S transcription is also repressed using mitotic kinase that has been immunopurified with antibodies against cdc2, cyclin B1 or cyclin B2.[132] In contrast, MAP kinase from mitotic extracts has no effect on 5S expression.[132] The possibility remains that the influence of cdc2/cyclin B on pol III transcription could be indirect. However, since repression is observed when affinity-purified cdc2 is combined with affinity-purified TFIIIB, it is likely that one or more components of TFIIIB is a direct substrate for the mitotic kinase in *Xenopus*.

CDC25

The cdc25 tyrosine phosphatases are involved in activating cyclin-dependent kinases.[129,133] Reynolds[134] demonstrated that recombinant cdc25 can repress $tRNA^{TyrC}$ transcription when added to *Xenopus* oocyte extracts. This effect was blocked by sodium vanadate, an inhibitor of tyrosine phosphatase activity.[134] Expression of a $tRNA^{TyrD}$ gene was unaffected under the same conditions.[134] Repression of $tRNA^{TyrC}$ by cdc25 was relieved by the addition of TFIIIC but not TFIIIB fractions.[134] It remains to be determined whether cdc25 acts directly on the pol III transcription apparatus or has an indirect effect, perhaps involving cyclin-dependent kinases.

Casein Kinase II

Casein kinase II (CKII) is a highly conserved serine/threonine protein kinase that has been implicated in growth and cell cycle control.[135] It is expressed in both the nucleus and cytoplasm of most metazoan and yeast cells. The yeast strain $cka2^{ts}$ carries a lesion in the catalytic α' subunit of CKII.[136] This mutation causes a decrease in tRNA synthesis at the restrictive temperature, whereas production of large rRNA is unaffected.[136] Extracts prepared from $cka2^{ts}$ cells are compromised for transcription of tRNA and 5S genes, whereas pol I transcription and basal pol II transcription are unchanged by the mutation.[136] Purified CKII stimulated tRNA and 5S transcription when added to the $cka2^{ts}$ extract, but had no effect on a wild-type extract.[136] This suggests that CKII activity is saturating for pol III transcription in wild-type cells. The CKII inhibitor 2,3-diphosphoglycerate suppresses tRNA synthesis in extracts from wild-type cells.[136] It remains to be determined whether CKII phosphorylates the pol III transcription apparatus directly. Candidate targets include TBP, the ABC23 and AC19 subunits of pol III, and the τ_{138}, τ_{131} and τ_{95} subunits of TFIIIC, since each of these is a phosphoprotein and each contains a consensus CKII target motif.[136]

In contrast to the evidence that CKII activates pol III transcription in yeast, Fan et al[137] demonstrated that CKII can inhibit the ability of La to stimulate recycling of the human pol III transcription complex. In vitro, CKII phosphorylates human La at serine 366 specifically.[137] This residue is part of a conserved CKII consensus sequence and has been shown to be phosphorylated in vivo.[137] La can stimulate transcription from isolated complexes preassembled on the VA_I promoter.[138] It remains to be determined whether inhibition of this activity by CKII is ever responsible for limiting the rate of pol III transcription under physiological conditions. In yeast, La is not required for viability.[139]

Protein Kinase C

Addition of phosphatidyl serine (PS; 200 μg/ml) to whole HeLa cell extracts activates protein kinase C (PKC) and inhibits transcription of the VA_I gene, whereas phosphatidyl choline and phosphatidyl ethanolamine affect neither.[140] Inhibitors of PKC reduce the ability of PS to repress pol III.[140] A peptide inhibitor of PKC can also raise the level of VA_I expression in the absence of added PS.[140] PS does not affect transcription if added after initiation complexes have assembled on the VA_I promoter.[140] PKC inhibitors have also been shown to stimulate pol III transcription when added to cytoplasmic extracts of *Drosophila* S-2 cells.[141]

Protein Phosphatase 2A

Strains of *S. cerevisiae* that lack the TPD3 gene are severely compromised for growth.[142] The TPD3 gene encodes the regulatory subunit A of protein phosphatase 2A (PP2A).[142] The regulatory subunits of PP2A are believed to define its substrate specificity.[143] A temperature-sensitive tpd3 strain was found to stop synthesizing tRNA at the nonpermissive temperature.[142] This defect could be reproduced in vitro using extracts prepared from tpd3 cells.[142] When grown at the nonpermissive temperature, the tpd3 strain gave extracts that were unable to transcribe tRNA genes.[142] When grown at the permissive temperature, the tpd3 extracts were 2- to 4-fold less active than wild-type.[142] In the latter case, transcription could be fully restored by the addition of partially purified TFIIIB, whereas pol III was partially stimulatory and TFIIIC had no effect.[142] However, fractionation revealed that the tpd3 extracts contained TFIIIB and pol III that was equivalent to wild-type in both activity and chromatographic properties.[142] Mixing experiments established that the defect in tpd3 is due to an inhibitory activity rather than the loss of a transcription component.[142] These results suggest that PP2A stimulates TFIIIB and, to a lesser extent, pol III in yeast cells. The inhibitory influence that prevents transcription in tpd3 mutants might be a kinase that represses TFIIIB and is normally balanced by PP2A. Experiments carried out in my laboratory suggest that PP2A is responsible for reversing the mitotic repression of TFIIIB in HeLa cells.[144]

References

1. Inostroza JA, Mermelstein FH, Ha I et al. Dr1, a TATA-binding protein-associated phosphoprotein and inhibitor of class II gene transcription. Cell 1992; 70:477-489.
2. Kuromori T, Yamamoto M. Cloning of cDNAs from *Arabidopsis thaliana* that encode putative protein phosphatase 2C and a human Dr1-like protein by transformation of a fission yeast mutant. Nucleic Acids Res 1994; 22:5296-5301.
3. Kim S, Na JG, Hampsey M et al. The Dr1/DRAP1 heterodimer is a global repressor of transcription in vivo. Proc Natl Acad Sci USA 1997; 94:820-825.
4. Mermelstein F, Yeung K, Cao J et al. Requirement of a corepressor for Dr1-mediated repression of transcription. Genes Dev 1996; 10:1033-1048.
5. Goppelt A, Stelzer G, Lottspeich F et al. A mechanism for repression of class II gene transcription through specific binding of NC2 to TBP-promoter complexes via heterodimeric histone fold domains. EMBO J 1996; 15:3105-3116.
6. Yeung KC, Inostroza JA, Mermelstein FH et al. Structure-function analysis of the TBP-binding protein Dr1 reveals a mechanism for repression of class II gene transcription. Genes Dev 1994; 8:2097-2109.
7. Kim TK, Zhao Y, Ge H et al. TATA-binding protein residues implicated in a functional interplay between negative cofactor NC2 (Dr1) and general factors TFIIA and TFIIB. J Biol Chem 1995; 270:10976-10981.
8. White RJ, Khoo BC-E, Inostroza JA et al. The TBP-binding repressor Dr1 differentially regulates RNA polymerases I, II and III. Science 1994; 266:448-450.
9. Khoo B, Brophy B, Jackson SP. Conserved functional domains of the RNA polymerase III general transcription factor BRF. Genes Dev 1994; 8:2879-2890.
10. Chaussivert N, Conesa C, Shaaban S et al. Complex interactions between yeast TFIIIB and TFIIIC. J Biol Chem 1995; 270:15353-15358.

11. Wang Z, Roeder RG. Structure and function of a human transcription factor TFIIIB subunit that is evolutionarily conserved and contains both TFIIB- and high-mobility-group protein 2-related domains. Proc Natl Acad Sci USA 1995; 92:7026-7030.
12. Buratowski S, Zhou H. A suppressor of TBP mutations encodes an RNA polymerase III transcription factor with homology to TFIIB. Cell 1992; 71:221-230.
13. Weinberg RA. The retinoblastoma protein and cell cycle control. Cell 1995; 81:323-330.
14. Whyte P. The retinoblastoma protein and its relatives. Seminars in Cancer Biology 1995; 6:83-90.
15. Friend SH, Bernards R, Rogelj S et al. A human DNA segment with properties of the gene that predisposes to retinoblastoma and osteosarcoma. Nature 1986; 323:643-646.
16. Huang H-JS, Yee J-K, Shew J-Y et al. Suppression of the neoplastic phenotype by replacement of the RB gene in human cancer cells. Science 1988; 242:1563-1566.
17. Bookstein R, Shew J, Chen P et al. Suppression of tumorigenicity of human prostate carcinoma cells by replacing a mutated RB gene. Science 1990; 247:712-715.
18. Qin X, Chittenden T, Livingston DM et al. Identification of a growth suppression domain within the retinoblastoma gene product. Genes Dev 1992; 6:953-964.
19. Clarke AR, Maandag ER, van Roon M et al. Requirement for a functional *Rb-1* gene in murine development. Nature 1992; 359:328-330.
20. Jacks T, Fazeli A, Schmitt EM et al. Effects of an *Rb* mutation in the mouse. Nature 1992; 359:295-300.
21. Lee EY-HP, Chang C-Y, Hu N et al. Mice deficient for Rb are nonviable and show defects in neurogenesis and haematopoiesis. Nature 1992; 359:288-294.
22. Hu N, Gutsmann A, Herbert DC et al. Heterozygous $Rb\text{-}1^{\Delta 20/+}$ mice are predisposed to tumors of the pituitary gland with a nearly complete penetrance. Oncogene 1994; 9:1021-1027.
23. Maandag ECR, van der Valk M, Vlaar M et al. Developmental rescue of an embryonic-lethal mutation in the retinoblastoma gene in chimeric mice. EMBO J 1994; 13:4260-4268.
24. Williams BO, Schmitt EM, Remington L et al. Extensive contribution of *Rb*-deficient cells to adult chimeric mice with limited histopathological consequences. EMBO J 1994; 13:4251-4259.
25. Nikitin AY, Lee W-H. Early loss of the retinoblastoma gene is associated with impaired growth inhibitory innervation during melanotroph carcinogenesis in $Rb^{+/-}$ mice. Genes Dev 1996; 10:1870-1879.
26. Sherr CJ. G1 phase progression: cycling on cue. Cell 1994; 79:551-555.
27. White RJ. Regulation of RNA polymerases I and III by the retinoblastoma protein: a mechanism for growth control? Trends Biochem Sci 1997; 22:77-80.
28. Larminie CGC, Alzuherri HM, Cairns CA et al. Transcription by RNA polymerases I and III: a potential link between cell growth, protein synthesis and the retinoblastoma protein. J Mol Med 1998; 76:94-103; in press.
29. White RJ, Trouche D, Martin K et al. Repression of RNA polymerase III transcription by the retinoblastoma protein. Nature 1996; 382:88-90.
30. Larminie CGC, Cairns CA, Mital R et al. Mechanistic analysis of RNA polymerase III regulation by the retinoblastoma protein. EMBO J 1997; 16:2061-2071.
31. White RJ, Gottlieb TM, Downes CS et al. Cell cycle regulation of RNA polymerase III transcription. Mol Cell Biol 1995; 15:6653-6662.
32. Chu W-M, Wang Z, Roeder RG et al. RNA polymerase III transcription repressed by Rb through its interactions with TFIIIB and TFIIIC2. J Biol Chem 1997; 272:14755-14761.

33. Voit R, Schafer K, Grummt I. Mechanism of repression of RNA polymerase I transcription by the retinoblastoma protein. Mol Cell Biol 1997; 17:4230-4237.
34. Hagemeier C, Bannister AJ, Cook A et al. The activation domain of transcription factor PU.1 binds the retinoblastoma (RB) protein and the transcription factor TFIID in vitro: RB shows sequence similarity to TFIID and TFIIB. Proc Natl Acad Sci USA 1993; 90:1580-1584.
35. Kim H-Y, Cho Y. Structural similarity between the pocket region of retinoblastoma tumor suppressor and the cyclin-box. Nature Structural Biol 1997; 4:390-395.
36. Chernova OB, Chernov MV, Agarwal ML et al. The role of p53 in regulating genomic stability when DNA and RNA synthesis are inhibited. Trends Biochem Sci 1995; 20:431-434.
37. Cox LS, Lane DP. Tumor suppressors, kinases and clamps: how p53 regulates the cell cycle in response to DNA damage. Bioessays 1995; 17:501-508.
38. Haffner R, Oren M. Biochemical properties and biological effects of p53. Curr Opin Genetics Dev 1995; 5:84-90.
39. Ko LJ, Prives C. p53: puzzle and paradigm. Genes Dev 1996; 10:1054-1072.
40. Hollstein M, Sidransky D, Vogelstein B et al. p53 mutations in human cancers. Science 1991; 253:49-53.
41. Mercer WE, Amin M, Sauve GJ et al. Wild-type human p53 is antiproliferative in SV40-transformed hamster cells. Oncogene 1990; 5:973-980.
42. Michalovitz D, Halevy O, Oren M. Conditional inhibition of transformation and of cell proliferation by a temperature-sensitive mutant of p53. Cell 1990; 62:671-680.
43. Martinez J, Georgoff I, Martinez J et al. Cellular localization and cell cycle regulation by a temperature-sensitive p53 protein. Genes Dev 1991; 5:151-159.
44. Crook T, Marston NJ, Sara EA et al. Transcriptional activation by p53 correlates with suppression of growth but not transformation. Cell 1994; 79:817-827.
45. Donehower LA, Harvey M, Slagle BL et al. Mice deficient for p53 are developmentally normal but susceptible to spontaneous tumors. Nature 1992; 356:215-221.
46. Bargonetti J, Friedman P, Kern S et al. Wild-type but not mutant p53 immunopurified proteins bind to sequences adjacent to the SV40 origin of replication. Cell 1991; 65:1083-1091.
47. Kern S, Kinzler K, Bruskin A et al. Identification of p53 as a sequence-specific DNA-binding protein. Science 1991; 252:1708-1711.
48. Farmer G, Bargonetti J, Zhu H et al. Wild-type p53 activates transcription in vitro. Nature 1992; 358:83-86.
49. Zambetti G, Bargonetti J, Walker K et al. Wild-type p53 mediates positive regulation of gene expression through a specific DNA sequence element. Genes Dev 1992; 6:1143-1152.
50. Seto E, Usheva A, Zambetti GP et al. Wild-type p53 binds to the TATA-binding protein and represses transcription. Proc Natl Acad Sci USA 1992; 89:12028-12032.
51. Liu X, Miller C, Koeffler P et al. The p53 activation domain binds the TATA box-binding polypeptide in Holo-TFIID, and a neighboring p53 domain inhibits transcription. Mol Cell Biol 1993; 13:3291-3300.
52. Martin DW, Munoz RM, Subler MA et al. p53 binds to the TATA-binding protein-TATA complex. J Biol Chem 1993; 268:13062-13067.
53. Truant R, Xiao H, Ingles C et al. Direct interaction between the transcriptional activation domain of human p53 and the TATA-box-binding protein. J Biol Chem 1993; 268:2284-2287.
54. Horikoshi N, Usheva A, Chen J et al. Two domains of p53 interact with the TATA-binding protein, and the adenovirus 13S E1A protein disrupts the association, relieving p53-mediated transcriptional repression. Mol Cell Biol 1995; 15:227-234.

55. Lu H, Levine AJ. Human TAF31 protein is a transcriptional coactivator of the p53 protein. Proc Natl Acad Sci USA 1995; 92:5154-5158.
56. Thut C, Chen JL, Klemm R et al. p53 transcriptional activation mediated by coactivators TAFII40 and TAFII60. Science 1995; 267:100-104.
57. Farmer G, Colgan J, Nakatani Y et al. Functional interaction between p53, the TATA-binding protein (TBP), and TBP-associated factors in vivo. Mol Cell Biol 1996; 16:4295-4304.
58. Chen X, Farmer G, Zhu H et al. Cooperative DNA binding of p53 with TFIID (TBP): a possible mechanism for transcriptional activation. Genes Dev 1993; 7:1837-1849.
59. Ginsberg D, Mechta F, Yaniv M et al. Wild-type p53 can down-modulate the activity of various promoters. Proc Natl Acad Sci USA 1991; 88:9979-9983.
60. Mercer W, Shields M, Lin D et al. Growth suppression induced by wild-type p53 protein is accompanied by selective downregulation of proliferating-cell nuclear antigen expression. Proc Natl Acad Sci USA 1991; 88:1958-1962.
61. Santhanam U, Ray A, Sehgal P. Repression of the interleukin 6 gene promoter by p53 and the retinoblastoma susceptibility gene product. Proc Natl Acad Sci USA 1991; 88:7605-7609.
62. Shiio Y, Yamamoto T, Yamaguchi N. Negative regulation of RB expression by the p53 gene product. Proc Natl Acad Sci USA 1992; 89:5206-5210.
63. Subler M, Martin D, Deb S. Inhibition of viral and cellular promoters by human wild-type p53. J Virol 1992; 66:4757-4762.
64. Ragimov N, Krauskopf A, Navot N et al. Wild-type but not mutant p53 can repress transcription initiation in vitro by interfering with the binding of basal transcription factors to the TATA motif. Oncogene 1993; 8:1183-1193.
65. Chesnokov I, Chu W-M, Botchan MR et al. p53 inhibits RNA polymerase III-directed transcription in a promoter-dependent manner. Mol Cell Biol 1996; 16:7084-7088.
66. Kley N, Chung RY, Fay S et al. Repression of the basal c-fos promoter by wild-type p53. Nucleic Acids Res 1992; 20:4083-4087.
67. Jackson P, Ridgway P, Rayner J et al. Transcriptional regulation of the PCNA promoter by p53. Biochem Biophys Res Commun 1994; 203:133-140.
68. Yamamoto M, Yoshida M, Ono K et al. Effect of tumor suppressors on cell cycle-regulatory genes: RB suppresses $p34^{cdc2}$ expression and normal p53 suppresses cyclin A expression. Exp Cell Res 1994; 210:94-101.
69. Cairns CA, White RJ. Unpublished observations.
70. Sturges MR, Peck LJ. Calcium-dependent inactivation of RNA polymerase III transcription. J Biol Chem 1994; 269:5712-5719.
71. Di Segni G, McConaughy BL, Shapiro RA et al. *TAP1*, a yeast gene that activates the expression of a tRNA gene with a defective internal promoter. Mol Cell Biol 1993; 13:3424-3433.
72. Aldrich TL, Di Segni G, McConaughy BL et al. Structure of the yeast TAP1 protein: dependence of transcription activation on the DNA context of the target gene. Mol Cell Biol 1993; 13:3434-3444.
73. Amberg DC, Goldstein AL, Cole CN. Isolation and characterization of *RAT1*: an essential gene of *Saccharomyces cerevisiae* required for the efficient nucleocytoplasmic trafficking of mRNA. Genes Dev 1992; 6:1173-1189.
74. Hermann-Le Denmat S, Sipiczki M, Thuriaux P. Suppression of yeast RNA polymerase III mutations by the *URP2* gene encoding a protein homologous to the mammalian ribosomal protein S20. J Mol Biol 1994; 240:1-7.
75. Hermann-Le Denmat S, Werner M, Sentenac A et al. Suppression of yeast RNA polymerase III mutations by *FHL1*, a gene coding for a *fork head* protein involved in rRNA processing. Mol Cell Biol 1994; 14:2905-2913.

76. Young LS, Dunstan HM, Witte PR et al. A class III transcription factor composed of RNA. Science 1991; 252:542-546.
77. Dunstan HM, Young LS, Sprague KU. TFIIIR is an isoleucine tRNA. Mol Cell Biol 1994; 14:3588-3595.
78. Dunstan HM, Young LS, Sprague KU. $tRNA^{IleIAU}$ (TFIIIR) plays an indirect role in silkworm class III transcription in vitro and inhibits low-frequency DNA cleavage. Mol Cell Biol 1994; 14:3596-3603.
79. Kassavetis GA, Blanco JA, Johnson TE et al. Formation of open and elongating transcription complexes by RNA polymerase III. J Mol Biol 1992; 226:47-58.
80. Smith TPL, Young LS, Bender LB et al. Silkworm TFIIIA requires additional class III factors for committment to transcription complex assembly on a 5S RNA gene. Nucleic Acids Res 1995; 23:1244-1251.
81. Waldschmidt R, Seifart KH. TFIIA is required for in vitro transcription of mammalian U6 genes by RNA polymerase III. J Biol Chem 1992; 267:16359-16364.
82. Meißner W, Holland R, Waldschmidt R et al. Transcription factor IIA stimulates the expression of classical polIII genes. Nucleic Acids Res 1993; 21:1013-1018.
83. Meißner W, Wanandi I, Carbon P et al. Transcription factors required for the expression of *Xenopus laevis* selenocysteine tRNA in vitro. Nucleic Acids Res 1994; 22:553-559.
84. Ranish JA, Hahn S. The yeast general transcription factor TFIIA is composed of two polypeptide subunits. J Biol Chem 1991; 266:19320-19327.
85. Usuda Y, Kubota A, Berk AJ et al. Affinity purification of transcription factor IIA from HeLa cell nuclear extracts. EMBO J 1991; 10:2305-2310.
86. Cortes P, Flores O, Reinberg D. Factors involved in specific transcription by mammalian RNA polymerase II: purification and analysis of transcription factor IIA and identification of transcription factor IIJ. Mol Cell Biol 1992; 12:413-421.
87. Kang JJ, Auble DT, Ranish JA et al. Analysis of the yeast transcription factor TFIIA: distinct functional regions and a polymerase II-specific role in basal and activated transcription. Mol Cell Biol 1995; 15:1234-1243.
88. Roberts S, Miller SJ, Lane WS et al. Cloning and functional characterization of the gene encoding the TFIIIB90 subunit of RNA polymerase III transcription factor TFIIIB. J Biol Chem 1996; 271:14903-14909.
89. Schuster C, Myslinski E, Krol A et al. Staf, a novel zinc finger protein that activates the RNA polymerase III promoter of the selenocysteine tRNA gene. EMBO J 1995; 14:3777-3787.
90. Schaub M, Myslinski E, Schuster C et al. Staf, a promiscuous activator for enhanced transcription by RNA polymerases II and III. EMBO J 1997; 16:173-181.
91. Myslinski E, Schuster C, Krol A et al. Promoter strength and structure dictate module composition in RNA polymerase III transcriptional activator elements. J Mol Biol 1993; 234:311-318.
92. Oei SL, Pieler T. A transcription stimulatory factor binds to the upstream region of *Xenopus* 5 S RNA and tRNA genes. J Biol Chem 1990; 265:7485-7491.
93. Marschalek R, Dingermann T. Identification of a protein factor binding to the 5'-flanking region of a tRNA gene and being involved in modulation of tRNA gene transcription in vivo in *Saccharomyces cerevisiae*. Nucleic Acids Res 1988; 16:6737-6752.
94. Giardina CA, Wu CW. The identification of two antagonistic activities in a *Xenopus* oocyte extract that can modulate the in vitro transcription of RNA polymerase III genes. J Biol Chem 1990; 265:9121-9130.
95. Van Dyke MW, Roeder RG. Multiple proteins bind to VA RNA genes of adenovirus type 2. Mol Cell Biol 1987; 7:1021-1031.

96. Wright PJ, DeLucia A, Tegtmeyer P. Sequence-specific binding of simian virus 40 A protein to nonorigin and cellular DNA. Mol Cell Biol 1984; 4:2631-2638.
97. Krayev AS, Markusheva TV, Kramerov DA et al. Ubiquitous transposon-like repeats B1 and B2 of the mouse genome: B2 sequencing. Nucleic Acids Res 1982; 10:7461-7475.
98. Perelygina LM, Tomilin NV, Podgornaya OI. Alu-family repeat binding protein from HeLa cells which interacts with regulatory region of SV40 genome. Mol Biol Rep 1987; 12:111-116.
99. Podgornaya OI, Perelygina LM, Tomilin NV. Multi-site binding of human nuclear protein to the Alu-family repeated DNA. FEBS Lett 1988; 232:99-102.
100. Tomilin NV, Bozhkov VM. Human nuclear protein interacting with a conservative sequence motif of Alu-family DNA repeats. FEBS Lett 1989; 251:79-83.
101. Chesnokov IN, Schmid CW. Specific Alu binding protein from human sperm chromatin prevents DNA methylation. J Biol Chem 1995; 270:18539-18542.
102. Howe JG, Shu M-D. Epstein-Barr virus small RNA (EBER) genes: unique transcription units that combine RNA polymerase II and III promoter elements. Cell 1989; 57:825-834.
103. Sorensen PD, Frederiksen S. Characterization of human 5S rRNA genes. Nucleic Acids Res 1991; 19:4147-4151.
104. Lescure A, Tebb G, Mattaj IW et al. A factor with Sp1 DNA-binding specificity stimulates *Xenopus* U6 snRNA in vivo transcription by RNA polymerase III. J Mol Biol 1992; 228:387-394.
105. Das G, Hinkley CS, Herr W. Basal promoter elements as a selective determinant of transcriptional activator function. Nature 1995; 374:657-660.
106. Kingsley C, Winoto A. Cloning of GT box-binding proteins: a novel Sp1 multigene family regulating T-cell receptor gene expression. Mol Cell Biol 1992; 12:4251-4261.
107. Bredow S, Surig D, Muller J et al. Activating-transcription-factor (ATF) regulates human 7S L RNA transcription by RNA polymerase III in vivo and in vitro. Nucleic Acids Res 1990; 18:6779-6784.
108. Brindle PK, Montminy MR. The CREB family of transcription activators. Curr Opin Gen Dev 1992; 2:199-204.
109. Piras G, Dittmer J, Radonovich MF et al. Human T-cell leukemia virus type I Tax protein transactivates RNA polymerase III promoter in vitro and in vivo. J Biol Chem 1996; 271:20501-20506.
110. Li J-M, Parsons RA, Marzluff WF. Transcription of the sea urchin U6 gene in vitro requires a TATA-like box, a proximal sequence element, and sea urchin USF, which binds an essential E box. Mol Cell Biol 1994; 14:2191-2200.
111. Chesnokov I, Schmid CW. Flanking sequences of an Alu source stimulate transcription in vitro by interacting with sequence-specific transcription factors. J Mol Evol 1996; 42:30-36.
112. Kurose K, Hata K, Hattori M et al. RNA polymerase III dependence of the human L1 promoter and possible participation of the RNA polymerase II factor YY1 in the RNA polymerase III transcription system. Nucleic Acids Res 1995; 23:3704-3709.
113. Roberts S, Colbert T, Hahn S. TFIIIC determines RNA polymerase III specificity at the TATA-containing yeast U6 promoter. Genes Dev 1995; 9:832-842.
114. Herschbach BM, Johnson AD. The yeast α2 protein can repress transcription by RNA polymerases I and II but not III. Mol Cell Biol 1993; 13:4029-4038.
115. Schreck R, Carey MF, Grummt I. Transcriptional enhancement by upstream activators is brought about by different molecular mechanisms for class I and II RNA polymerase genes. EMBO J 1989; 8:3011-3017.

116. Berger SL, Folk WR. Differential activation of RNA polymerase III-transcribed genes by the polyomavirus enhancer and the adenovirus E1A gene products. Nucleic Acids Res 1985; 13:1413-1428.
117. Leveillard T, Kassavetis GA, Geiduschek EP. Repression and redirection of *Saccharomyces cerevisiae* tRNA synthesis from upstream of the transcriptional start site. J Biol Chem 1993; 268:3594-3603.
118. Gill G. Complexes with a common core. Current Biology 1992; 2:565-567.
119. Pugh BF, Tjian R. Diverse transcriptional functions of the multisubunit eukaryotic TFIID complex. J Biol Chem 1992; 267:679-682.
120. Sharp PA. TATA-binding protein is a classless factor. Cell 1992; 68:819-821.
121. Hernandez N. TBP, a universal eukaryotic transcription factor? Genes Dev 1993; 7:1291-1308.
122. Struhl K. Duality of TBP, the universal transcription factor. Science 1994; 263:1103-1104.
123. Tansey WP, Herr W. TAFs: guilt by association? Cell 1997; 88:729-732.
124. Goodrich JA, Hoey T, Thut CJ et al. Drosophila $TAF_{II}40$ interacts with both a VP16 activation domain and the basal transcription factor TFIIB. Cell 1993; 75:519-530.
125. Hoey T, Weinzierl ROJ, Gill G et al. Molecular cloning and functional analysis of *Drosophila* TAF110 reveal properties expected of coactivators. Cell 1993; 72:247-260.
126. Weinzierl ROJ, Dynlacht BD, Tjian R. Largest subunit of *Drosophila* transcription factor IID directs assembly of a complex containing TBP and a coactivator. Nature 1993; 362:511-517.
127. Chao DM, Young RA. Activation without a vital ingredient. Nature 1996; 383:119-120.
128. Pines J. Cyclins and cyclin-dependent kinases: take your partners. Trends Biochem Sci 1993; 18:195-197.
129. Morgan DO. Principles of CDK regulation. Nature 1995; 374:131-134.
130. Pines J. Cyclins, CDKs and cancer. Seminars in Cancer Biology 1995; 6:63-72.
131. Gottesfeld JM, Wolf VJ, Dang T et al. Mitotic repression of RNA polymerase III transcription in vitro mediated by phosphorylation of a TFIIIB component. Science 1994; 263:81-84.
132. Wolf VJ, Dang T, Hartl P et al. Role of maturation-promoting factor ($p34^{cdc2}$-cyclin B) in differential expression of the *Xenopus* oocyte and somatic-type 5S RNA genes. Mol Cell Biol 1994; 14:4704-4711.
133. Millar JBA, Russell P. The cdc25 M-phase inducer: an unconventional protein phosphatase. Cell 1992; 68:407-410.
134. Reynolds WF. The tyrosine phosphatase cdc25 selectively inhibits transcription of the *Xenopus* oocyte-type tRNAtyrC gene. Nucleic Acids Res 1993; 21:4372-4377.
135. Pepperkok R, Lorenz P, Ansorge W et al. Casein kinase II is required for transition of G_0/G_1, early G_1, and G_1/S phases of the cell cycle. J Biol Chem 1994; 269:6986-6991.
136. Hockman DJ, Schultz MC. Casein kinase II is required for efficient transcription by RNA polymerase III. Mol Cell Biol 1996; 16:892-898.
137. Fan H, Sakulich AL, Goodier JL et al. Phosphorylation of the human La antigen on serine 366 can regulate recycling of RNA polymerase III transcription complexes. Cell 1997; 88:707-715.
138. Maraia RJ. Transcription termination factor La is also an initiation factor for RNA polymerase III. Proc Natl Acad Sci USA 1996; 93:3383-3387.
139. Yoo CJ, Wolin SL. La proteins from Drosophila melanogaster and Saccharomyces cerevisiae: a yeast homolog of the La autoantigen is dispensable for growth. Mol Cell Biol 1994; 14:5412-5424.

140. James CBL, Carter TH. Activation of protein kinase C inhibits adenovirus VA gene transcription in vitro. J Gen Virol 1992; 73:3133-3139.
141. Wang H-D, Yuh C-H, Dang CV et al. The hepatitis B virus X protein increases the cellular level of TATA-binding protein, which mediates transactivation of RNA polymerase III genes. Mol Cell Biol 1995; 15:6720-6728.
142. Van Zyl W, Huang W, Sneddon AA et al. Inactivation of the protein phosphatase 2A regulatory subunit A results in morphological and transcriptional defects in *Saccharomyces cerevisiae*. Mol Cell Biol 1992; 12:4946-4959.
143. Cohen PTW. Novel protein serine/threonine phosphatases: variety is the spice of life. Trends Biochem Sci 1997; 22:245-251.
144. McLees A, White RJ. Unpublished observations.

CHAPTER 9

Regulation of RNA Polymerase III Transcription

Developmental Regulation

Regulation During Xenopus *Embryogenesis*

There are two families of active 5S genes in *Xenopus laevis*. One consists of the somatic 5S genes, of which there are 400 copies per haploid genome, organized in a single cluster.[1,2] The other is divided into two classes, the 20,000 major oocyte and the 1,300 trace oocyte 5S genes.[3,4] There are only 6 nucleotides different between the 120 bp coding regions of the somatic and major oocyte types, but the flanking sequences are completely divergent.[1,3,4]

Each of the ribosomal RNAs is synthesized at greatly elevated rates in oocytes, but this is achieved by differing mechanisms. The genes encoding 18S and 28S rRNAs are amplified specifically in oocytes.[5,6] In contrast, the large auxiliary oocyte 5S rRNA gene family is present in all cell types but is active only in oocytes.[7-9] Both somatic and oocyte 5S genes are actively expressed during oogenesis, resulting in a massive accumulation of 5S rRNA for subsequent incorporation into ribosomes.[10] Following fertilization and meiosis there is a generalized repression of transcription that continues through the first 12 cleavage divisions of the fertilized egg, until the mid-blastula transition (MBT). This repression appears to result primarily from a large excess of chromatin proteins.[11-13] The oocyte accumulates sufficient core histones to assemble 13,000-16,000 nuclei.[14] These gain access to the chromosomes when the nucleus (germinal vesicle) breaks down during oocyte maturation. Titrating away the excess of histones by injecting nonspecific DNA at a concentration comparable to the normal DNA mass present at the MBT is sufficient to activate tRNA expression precociously.[11,13] It therefore seems that the class III transcription apparatus remains functional prior to MBT, but cannot gain access to appropriate templates. When transcription resumes at the MBT, approximately equal amounts of somatic and oocyte 5S RNA are produced, representing a 50-fold transcriptional preference for somatic over oocyte 5S genes.[10,15] Two or three cell divisions later, the final state of differential gene expression is established as the oocyte 5S genes are inactivated and the transcriptional preference reaches 1000.[10,15,16] This process is not irreversible, since the oocyte 5S genes can be reactivated if somatic cell nuclei are transferred into oocytes.[16]

RNA Polymerase III Transcription, Second Edition, by Robert J. White.

Considerable effort has been invested in trying to understand the differential regulation of the two 5S gene families.[17-20] The system provides an important paradigm for what may be a common developmental mechanism, where two or more gene families have similar but not identical cis-acting sequences recognized by the same transcription factors, but are nonetheless controlled differentially.

Following the discovery that the trace oocyte 5S genes have a 4-fold lower competitive strength towards TFIIIA binding than the somatic 5S genes,[21,22] models were proposed for the differential regulation of these genes based on the premise that TFIIIA binding is the critical factor in determining transcriptional activity.[18,23-25] The rationale for these ideas was to some extent compromised by the subsequent finding that TFIIIA binds to the major oocyte and somatic 5S genes with a similar affinity.[26-28] Nevertheless, the models were supported by the developmental profile of TFIIIA levels.[25,27,29,30] Shastry et al[29] showed that the *Xenopus* egg has 70,000 TFIIIA molecules per 5S gene and that this absolute amount remains constant following fertilization. However, the cellular TFIIIA level decreases dramatically during embryogenesis as cells divide rapidly, until by the gastrula stage, 2-3 cell divisions after the MBT, there is only one TFIIIA molecule per 5S gene.[29] As development continues, the ratio declines further to the value of 0.2 that is seen in adult cells.[29] Therefore, as the concentration of TFIIIA drops, a higher affinity for the somatic 5S genes may contribute to the exclusion of the oocytic genes. In this simple model of gene discrimination, the concentration of TFIIIA relative to the number of genes available and their binding affinities are the primary determinants of developmental control. High TFIIIA levels during oogenesis would allow expression of all types of 5S genes; lower concentrations in gastrulating embryos would result in the progressive inactivation of those genes which bind TFIIIA more weakly.

The model predicts that raising the concentration of TFIIIA at the blastula stage will retard the inactivation of oocyte 5S genes. This prediction was tested directly by Brown and Schlissel,[24] who injected purified TFIIIA protein into embryos. They were obliged to use syncytial embryos in order to overcome the toxic effects of injecting large amounts of protein and were restricted by the failure of such embryos to develop much beyond the blastula stage.[24] Nevertheless, they found that raising TFIIIA to 500 times its normal level causes a dramatic increase in 5S rRNA at the MBT, and that this is primarily due to activation of the oocyte 5S genes,[24] in keeping with the model. The effect was seen even if DNA replication was inhibited.[24] They concluded that TFIIIA is limiting for 5S transcription during embryogenesis, and that this limitation contributes to the inactivation of 5S genes.[24]

Andrews and Brown[25] avoided the problem of toxicity by injecting TFIIIA mRNA, which allowed embryos to develop normally. This resulted in a 15-fold increase in the level of TFIIIA protein that persisted into the swimming tadpole stage.[25] Chromatin templates isolated from TFIIIA-enhanced gastrulae synthesized 20-fold more 5S rRNA than chromatin from control embryos at the same stage.[25] Much of this increase was due to transcription of oocyte 5S genes.[25] This shows that TFIIIA is limiting for 5S rRNA synthesis in vivo just after the MBT.[25] However, this elevation fell towards the late gastrula stage, and by neurulation and beyond 5S expression occurred at about the same level with chromatin from injected or control embryos.[25] Embryos with high TFIIIA were found to inactivate oocyte 5S genes and impose the 1000:1 somatic:oocyte (S:O) 5S gene transcriptional ratio in the same time frame as uninjected embryos.[25] This effect was independent of DNA replication.[25] Assuming that the elevated TFIIIA remained functional, then either a different factor may become limiting or the oocyte 5S genes may become inaccessible to even high levels of transcription factors, perhaps due to changes in chromatin structure. tRNA and satellite 1 transcription from chromatin decreases at the same time as that of oocyte 5S genes.[25,31,32] This would suggest that one of the general class III factors becomes limiting. In cleavage-arrested embryos with a reduced DNA content, the period of satellite I expression is prolonged, consistent with the idea that inactivation is due to an excess of templates competing for a limiting factor.[32]

The "replication-expression" model suggested that somatic 5S genes replicate earlier in the cell cycle than do oocyte 5S genes and therefore have a competitive advantage in binding limiting amounts of TFIIIA.[23] Somatic 5S genes replicate early and oocyte 5S genes late in S phase in a *Xenopus* cell line expressing only somatic 5S rRNA.[33-35] It is difficult to establish cause and effect in this case. However, in one kidney cell line a block of oocyte 5S genes is translocated from the heterochromatic and late-replicating telomeres to a site adjacent to the nucleolar organizer; 10-20% of 5S rRNA in this line is of the oocytic type and some oocyte 5S genes are replicated early,[35] although it remains to be shown that these are the ones that are transcribed.

Wolffe and Brown[36] proposed that selective inactivation of oocyte 5S genes involves the preferential destabilization of their transcription complexes in somatic cells. These workers demonstrated a differential stability using extracts from activated eggs.[36] In an oocyte nuclear extract, somatic and oocyte 5S genes are transcribed at approximately equal rates and both form stable transcription complexes.[36] However, the addition of an activated egg extract leads to the rapid inactivation of oocyte 5S genes and generates an S:O ratio of 50.[36] This inactivation correlates with a rapid destabilization of the oocyte 5S transcription complexes, as assayed either by footprinting or by template challenge.[36] Whereas in the oocyte nuclear extract the S:O ratio is insensitive to TFIIIA concentration, in the activated egg extract it is very sensitive: addition of excess TFIIIA reduces the ratio from 50 to 5, whereas immunodepletion of TFIIIA raises it to 400; this is due to variation in the rate of transcription of the oocyte 5S genes.[36] Thus, transcription directed by unstable complexes is especially dependent upon the concentration of the limiting factor, because factors in such complexes are in equilibrium with free factors.

The involvement of factors other than TFIIIA in the differential regulation of somatic and oocyte 5S genes is suggested by the fact that sequences outside the TFIIIA binding site have been clearly shown to influence the relative levels of expression of these two gene types.[37-39] In a reconstituted system, the S:O ratio can be exaggerated by lowering the concentration of limiting amounts of TFIIIB, and this has been shown to reflect sequence differences between -32 and +37.[39] The difference in complex stability between the somatic and major oocyte 5S genes in the presence of an activated egg extract is due to three changes in nucleotides in the upstream part of the ICR that contribute little to TFIIIA binding.[36] As a result of these differences, TFIIIC has a 5-fold higher affinity for the TFIIIA/somatic 5S complex than it does for the TFIIIA/major oocyte 5S complex.[40] Whole oocyte extracts (S150s) give the same S:O ratio as occurs at MBT, viz 50.[41] Since an oocyte nuclear extract gives an S:O characteristic of oogenesis, viz 5, this may reflect the storage of regulatory substances in the oocyte cytoplasm for use in later development. In oocyte S150s the S:O ratio is dependent upon TFIIIC and addition of a TFIIIC fraction lowers the S:O ratio.[42,43] Unlike the case in nuclear extracts, TFIIIA is in relative excess in oocyte S150s, but again, complexes on oocyte 5S genes are selectively destabilized.[42] The limiting TFIIIC stabilizes the binding of TFIIIA to a somatic 5S gene much more efficiently than to an oocyte 5S gene.[42,43] Therefore under conditions of preferential destabilization of the oocyte 5S complex the limitation of any required transcription factor may exaggerate the S:O differential in favor of the somatic 5S genes. Such a combinatorial effect of separate factors on gene activity could provide a simple mechanism for amplifying individual differences in affinity.

Contrary to Wolffe,[42] Millstein et al[44] found that neither TFIIIA, TFIIIB, nor TFIIIC fractions could change the selective transcription of somatic over oocytic 5S genes in an oocyte S150. These workers preincubated a 50:1 mix of oocyte and somatic genes in an oocyte S150 and then injected it into live oocyte nuclei.[44] They found that only the somatic 5S genes were expressed.[44] Predominantly oocytic 5S rRNA was obtained after injecting the same mix of naked DNA.[44] The effect was not due to the loss of the oocytic genes, since deproteinization after preincubation in the S150 and prior to injection again gave mainly oocyte-type 5S rRNA.[44] A possible explanation is that the oocyte S150 assembles stable inactive complexes on oocyte 5S genes and stable active complexes on somatic 5S genes.[44] An alternative answer was provided by the observation that oocyte 5S genes take much longer to assemble a stable (preemptive) complex than somatic 5S genes when incubated in oocyte S150 extracts.[45] It is therefore plausible that the oocyte 5S templates had not formed stable complexes prior to injection. Although footprinting suggested that most oocyte 5S promoters are, like the somatic 5S genes, bound by TFIIIA in the oocyte S150,[44] stable complex formation requires the additional presence of TFIIIC.[46]

A candidate differential repressor was provided by Blanco et al,[27] who detected a 42 kD protein that is immunologically related to TFIIIA and has a similar but distinct peptide map. They observed that this protein binds to somatic and oocyte 5S genes with comparable affinities and gives footprints indistinguishable from those of TFIIIA.[27] However, it will only support transcription of the somatic 5S genes.[27] Since the abundance of this protein increases dramatically during oogenesis, the authors suggested that it serves as an activator of somatic 5S transcription and as a repressor of oocyte 5S transcription during early embryogenesis, contributing to the differential regulation of these gene families.[27] However, Brown[47] was unable to detect any activity of the 42 kD protein, either as an activator of somatic 5S genes or as a specific repressor of oocyte 5S genes. He suggested[47] that the results of Blanco et al[27] reflect the contamination of their protein fractions with TFIIIA, a possibility conceded by the authors.

Although raising the TFIIIA level following the MBT causes a transient increase in the relative expression of oocyte 5S genes, this increase is soon replaced by the situation found in adult somatic cells, in which somatic 5S genes exist in active stable transcription complexes and oocyte 5S genes are repressed.[24,25] The packaging of DNA into chromatin is likely to play a major role in the differential expression of the 5S gene families in the somatic cells of *Xenopus*.[16,19,23,48] The number of 5S transcription complexes that are accessible to pol III in isolated chromatin declines rapidly during the stages in which the oocyte 5S genes become suppressed.[31] This is also true of satellite I genes and of certain tRNA genes.[31] In contrast, the number or accessibility of transcription complexes on somatic 5S genes shows a slight increase during this period.[31] Clear differences have been detected between the chromatin structures of somatic and oocyte 5S genes in somatic cells.[49,50]

Korn and Gurdon[16] found that oocyte 5S genes can remain repressed following the microinjection of intact somatic nuclei into certain types of mature oocytes unless the nuclei are pretreated with high salt concentrations. Whereas both the oocyte and somatic 5S genes are actively transcribed from naked DNA, only the latter are transcribed following the addition of pol III to chromatin or nuclei from somatic cells.[15,23,29,48,51,52] If, however, histone H1 is removed from such chromatin by salt washes or ion-exchange chromatography, then the oocyte 5S genes become accessible for transcription.[23,48,50,51] This requires the addition of all the transcription factors, showing that TFIIIA and the other factors are not associated with oocyte 5S genes in the repressed chromatin.[51] Purified histone H1, when added back at levels found in vivo to chromatin that had been derepressed by H1-removal, restored repression of the oocyte 5S genes without affecting the somatic genes.[51] The other histones could also repress oocyte 5S genes, but only at higher concentrations.[51] However, preincubation of derepressed chromatin with an oocyte nuclear extract in order to form transcription complexes on oocyte 5S genes prior to the re-addition of histone H1 resulted in both types of 5S gene retaining their activity after H1 addition; thus, the developmental state of somatic cell chromatin was transformed to that of oocyte chromatin.[51] Although histone H1 plays a cardinal role in the compaction of chromatin into higher order structures, the level of chromatin structure that is responsible for the silencing of oocyte 5S genes in somatic cells can occur within fragments the size of individual genes.[53]

Wolffe and Brown[19] described a model for developmental regulation based on the progressive limitation of transcription factors during embryogenesis coupled with the difference in the stability of complexes on somatic and oocyte 5S genes which they had demonstrated in the oocyte S150 and activated egg extracts. According to this model,[19] when an oocyte 5S gene becomes unoccupied of factors due to their unstable interaction, a repressed chromatin structure forms that can exclude factor binding. This repressive structure is opportunistic and depends on the interaction of histone H1 with nucleosomes. Compaction of H1-containing nucleosomes produces more inaccessible and repressed structures in which H1 is much less likely to exchange. Active transcription complexes interrupt the repetitive positioning of nucleosomes that is necessary for compacted structures. The stability of repression dictates the factor concentration required for reactivation. Since transcription factors remain associated with somatic 5S genes, these genes resist the formation of repressed chromatin structures throughout embryogenesis. As the cell cycle lengthens after MBT, regions of interphase chromatin may become more compacted, much less accessible to transcription factors and less available for H1 exchange. However, competition to activate or repress a gene must occur each time it is replicated, because replication forks remove even stable transcription complexes, making genes susceptible to either activation by transcription factors or to compaction into repressed chromatin.[19]

Wolffe[54] found that chromatin isolated at different embryonic stages maintains its developmental programming. Transcription complexes are absent from 5S genes in chromatin from morulae, but can be formed by the addition of an oocyte nuclear extract.[54] The important novel observation was that addition of purified somatic histone H1 to chromatin treated in this way produces a selective inactivation of oocyte 5S genes, and also of the developmentally regulated satellite I gene family, while somatic 5S and tRNA genes are relatively unaffected.[54] This observation contrasts with that of Schlissel and Brown,[51] who found that complexes formed on oocyte 5S genes in tissue culture cell chromatin are resistant to histone H1-mediated repression. Linker histone synthesis is developmentally regulated in *Xenopus*; early stages have an embryonic variant (B4 or H1M) and the somatic H1 forms only accumulate after the MBT.[54-56] Interaction of histone B4 with nucleosomes elicits almost identical structural transitions to those produced by histone H1.[57,58] However, B4 is much more acidic than H1 in the C-terminal tail.[56,57] This may weaken its interaction with DNA and decrease its ability to compete effectively with transcription factors. Early embryonic chromatin is also enriched in HMG1.[57] Although HMG1 and histone B4 can both repress 5S transcription, neither is as efficient in this regard as somatic H1.[58] This difference correlates with their relative affinities for nucleosomes assembled on 5S genes; thus, H1 (Kd, 7.4 nM) binds 6-fold more tightly than B4 (Kd, 45 nM) and 40-fold more tightly than HMG1 (Kd, 300 nM).[58] Since the accumulation of somatic histone H1 correlates with the decreased accessibility and transcription of oocyte 5S genes,[55] Wolffe[54,55] proposed that the presence of somatic H1 in chromatin may be important for the repression of oocyte 5S and satellite I genes: early embryonic chromatin is deficient in somatic H1, is less compacted and expresses these genes when excess transcription factors are provided, whereas gastrula chromatin has significant amounts of somatic H1, is more compacted and contains these genes in a largely repressed state.[54,55] Thus, the appearance of somatic histone H1 in embryonic chromatin coincides with the establishment of repression on oocyte 5S genes and satellite I genes, leading Wolffe[54,55] to suggest that H1-dependent changes in chromatin structure may have a dominant role in class III gene regulation during *Xenopus* development.

This idea was corroborated by two elegant studies that manipulated the levels of H1 in vivo.[59,60] Overexpression of histone H1 from microinjected mRNA repressed transcription of oocyte 5S genes without affecting the expression of somatic 5S, tRNA, U1 or U2 genes.[59] This was the case even if TFIIIA was overexpressed in parallel.[59] When H1 was specifically depleted using a hammerhead ribozyme, transcription of oocyte 5S genes was selectively raised.[59,60] The effect of H1 depletion increased as gastrulation progressed, in parallel with the transition from the cleavage linker histone B4 to the adult H1 forms.[59] The response of oocyte 5S promoters to raising the level of TFIIIA was increased in H1-depleted embryos.[59] Although these results prove the involvement of H1 in controlling 5S template activity, a substantial S:O differential was nevertheless maintained in gastrulating embryos following H1 depletion.[59] This reinforces the idea that multiple factors are likely to be important for the developmental regulation of 5S genes.

The question remains as to what feature of oocyte 5S and satellite I genes allows them to be repressed while somatic 5S and tRNA genes remain active. The propensity for being repressed by somatic histone H1 does not correlate with a simple inability to bind transcription factors stably, as measured in template competition assays, since oocyte 5S genes can form stable preemptive transcription complexes in the oocyte nuclear extracts used for these experiments.[36,54] Satellite I genes too can form preemptive transcription complexes and have equivalent competitive strength to tRNA genes in an oocyte nuclear extract.[54,61,62] In an egg extract, satellite I DNA competes as efficiently as somatic 5S DNA for common factors and, once assembled, transcription complexes have comparable stabilities to template challenge on these two types of genes.[62] Indeed, complexes formed on satellite I DNA are much more stable to repression by core histones in an egg extract than are complexes formed on somatic 5S genes.[62,63] However, the normal somatic form of histone H1 is deficient in *Xenopus* egg and oocyte extracts[55,62] and satellite I and oocyte 5S genes are clearly less stable to this form of histone H1.[54,59] The dominant repression of oocyte 5S genes and satellite I genes by somatic histone H1 can occur in the presence of excess transcription factors and requires some feature of chromatin structure since it does not occur with naked genomic DNA.[54] Although the H1 response is sensitive to chromatin structure, this must be on a local scale since fragments smaller than 600 bp maintain differential repression.[53] One candidate feature is the spacer sequence, which has an A/T content of 76% for major oocyte 5S genes, 60% for minor oocyte 5S genes, and only 43% for somatic 5S genes.[16] A/T-richness strongly promotes histone H1 binding; for example, DNA fragments that are 57% A/T bind more than 10 times as much H1 as DNA of equal size that is 50% A/T.[64] This sequence preference may favor H1 binding to oocyte 5S genes and contribute to the selectivity of repression during embryogenesis. Indeed, the flanking sequences have been found to confer differential inhibition of the major oocyte and somatic 5S genes by histone H1 in vitro.[65] The preferential association of histone H1 with the oocyte 5S repeat in the linker regions between nucleosomes may therefore facilitate the differential repression that occurs during embryogenesis.[54,66]

The developmental regulation of 5S genes in *Xenopus* appears a relatively simple problem. Yet considerable effort has been required in order to explore the many facets of this system. A variety of mechanisms for controlling 5S transcription have been described, some of which are illustrated in Figure 9.1. None of these is sufficient individually to account for the overall pattern of expression. It is likely, however, that each model has provided some aspect of the answer, and that a multiplicity of considerations contribute to the biological phenomenon. The more rapid and stable interaction of transcription factors with somatic 5S genes, together with decreasing factor abundance and the increasing amounts of somatic histone H1, which binds preferentially to oocyte 5S genes, may be sufficient to account for the 50-fold transcriptional preference observed soon after the MBT. More strongly repressing chromatin structures and the prior replication of somatic 5S genes when transcription factors are severely limiting may then combine with these considerations to establish the S:O ratio of 1000 that is observed in somatic cells.

Regulation During the Maturation of Xenopus *Oocytes*

Satellite I genes and a number of multicopy oocytic tRNA genes, such as $tRNA^{TyrC}$ and $tRNA^{Met1}$, are expressed during the early stages of *Xenopus* oogenesis but become inactivated as the oocytes approach maturity.[67] The $tRNA^{TyrC}$, $tRNA^{Met1}$ and satellite I genes are actively transcribed in extracts of immature oocytes but not in extracts of mature oocytes.[68,69] Reynolds and Johnson[68] attributed this to a change in TFIIIC that occurs during oocyte maturation. A PC-C fraction from mature oocytes is unable to activate the $tRNA^{Met1}$ and satellite I genes whereas the corresponding fraction from immature oocytes functions 50- to 100-fold more efficiently.[68] In contrast, PC-B fractions from mature and immature oocytes display comparable activities.[68] The levels of reconstituted 5S transcription are not affected by the maturity of the oocytes from which the PC-C is derived.[68]

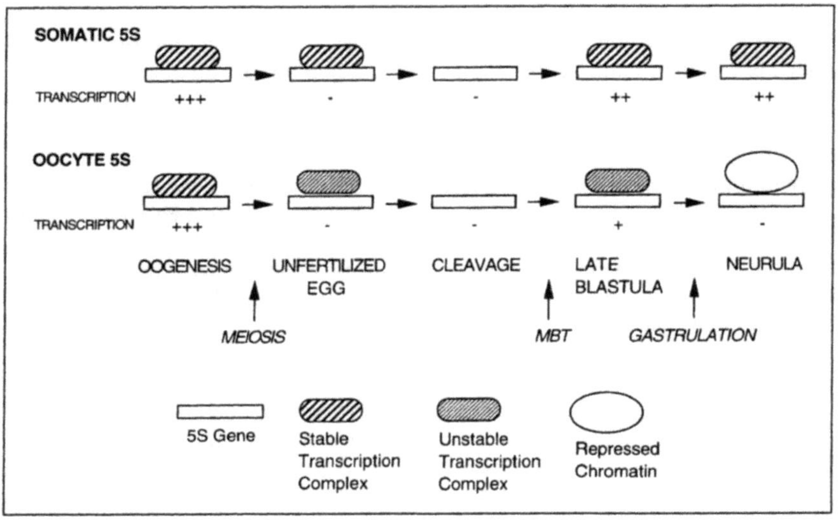

Fig. 9.1. Model illustrating some aspects of the developmental regulation of 5S genes in *Xenopus*. In immature oocytes both types of 5S gene are associated with stable transcription complexes and are actively transcribed (S:O ratio of 4). Endogenous genes are repressed in eggs following meiosis. Transcription factors are still abundant in unfertilized eggs, but their function is impeded by the presence of unidentified inhibitors. High concentrations of DNA can overcome this inhibition and allow 5S transcription. Under these conditions the oocyte 5S genes are transcribed much less efficiently than the somatic 5S genes (S:O ratio of 50). This differential is caused by the relative instability of transcription complexes on oocyte 5S genes. Following fertilization, the egg undergoes twelve rapid cycles of cell division. During this period endogenous genes are silent, probably because of the presence of a large excess of inhibitory chromatin proteins. When transcription resumes at the MBT, oocyte 5S genes are expressed much less efficiently than somatic 5S genes (S:O ratio of 50). Several considerations may contribute to this differential expression. One is the decreasing level of TFIIIA. Another is the relative instability of transcription complexes on oocyte 5S genes. A third is that increasing numbers of oocyte 5S genes become assembled into histone H1-containing chromatin. By early neurulation the repression of oocyte 5S genes is nearly complete (S:O ratio of 1000). Transcription factors other than TFIIIA have become rate-limiting, despite the continued decline in TFIIIA levels. The majority of oocyte 5S genes are assembled into repressive chromatin structures. These genes are replicated late in S phase relative to the active somatic 5S genes.

The $tRNA^{Met1}$ and satellite I genes can form stable transcription complexes in the presence of PC-C from immature oocytes or HeLa cells but cannot form stable complexes with a mature oocyte PC-C.[68] The activity responsible for the differential expression of $tRNA^{Met1}$ and satellite I genes copurifies with TFIIIC during gradient elution of phosphocellulose and during B-block affinity chromatography.[68] Thus, oocyte maturation appears to be accompanied by a change in TFIIIC which reduces its ability to form a stable complex with certain promoters, whereas other promoters are unaffected.

Mixtures of immature and mature oocyte extracts display the selective repression of oocytic tRNA genes that characterizes mature extracts.[69] Preincubation with the tyrosine phosphatase inhibitor sodium vanadate prevents inactivation of $tRNA^{TyrC}$.[69] The effect of sodium vanadate is only observed in the mixed extracts and does not occur with mature or immature extract alone.[69] These observations suggest that during oocyte maturation a tyrosine phosphatase activity emerges which modifies a factor that only supports $tRNA^{TyrC}$ expression in immature oocytes. TFIIIC was again implicated, since sodium vanadate enhances the ability of TFIIIC from immature extracts to stimulate $tRNA^{TyrC}$ transcription in mature extracts.[69] The activity of the tyrosine phosphatase cdc25 increases substantially during oocyte maturation. When added to extracts of immature oocytes, recombinant cdc25 represses $tRNA^{TyrC}$ without affecting expression of the somatic $tRNA^{TyrD}$ gene.[69] This effect is blocked by sodium vanadate.[69] Repression of $tRNA^{TyrC}$ by cdc25 can be relieved by TFIIIC but not TFIIIB fractions.[69] It remains to be determined whether the effect of cdc25 on the class III transcription apparatus is direct. Since the differential expression of $tRNA^{TyrC}$ and $tRNA^{TyrD}$ is attributable to sequences in the 5'-flanking regions,[70] Reynolds[69] proposed that during oocyte maturation cdc25 influences TFIIIC so that it loses the ability to recruit TFIIIB into a productive complex on the $tRNA^{TyrC}$ promoter.

Regulation During Early Mouse Development

Significant amounts of B1 and B2 transcripts are present in the oocytes and unfertilized eggs of mice.[71-74] Following fertilization, the steady-state levels of B1 and B2 RNA increase several-fold,[72,73] such that early blastocysts each contain about two million copies of both B1 and B2 transcripts, representing 5-6% of the total amount of $poly(A)^+$ RNA present at this stage.[72] Injection of VA genes into single cell embryos has demonstrated that the pol III machinery is active even prior to cell cleavage.[75] The majority of B1 and B2 transcription during early stages is likely to be by pol III, since the transcripts are generally short and discrete,[72,73] unlike the heterogeneous B2 RNA that is produced by pol II readthrough.[76] The relative abundance of small B2 transcripts subsequently declines, until by 12 days postcoitum (dpc) the level is similar to that found in adult brain.[73] In contrast, the amount of B2 sequence-containing heterogeneous high molecular weight RNA rises significantly during this period, reaching maximal levels at 9.5 to 11 dpc, a time when the embryo is growing rapidly and organogenesis is beginning.[73,77]

A detailed analysis of the spatial distribution of B2 transcripts during early stages of development has been provided by in situ hybridization.[78] B2 transcripts are detected in unfertilized eggs and rise in abundance following fertilization.[78] Increased amounts of B2 RNA in the pronuclei and polar bodies are consistent with de novo transcription at this very early stage.[78] The blastomeres of two- and four-cell morulae and of eight-cell blastocysts are heavily labeled.[78] Hybridization decreases in the trophectoderm cells of expanding blastocysts while the inner cell mass remains heavily labeled.[78] By the late primitive streak stage (7.5 dpc) hybridization is restricted to the ectoderm and mesoderm, whereas the embryonic and extra-embryonic endoderm is negative. This approach does not distinguish between the products of pols II and III. However, the B2 sense strand is the most abundant at all stages studied.[78] B2 sequences are present in both orientations in approximately equal amounts in nuclear RNA transcribed by pol II,[79,80] consistent with their random integration into transcription units. Therefore the high levels of sense relative to antisense B2 sequences that are detected in nuclei of early embryonic stages by in situ hybridization can be ascribed to promoter-directed transcription by pol III. This conclusion is supported by the short and discrete nature of the major B2 species detected on northern blots of RNA from the egg to the blastocyst stages.[73]

The strong developmental regulation of B2 expression makes this an interesting model system for investigating the mechanisms of gene regulation during early mouse embryogenesis. However, its small size and the variety of cell types present preclude the use of the early mouse embryo as a source of the relatively large quantities of pure cell populations required for biochemical studies. An alternative approach involves the use of embryonal carcinoma (EC) and embryonic stem (ES) cell culture systems which mimic the differentiation events that occur within the developing embryo. An example is provided by the F9 EC cell line, that was derived from a teratocarcinoma produced by grafting a 6 dpc male embryo to a host testis.[81] Suspension culture of F9 cells in the presence of retinoic acid results in the formation of embryoid bodies, the surface cells of which resemble visceral endoderm (VE) by both morphological and biochemical criteria.[82] In contrast, treatment of F9 monolayers with retinoic acid plus cAMP generates cells resembling parietal endoderm (PE).[83]

High levels of small B2 RNA are expressed in a variety of EC cell lines, including PCC4, PC13 and F9.[77,84-87] In contrast, very little small B2 RNA is present in PE and VE cells derived from EC cells[77,86] or in the trophoblastoma line TDM-1.[85] The level of these transcripts decreases dramatically following the differentiation of F9 EC cells to VE-like or PE-like cells.[77,86] This is not a secondary response to a slowing of growth rate, since F9 EC cells made quiescent by serum deprivation continue to express substantially more small B2 RNA than do F9 PE cells.[86] Differentiation of PC13 EC cells to a VE phenotype is also accompanied by loss of low molecular weight B2 RNA.[84] However, differentiation of the ES cell lines EKB2B2 and CCE to give a heterogeneous population containing both pluripotential and endoderm-like cell types does not result in the disappearance of small B2 RNA.[77,86] These EC and ES cell culture systems therefore accurately mimic the patterns of B2 gene expression observed in vivo.

F9 cell differentiation into parietal endoderm has been used as a model system in which to investigate the developmental regulation of B2 transcription.[86] Nuclear run-on experiments confirmed that the decline in B2 transcript levels that accompanies the differentiation of F9 EC into PE cells is due to a decrease in pol III transcription.[86] Transcription of B1 and tRNA genes by pol III is also downregulated under these conditions, but pol II transcription does not show a similar response.[86] Although tRNA transcription decreases 9-fold as F9 EC cells differentiate, the steady-state level of tRNA transcripts does not change.[86] This implies that the reduction in transcription is compensated for by a similar fall in turnover rate. Other situations have also been reported in which the abundance of tRNA molecules does not reflect changes in the rate of tRNA transcription.[88,89] Steady-state levels of U6 snRNA and 5S rRNA decrease 2- to 4-fold following F9 cell differentiation.[90]

Whole cell extracts prepared from F9 EC and PE cells accurately mimic the transcriptional status of the cells from which they derive, with active pol II transcription in both types of extracts and active pol III transcription in F9 EC but not PE extracts.[86] Mixing experiments indicated that the inactivity of F9 extracts for

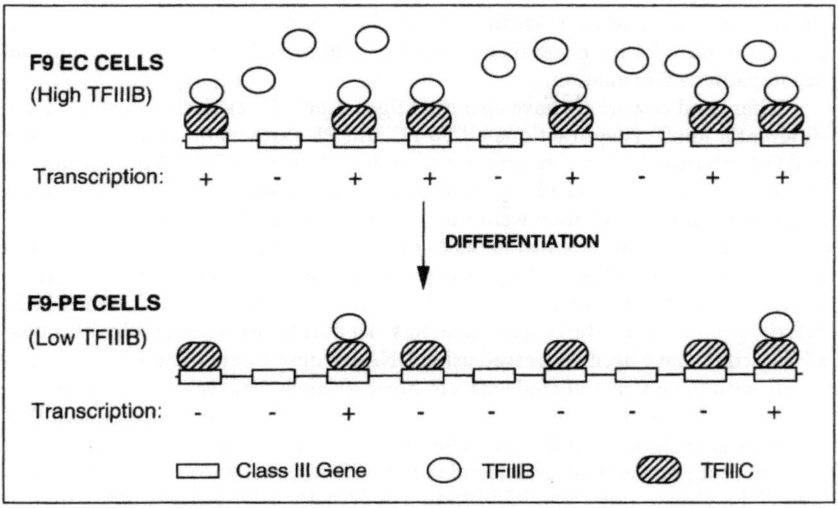

Fig. 9.2. Model for class III gene regulation during the differentiation of F9 EC cells into PE cells. The decrease in pol III transcription is due to a specific decrease in TFIIIB activity, whereas TFIIIC activity does not diminish. Although TFIIIC is limiting in F9 EC cells, TFIIIB becomes severely limiting in differentiated PE cells.

pol III transcription is due to the lack of positive factors rather than the presence of an inhibitor.[86] Since all class III genes tested are affected, both in vivo and in vitro, this positive component is likely to be a general class III factor. Complementation experiments demonstrated that a PC-B fraction is sufficient to restore pol III transcription in PE extracts to levels occurring in EC extracts, whereas a PC-C fraction does not have this effect.[86] This implicated TFIIIB as the factor which is downregulated during differentiation of F9 cells, since none of the other known class III general factors are specific to PC-B. Direct measurement of the activities of TFIIIC and pol III enzyme showed similar levels of these components in EC and PE extracts.[86] Complementation assays demonstrated that TFIIIC activity is limiting for class III gene transcription in F9 EC extracts, whereas an activity specific to PC-B is rate-limiting in PE extracts.[86] On the basis of these data, White et al[86] proposed that TFIIIB activity decreases and becomes limiting for pol III transcription as F9 cells differentiate, resulting in the downregulation of class III gene expression (Fig. 9.2).

A recent analysis has extended this investigation.[91] Mono Q-purified TFIIIB has been found to be sufficient to raise transcription in PE extracts to levels occurring in EC extracts.[91] This confirms that TFIIIB is the only general pol III factor that is substantially deficient in differentiated F9 cells. TFIIIB has been partially purified from EC and PE cells and shown to be much more active in the former.[91] The overall level of TBP changes little when F9 cells differentiate.[91] However, significantly less TBP is found in TFIIIB fractions.[91] Furthermore, direct assays demonstrate that TFIIIB TAF activity decreases approximately 8-fold.[91] It is the TAF component of TFIIIB that is limiting for VA_I transcription in PE extracts.[91] Western blotting reveals that BRF is approximately 5-fold less abundant following differentiation.[91] This decrease in the level of an essential pol III TAF may be sufficient to account for the general downregulation of class III gene transcription that accompanies PE formation.

Seifart and coworkers have also investigated pol III regulation during F9 cell differentiation.[90] They fractionated F9 EC and PE extracts on phosphocellulose and Q-Sepharose.[90] Complementation experiments to assay TFIIIB produced variable results; in some cases PE extracts were found to have reduced TFIIIB activity, whereas in other cases they were not.[90] In contrast, a PC-C fraction from a PE extract was found to have lost all ability to support VA_I expression.[90] This result was surprising, since the starting PE extract was only 5-fold less active in VA_I transcription than the corresponding EC extract.[90] A likely explanation is that the TFIIIC activity present in the starting material was lost during chromatography. This possibility could have been addressed using DNA-binding assays or add-back experiments with the unfractionated extracts, however, such data were not presented.

Band-shift and footprinting assays with fractionated extracts suggested that PSE-binding activity also decreases following F9 cell differentiation.[90] Two other DNA-binding proteins of undetermined specificity were shown to remain constant.[90] Fractions containing PSE-binding activity from EC or HeLa extracts were found to stimulate U6 transcription when added to PE extract.[90] This may have been a nonspecific effect, since every fraction tested in this way increased U6 expression.[90]

Encystment of Acanthamoeba castellanii

When starved of essential nutrients, *Acanthamoeba castellanii* differentiates into a dormant cyst. This is accompanied by a rapid and almost coordinate decrease in the synthesis of large rRNA by pol I and 5S rRNA by pol III.[92] Extracts prepared from *Acanthamoeba* that have encysted for 10 hours transcribe a 5S gene with only 12% of the efficiency of extracts from undifferentiated trophozoites.[92] Whereas the activities of TFIIIB, TFIIIC and pol III are unchanged at this time, both the transcriptional and DNA-binding activities of TFIIIA are substantially diminished.[92] Pol III and the general class III factors are downregulated more gradually during differentiation.[92]

Tissue-Specific Regulation

Silk Glands

Differentiated cells that synthesize high proportions of particular proteins sometimes adapt their tRNA synthesis to suit their specialized translational requirements. For example, when reticulocytes differentiate the relative levels of individual tRNA species change so as to be optimal for high levels of globin production.[93] An extreme example of this behavior is provided by the terminally differentiated silk gland cells of the silkworm *Bombyx mori*, which synthesize little protein other than silk. The distribution of tRNAs in mature silk glands reflects the amino acid composition of fibroin, the major constituent of silk, which is extremely rich in glycine (46%) and alanine (26%). Sprague and coworkers have studied how the relative abundance of $tRNA^{Ala}$ is increased in silk glands.

Silkworms produce two types of $tRNA^{Ala}$: a constitutive type ($tRNA^{AlaC}$) that is found in all cells, and a second type ($tRNA^{AlaSG}$) that is restricted to silk glands.[94] These tRNAs have different primary sequences and each is encoded by a distinct set of genes, both of which are present in approximately 20 clustered copies per haploid genome.[95] In silk glands, transcription of the $tRNA^{AlaSG}$ genes adds to that of the $tRNA^{AlaC}$ genes in order to achieve high levels of $tRNA^{Ala}$. The transcriptional properties of these gene families in vitro are consistent with their pattern of expression in vivo. Thus, $tRNA^{AlaC}$ genes are transcribed efficiently under a variety of conditions, whereas $tRNA^{AlaSG}$ genes are actively transcribed only in concentrated extracts of silk glands.[96] This difference in transcriptional efficiency is due to sequences between -30 and the initiation site.[96] Single-round transcription assays demonstrated that fewer active initiation complexes assemble on the $tRNA^{AlaSG}$ genes, although once formed these complexes function efficiently.[97] The $tRNA^{AlaSG}$ promoter is 3-fold less capable than the $tRNA^{AlaC}$ promoter at competing for TFIIIC or TFIIID, but it is 50-fold less efficient at recruiting TFIIIB or pol III.[97] The low affinity for TFIIIB may account for its difficulty in assembling pol III. Although high concentrations of TFIIIB can drive recruitment to the $tRNA^{AlaSG}$ genes, the resulting complexes differ qualitatively from those formed on the $tRNA^{AlaC}$ promoters; the upstream flanking region is not protected from DNase I digestion and only a minority of the complexes are transcriptionally active.[98] These results can explain why the $tRNA^{AlaSG}$ genes are poorly expressed in most tissues, but the question remains as to how they become activated in silk glands. Silk glands are reported to have especially high levels of TFIIIB activity[97] and this may be sufficient to drive transcription complex assembly on the $tRNA^{AlaSG}$ promoters. It is also possible that an uncharacterized silk gland-specific factor helps TFIIIB to function on the $tRNA^{AlaSG}$ genes.

Regulation in Response to Growth Conditions

Growth into Stationary Phase

In yeast maintained under balanced growth conditions, the rate of pol III transcription is proportional to the rate of growth.[99] However, not all treatments that slow or stop growth also downregulate pol III activity.[99] For example, energy starvation can stop yeast from growing without inhibiting their pol III transcriptional capacity.[99] When grown to high density, yeast undergo a diauxic shift in which the nutrients required for fermentative growth become exhausted and the cells switch to respiratory metabolism. During this transition, tRNA and 5S gene transcription is downregulated.[99,100] When fresh medium is added to these dense stationary phase cells, pol III transcription resumes although growth does not.[99] This reactivation of pol III can be blocked with cycloheximide.[99] The decrease in pol III transcription that accompanies the transition from logarithmic to stationary phase growth can be reproduced in vitro using whole cell extracts.[99,100] Extracts prepared from postdiauxic cells are 3-fold less active than those prepared from log phase cells.[100] This difference was exaggerated to 20-fold in a pcf1-4 mutant strain that carries a substitution at residue 728 of τ_{131}.[100] Mixing experiments did not reveal any diffusible inhibitors in the postdiauxic phase extracts.[100] Add-back experiments demonstrated that BRF is the limiting component of the pol III transcription apparatus in extracts prepared from either logarithmic or stationary phase cells.[100] This is consistent with the observation that BRF is the limiting factor in vivo.[101] BRF is sufficient to raise transcription in postdiauxic phase extracts to levels obtained in extracts prepared from logarithmically growing cells.[100] Western blotting showed that a 3-fold decrease in BRF levels accompanies the transition into early stationary phase, whereas no difference was detected in the abundance of τ_{131}.[100] A 3-fold difference in the abundance of BRF correlates well with the 3-fold reduction in transcription.[100] However, in vivo footprinting of the SUP53 tRNA gene revealed no decrease in the occupancy of the TFIIIB binding site (-40 to -10) when growing and saturated cultures were compared.[102] In contrast, the region between -10 and +15 showed reduced footprinting in stationary phase cells.[102] These data suggest a reduced promoter occupancy by pol III. A decrease in pol III recruitment might be caused by changes in the availability or activity of TFIIIB.

Upon nutritional starvation of cells grown to high density, 5S transcription decreases more rapidly than tRNA transcription.[99] Mixing experiments suggested that the differential response of 5S and tRNA genes is due to a proteinaceous 5S-specific inhibitor.[99] These data suggest a biphasic regulation during growth of yeast into stationary phase: the rapid appearance of an inhibitor that is specific for 5S genes and the subsequent inactivation of the general pol III transcription apparatus.

More transcription complexes are found on B2 genes in chromatin from growing 3T3 cells than in chromatin from confluent cells.[103] However, the majority of tRNA genes are occupied by transcription complexes, even in chromatin from confluent 3T3 cells.[103] Extracts made from log-phase mouse cells transcribe class III genes about 5-fold more efficiently than extracts prepared from cells grown into stationary-phase.[103,104] Mixing experiments demonstrated that inhibitors are not responsible for the low activity of the stationary-phase extracts.[103,104] TFIIIC activity was found by complementation to be 5-fold lower in extracts of confluent 3T3 cells, whereas pol III activity decreased by less than 2-fold and TFIIIB activity showed little change.[103] However, reconstitution assays showed that stationary-phase extracts from Ehrlich ascites cells have 40% of the TFIIIB, 75% of the TFIIIC, and 80% of the pol III activities of the corresponding log-phase extracts.[104] This resembles the situation in yeast, where the downregulation of pol III transcription in stationary phase is primarily due to a decrease in TFIIIB activity.[100] TFIIIB appears to be rate-limiting in stationary-phase extracts from Ehrlich ascites cells, since transcription is stimulated by PC-B, but not by PC-C or pol III fractions.[104]

Serum Stimulation

tRNA levels are regulated in response to growth conditions. A modest increase is seen during the first 10-12 hours after serum stimulation of resting 3T6 fibroblasts.[105,106] Then, as cells reach S phase, tRNA synthesis rises sharply.[105,106] Differential regulation is observed in some cases. For example, expression of $tRNA^{Lys4}$ increases markedly with proliferation rate, but other $tRNA^{Lys}$ genes show much less effect.[107] The rise in tRNA levels is primarily due to increased transcription, although reduced turnover is also involved.[108] For example, the half-life of tRNA is 36 hours in resting 3T3 and 3T6 cells, and 60 hours in growing 3T6 cells.[109] However, the opposite situation is found in SV3T3 cells, where turnover increases with feeding.[110]

The levels of small B2 RNA also increase dramatically following serum stimulation of growth-arrested cells.[111,112] Nuclear run-on assays have shown that increased transcription is at least partly responsible.[112] B2 expression only reaches maximum levels as the cells move into S phase.[111,112] 5S transcription increases in parallel with that of B2 genes, although the effect is much less marked.[112] The fact that the major increase in tRNA and B2 expression coincides with S phase entry[106,111,112] suggests that the response may require the recruitment of previously repressed genes into active transcription complexes and that replication may be necessary to disrupt chromatin structures. Alternatively, maximal pol III transcription may require the inactivation of RB at the G1/S transition.

Nuclear extracts made from HeLa cells grown in high (5-10%) serum transcribe VA1, 5S and tRNA genes 3.5- to 10-fold more efficiently than extracts of cells kept in low (0.5%) serum.[113,114] The increased transcriptional capacity of the high serum extracts correlates with an increase in the proportion of TFIIIC2 in the transcriptionally active TFIIIC2a form.[113] Immunoblot analysis revealed that the serum concentration had no effect on TFIIICα levels, but that the TFIIICβ subunit was more abundant in extracts of HeLa cells grown at high serum concentrations.[114] TFIIICα is present in both the active TFIIIC2a factor and the inactive TFIIIC2b factor, whereas TFIIICβ is unique to the former.[114] Thus, a serum-induced increase in TFIIICβ may convert pre-existing TFIIIC2b into transcriptionally active TFIIIC2a. It remains to be determined whether TFIIIC2b that is bound to a promoter can be activated directly by recruiting TFIIICβ.

Phorbol Esters

Treatment of *Drosophila* Schneider S-2 cells with the tumor-promoting phorbol ester 12-O-tetradecanoylphorbol-13-acetate (TPA) induces a 3- to 5-fold increase in tRNA synthesis.[115] 5S transcription displays a parallel response, whereas pol II transcription of the actin 5C gene is unaffected.[115] The tRNA response is rapid and transient, peaking after 45 minutes and dissipating after an hour.[115] The effect can be mimicked in vitro. Extracts of cells treated with TPA for 15-45 minutes transcribe tRNA, 5S and U6 genes ~10-fold more efficiently than extracts of untreated cells.[115,116] Mixing experiments suggested that this is due to the increased activity of a positive factor rather than the downregulation of a repressor.[115] Single round transcription assays demonstrated that the number of functional initiation complexes increases in response to TPA.[116] Add-back experiments suggested that TFIIIB is limiting in S100 extracts of untreated or TPA-treated S-2 cells.[116] However, TFIIIB activity is substantially elevated following TPA treatment, while pol III and TFIIIC show little or no change.[115,116] Western analysis demonstrated that the abundance of TBP in crude extracts or PC-B fractions increases 4- to 10-fold following TPA stimulation.[116] This is a specific effect, since levels of eIF-2α change little.[116] Overexpression of TBP in Schneider S-2 cells by transient or stable transfection is sufficient to stimulate transcription of tRNA and U6 genes.[117] This is likely to be a direct effect, rather than an indirect consequence of elevated pol II activity, since overexpressing a TBP mutant that supports pol II but not pol III transcription has no effect upon tRNA or U6 expression.[117] These results indicate that TBP is limiting for the synthesis of tRNA and U6 RNA in S-2 cells. It is therefore likely that the increased TBP levels in response to TPA are sufficient to account for the stimulation of TFIIIB activity. It remains to be determined how TPA can produce such a rapid and dramatic increase in TBP abundance.

Mitosis

Nuclear transcription is repressed during mitosis[118-121] (reviewed by Gottesfeld and Forbes[122]). Hartl et al[123] found that pol III transcription is also repressed when *Xenopus* egg extracts are shifted to the mitotic state by the addition of a proteolysis-resistant form of cyclin B. The degree of repression was 4.5-fold for a *Xenopus* somatic 5S gene and 15-fold for a yeast $tRNA^{Leu3}$ gene.[123] Oocyte 5S and $tRNA^{Met1}$

genes are also inhibited more substantially than somatic 5S genes.[124] Sequence differences within the coding regions account for the differential response of somatic and oocyte 5S genes.[124] Repression occurs even if transcription complexes are preassembled onto these templates, although prolonged preincubation can reduce the level of subsequent inhibition.[123,124] Preincubation of the extract with nonspecific DNA to titrate out histones and other chromatin-binding proteins made no difference to the extent of mitotic repression.[123] The topoisomerase II inhibitor VM26, which blocks mitotic chromosome condensation, also failed to prevent repression.[123] Therefore the inhibition of pol III transcription does not require the formation of nucleosomes or condensed chromatin nor the binding of nonspecific titratable repressors.

A fraction prepared from mitotic extracts inhibited expression when added to class III genes that had been preincubated with partially purified transcription factors.[123,124] This repression was prevented by the kinase inhibitor DMAP.[123,124] An equivalent fraction from interphase extracts did not inhibit transcription.[123] The fraction from mitotic extracts was enriched in cdc2/cyclin B kinase.[123] Indeed, identical results were obtained when cyclin-dependent kinases were affinity-purified from mitotic extracts using p13suc1 or antibodies to cdc2.[124-126] In contrast, MAP kinase that had been immunoprecipitated from mitotic extracts had no effect on pol III activity.[124] Gottesfeld et al[125] showed that cdc2 bound to recombinant cyclin B1 is sufficient to inhibit 5S transcription. However, removal of cdc2 from mitotic extract using p13suc1-beads did not prevent the DMAP-sensitive repression of a subsequently added tRNA gene.[123] Furthermore, cdc2-depleted mitotic extract could inhibit transcription when mixed with interphase extract.[123] These results suggest that mitotic extracts contain at least one other kinase that is capable of repressing pol III but is not part of the cdc2 family. This inhibition only occurred in the presence of the phosphatase inhibitor okadaic acid,[123] which implies that the action of mitotic kinase is overcome by a phosphatase found in interphase extracts. This phosphatase is also present in M phase extracts, as transcription can be reactivated if mitotic kinase activity is subsequently inhibited using DMAP.[126] Reactivation can be blocked with okadaic acid.[126]

Repression by cdc2 could be reversed by the addition of highly purified TFIIIB, whereas TFIIIA and TFIIIC had no effect.[125] Although TFIIIB from interphase extracts was very efficient in this respect, TFIIIB isolated from mitotic extracts was inactive.[125] The activity of mitotic TFIIIB was restored by treatment with alkaline phosphatase.[125] Affinity-purified interphase TFIIIB could be inactivated using cdc2 kinase, an effect which was sensitive to DMAP.[125] These results suggest that the repression of pol III transcription in metaphase frog eggs involves the inactivation of TFIIIB through the action of cdc2 and one or more additional mitotic kinase(s). Polypeptides of 90/92, 42, and 33/34 kD present in affinity-purified TFIIIB were found to be phosphorylated by mitotic kinase immobilized on p13 beads.[126] The 33/34 kD species comigrates with TBP.[126] Although recombinant *Xenopus* TBP is phosphorylated by p13-associated mitotic kinases, it is not a substrate for cdc2/cyclin B and does not contain consensus cdc2 sites.[126] The other polypeptides may represent TAF subunits of *Xenopus* TFIIIB.[126]

Whereas the cell cycle in early *Xenopus* embryos consists of a simple fluctuation between S and M phases, somatic cells undergo a more complex cycle that involves sequential passage through G1, S, G2 and M phases. Nevertheless, mitotic HeLa cells repress pol III transcription by a mechanism that appears to resemble the situation in frog eggs.[127] Thus, TFIIIB is inactivated specifically in extracts of HeLa cells that have been synchronized in M phase.[127] As a consequence, TFIIIB activity becomes limiting for pol III transcription during mitosis and expression of class III templates is substantially diminished.[127] Western blotting revealed that the abundance of TBP is unchanged at M phase.[127] However, TBP becomes hyperphosphorylated during this phase of the cell cycle.[127] These changes in TBP are not sufficient to account for the reduction in pol III transcription, since expression cannot be restored by the addition of recombinant TBP to mitotic extracts.[127] In contrast, affinity-purified TFIIIB TAFs reverse repression efficiently and can raise transcription in M phase extracts to levels obtained using protein extracted from asynchronous HeLa cells.[127] Direct assay confirmed that the TAF component of TFIIIB is inactivated during mitosis.[127] Western blotting has revealed that hBRF is hyperphosphorylated in mitotic extracts.[128]

Interphase

After cycling HeLa cells have left mitosis, pol III transcription remains low during early G1 phase.[129] Transcriptional activity increases gradually as cells progress through G1, and maximal expression is only achieved during S and G2 phases.[129] Nuclear run-on assays demonstrated that U6 and tRNA synthesis is 2- to 3-fold less active during early G1 than it is during S or G2.[129] This pattern of expression could be reproduced in vitro using whole cell extracts prepared from synchronized HeLa cells.[129]

Random polymerization assays revealed no consistent cell cycle variation in the catalytic activity of the pol III enzyme.[129] The B-block-binding activity of TFIIIC also remains constant.[129] Add-back experiments indicated that TFIIIB activity is limiting during early G1 phase.[129] Although TBP is hyperphosphorylated at mitosis,[127] this modification is rapidly reversed at the beginning of G1 and the abundance and electrophoretic mobility of TBP remains constant throughout interphase.[129] However, complementation assays revealed that the TAF component of TFIIIB is 6- to 8-fold less active in early G1 phase than it is in S or G2.[129] Addition of affinity-purified TFIIIB TAFs is sufficient to raise expression in early G1 phase extracts to levels occurring in S or G2 phases.[129] In contrast, by S and G2 phases TAF activity has increased to such an extent that TFIIIB is in relative excess.[129]

Although TFIIIB TAF activity is limiting for pol III transcription in both M and G1 phases,[127,129] different mechanisms appear to be responsible in each case. During mitosis, TFIIIB is repressed due to the hyperphosphorylation of multiple subunits. The general kinase inhibitor DMAP is sufficient to reactivate transcription in a mitotic extract by allowing the unbalanced action of endogenous PP2A.[128] Hyperphosphorylation of TBP and BRF is reversed at the beginning of interphase, and DMAP is unable to activate expression in a G1 phase extract.[128,129] The mechanism(s) responsible for repressing TFIIIB TAF activity during early G1 remains to be determined. One obvious possibility is that RB contributes to the silencing of TFIIIB at this stage of the cell cycle. RB has been shown to bind TFIIIB and repress its TAF component.[130] Furthermore, RB is known to function during early G1 and then become inactivated as cells progress into S phase due to the action of cyclin D- and E-dependent kinases.[131,132] Since TFIIIB TAF activity is targeted by RB and increases just as RB is switched off, it seems highly likely that RB

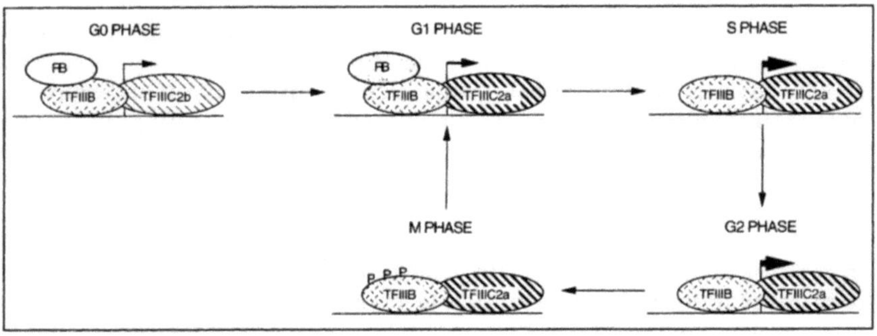

Fig. 9.3. Schematic model of the cell cycle regulation of TFIIIB and TFIIIC2 in mammalian cells. During Go and early G1 phase, TFIIIB is subject to repression by RB. TFIIIB activity is maximal during S and G2 phases. At mitosis, TFIIIB becomes repressed through hyperphosphorylation. TFIIIC2 is regulated as cells enter and exit the cycle; in stationary (Go) phase, the inactive TFIIIC2b form predominates, whereas the active TFIIIC2a form predominates in cycling cells. As a consequence of these changes, the rate of pol III transcription varies substantially through the cell cycle; it is low during Go and early G1, high during S and G2, and ceases at mitosis.

contributes to the cell cycle regulation of pol III transcription. However, it remains to be determined to what extent this is the case. Given the complexity of cell cycle control, it would not be surprising if multiple regulatory mechanisms are involved.

Following serum stimulation of quiescent cells, maximal rates of pol III transcription are only achieved once cells have entered S phase.[105,106,112] Thus, it appears that the pol III transcription apparatus becomes activated at the G1/S transition whether the starting point is the Go phase of growth-arrested cells or the M phase of proliferating cells. In the former case, activation seems to involve changes in both TFIIIB and TFIIIC,[104,113,114] whereas TFIIIB alone appears to be regulated once cells are cycling.[129] It may therefore be that when quiescent cells are stimulated, TFIIIC is activated during progress from Go into G1 and then TFIIIB is released from repression during passage into S phase (Fig. 9.3).

Cycloheximide Treatment

Gokal et al[88] found that synthesis of tRNA and 5S rRNA is rapidly inhibited after the treatment of murine lymphosarcoma cells with the protein synthesis inhibitors cycloheximide (CHX) or emetine. Measurements of synthesis in isolated nuclei showed that the effect occurs at the transcriptional level.[88] However, the steady-state levels of tRNA and 5S rRNA are unchanged.[88] S100 extracts mimic the response.[88] Thus, S100s prepared 2 hours after CHX treatment are 90% inhibited in 5S transcription and 75% inhibited in tRNA transcription.[88] A time course showed that 50% inhibition occurs after 20 minutes for 5S and 50-60 minutes for tRNA.[88] No inhibitory substances were detected by mixing experiments and pol III activity was unchanged.[88]

Tower and Sollner-Webb[104] found that extracts made 1.5 hours after CHX treatment exhibit a 5-fold reduction in class III transcriptional capacity.[104] This is associated with unchanged pol III activity, but the activities of TFIIIB and TFIIIC are

40% and 80%, respectively of control levels.[104] Therefore, inhibition of protein synthesis results in the rapid depletion of TFIIIB and, to a lesser extent, TFIIIC.[104] TFIIIB is the rate-limiting factor under these conditions[104] and is likely to mediate the downregulation of class III gene transcription in response to CHX.

In contrast to these results, CHX elicits a rapid and transient increase in the abundance of Alu RNA in HeLa or 293 cells.[133,134] Induction occurs within 20 minutes and results in a 20-fold increase after 3 hours.[134] By 7 hours, transcript abundance has decayed to uninduced levels.[134] Puromycin, which blocks protein synthesis at a different stage from CHX, also raises Alu RNA levels.[134] Similar effects are seen with B1 and B2 RNA, but not with 5S, U6, 7SL or 7SK RNAs.[134] This regulation has not been shown to be a transcriptional response; it may be that an unstable protein is involved in degrading SINE transcripts and that translational inhibitors increase RNA accumulation by reducing the rate of turnover.

CHX can also inhibit pol III transcription in yeast.[135] However, this may depend upon the strain or growth conditions employed, since Clarke et al[99] were able to inhibit cell proliferation using CHX without inhibiting pol III. In contrast, Dieci et al[135] found that the in vivo synthesis of tRNA and 5S rRNA decreases rapidly upon CHX administration. These workers observed that tRNA and 5S rRNA synthesis decreases with half-times of 20 and 10 minutes, respectively, and levels off after 2 hours at 15% and 10% of the rate seen in untreated cells.[135] However, the steady-state levels of tRNA and 5S rRNA are unchanged.[135] Although pols I and II are also sensitive to CHX, the rates of inhibition vary widely for the three polymerases. Pol III transcription decreases 4.5- to 9-fold more rapidly than pol II, but 4-fold more slowly than pol I.[135] Extracts prepared after 3 hours of CHX treatment have almost completely lost the ability to transcribe a class III gene.[135] Mixing experiments did not reveal any diffusible inhibitor in these extracts.[135] Direct assays demonstrated that TFIIIB activity is reduced by 10-fold, whereas the activities of TFIIIA, TFIIIC, TFIIIE and pol III are near normal.[135] Addition of TFIIIB specifically restored transcription to the extracts prepared from CHX-treated cells.[135] Although none of the individual subunits of TFIIIB could do this, a combination of BRF and B" restored activity.[135] Western analysis detected no change in TBP, but a substantial decrease in BRF levels in response to CHX.[135] The activities of BRF and B" were reduced by 10- and 7-fold respectively, whereas TBP activity was essentially unchanged.[135] BRF contains a degradation-promoting "PEST" sequence, which may account for its rapid turnover in CHX-treated cells.[135]

DNA Damage

Following UV-induced DNA damage, the transcribed strand of class II genes is repaired faster than the coding strand or flanking genomic DNA.[136,137] Such transcription-coupled repair is not seen in human class I genes.[137] Following UV irradiation of human fibroblasts, a $tRNA^{Val}$ gene and the $tRNA^{Sec}$ gene were repaired without strand preference and much more slowly than class II genes.[138] These observations suggest that the pol III transcription apparatus has no stimulatory effect on the nucleotide excision repair machinery. Both tRNA genes examined in this study were actively transcribed and expression did not decrease in response to UV treatment.[138] Preferential repair of class III genes may not be important because these genes are short and are generally present in multiple copies.

Heat Shock

Heat shock of mammalian cells causes a rapid increase in the levels of SINE transcripts.[134,139,140] In the case of B2 genes, this has been shown to result from elevated rates of transcription.[139] The response is extremely fast, with a 6-fold increase in small B2 RNA after 15 minutes, and a maximum 10- to 20-fold induction reached after 4 hours and then remaining constant for 24 hours.[139] Continued synthesis is required, since B2 RNA levels decline in the presence of actinomycin D with a half-life of one hour.[139] The induction of SINE transcripts precedes the increase in hsp70 mRNA[134] and B2 may be the most abundant transcript induced by heat shock in several rodent cell lines.[139] This may be a general stress response, since heat shock mimetics such as ethanol or arsenite produce the same effect.[134,139] However, it is not a nonspecific response to cellular injury, because no induction follows UV- or X-irradiation or treatment with the DNA damaging agent methylmethane sulphonate.[139,140] The steady-state levels of 7SL, 7SK, 5S and U6 RNAs do not change following heat shock.[134] No mechanism has been elucidated.

Regulation in Response to Viruses

Poliovirus Infection

Extracts prepared from HeLa cells 4 to 5 hours after infection with poliovirus have a reduced capacity to transcribe class III genes.[141] If guanidine is used to block viral replication, class III transcription is no longer inhibited, despite large-scale shutoff of host protein synthesis.[141] Fradkin et al[141] compared the activities of fractions from infected and mock-infected extracts. They found that pol III activity is unchanged following infection, TFIIIA activity is 1.4-fold less, TFIIIB activity decreases 4-fold, and TFIIIC activity decreases 8-fold.[141] TFIIIC was shown to be the rate-limiting factor both before and after infection,[141] implying that downregulation is mediated primarily through changes in this factor. However, the overall decrease in transcription is much greater than 8-fold,[141] suggesting that changes in several factors may amplify the effect or that transcription responds in a nonlinear fashion to changes in factor concentration. Despite its loss of transcriptional activity, TFIIIC displays little change in its ability to bind to the B-block of VA_I.[141-143] However, whereas TFIIIC2 from uninfected and mock-infected cells migrates as a doublet of bands in nondenaturing gels, the upper TFIIIC2a band is missing and the lower TFIIIC2b band is more abundant in extracts from poliovirus-infected cells.[142] Only the TFIIIC2a form has transcriptional activity.[113,114,142,144] It appears to contain subunits that are missing or modified in the TFIIIC2b form.[114,144] A partial conversion of TFIIIC2a to the inactive TFIIIC2b can be achieved by treatment with acid phosphatase.[113,142] In addition to TFIIIC2b, extracts from poliovirus-infected cells contain a much faster migrating TFIIIC2 species that is not seen in uninfected cells.[142] This form also has no transcriptional activity.[142] The loss of the active TFIIIC2a form, the increase in TFIIIC2b and the appearance of the fastest form correlate kinetically with the decrease in TFIIIC activity during the course of infection.[142] The inactive fast migrating species can be generated from the other forms by incubation with recombinant proteinase 3C, one of the two poliovirus proteases that are involved in processing the viral polyprotein precursor into capsid and noncapsid proteins.[143,145] A poliovirus with an inactivating point mutation in proteinase 3C fails to produce the fast form of TFIIIC2 and is limited in its ability to inhibit pol III transcription.[145] In contrast, the gene for proteinase 3C alone is suf-

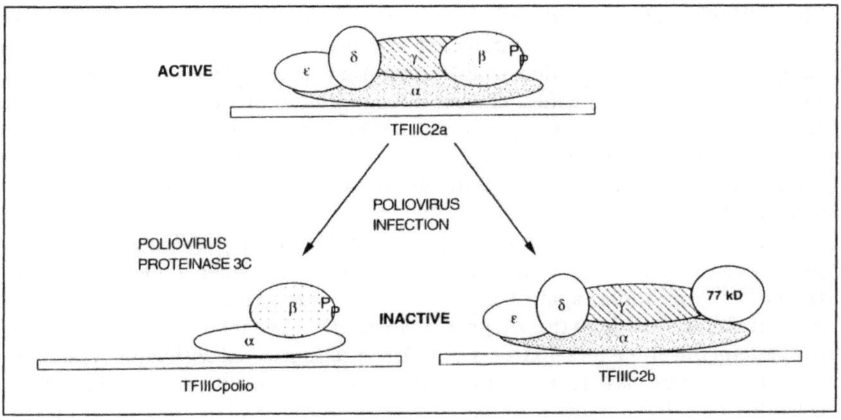

Fig. 9.4. Schematic illustration of how TFIIIC2 responds to poliovirus infection. Infected cells contain very little of the active TFIIIC2a form. As well as containing inactive TFIIIC2b, they also contain a proteolyzed complex in which the α subunit is truncated and the β,γ and δ subunits are missing. The poliovirus 3C protease is responsible for degrading TFIIIC2.

ficient to produce a significant decrease in TFIIIC activity and pol III transcription when transfected into HeLa cells.[145] Both the 240 kD α subunit and the 80 kD δ subunit of TFIIIC2 are cleaved by recombinant proteinase 3C in vitro.[143] TFIIICα is cleaved at residue 732 to generate an 83 kD N-terminal fragment that remains associated with TFIIICβ and is fully active for DNA binding; however, it no longer interacts with the γ, δ and ε subunits and it is unable to support transcription.[143] Similar digestion products are produced in poliovirus-infected cells[143] (Fig. 9.4).

Although the proteolysis induced by poliovirus infection is neither large-scale nor random, TBP is also attacked by proteinase 3C.[146] Proteinase 3C cleaves TBP near its N-terminus, at residue 12 and/or 18.[146] However, this is unlikely to be involved in the diminished expression of class III genes, since the N-terminal region of human TBP is not required for efficient transcription of most pol III templates in HeLa extracts.[147] Furthermore, the reduction in TFIIIB activity that accompanies poliovirus infection is not affected by the inactivating mutation in proteinase 3C.[145] Thus, poliovirus would appear to inhibit pol III factors by a combination of mechanisms.

Adenovirus Infection

Pol III transcription is stimulated strongly following adenovirus infection of human cells.[113,114,133,148-154] This enables the virus to express its VA_I and VA_{II} gene products at very high levels late in infection.[148] Adenovirus infected cells also overexpress transfected VA_I and tRNA genes[149-151] and endogenous Alu genes.[133,154] However, transcript levels from endogenous tRNA, 5S and 7SL genes show little or no response.[133,150,154] Induction occurs in the absence of viral DNA replication, suggesting that it is mediated by the early gene products.[133,150] Little or no VA activation is observed using mutant virus strains that lack the E1A protein.[114,133,149-151,153,155] Furthermore, a transfected E1A gene alone can be sufficient to trans-activate

VA_I,[156,157] although one study failed to observe this effect.[153] However, at high multiplicities of infection Alu induction can also occur in the absence of E1A.[133] This suggests that part of the effect of E1A upon Alu transcription is due to its activation of other viral early genes. The use of a range of viral mutants has shown that full induction of Alu genes requires the 58 kD E1B gene product and the products of the E4 open reading frames 3 and 6.[133] These E1b and E4 proteins are post-transcriptional regulators that may facilitate the processing or nuclear transport of viral late RNAs.[158]

Recombinant 289 amino acid E1A protein that has been purified to near homogeneity can stimulate VA_I transcription by up to 50-fold when added to HeLa extracts.[159,160] Antibodies against E1A can prevent this effect specifically.[159,160] E1A does not bind directly to the VA_I gene.[160] Instead, it appears to exert its effects via the general pol III factors. Extracts from HeLa cells infected with wild-type adenovirus transcribe class III genes 1.5- to 17-fold more efficiently than extracts from cells that have been mock-infected or infected with the dl312 mutant strain that lacks E1A.[113,114,151,152] TFIIIC is the rate-limiting factor in these extracts and the PC-C fraction is up to 5-fold more active in infected than in uninfected or mock-infected extracts.[151,152] Yoshinaga et al[152] used a template commitment assay to show that TFIIIC is 4- to 8-fold more abundant in extracts prepared after 30 hours of adenovirus infection. Using the same assay, Hoeffler et al[113] found little or no change in TFIIIC abundance 6 hours after productive infection. However, these workers demonstrated a clear increase in the proportion of TFIIIC molecules existing in the transcriptionally active TFIIIC2a form.[113] A similar change was reported following the addition of recombinant E1A to an uninfected HeLa extract.[160] This effect can be prevented if anti-E1A antibodies are added at the same time as E1A, but not if they are added 15 minutes later.[160] An increase in the number of active initiation complexes following adenovirus infection was confirmed using single-round transcription assays in the presence of 0.05% sarkosyl.[161] It may be that adenovirus acts initially by converting extant TFIIIC2b into TFIIIC2a, and then subsequently reinforces this effect by inducing an increase in TFIIIC2 abundance. Immunoblotting demonstrated that cells infected with wild-type adenovirus had ~10-fold higher TFIIICβ levels than cells infected with the E1A mutant dl312.[114] In contrast, TFIIICα levels were unchanged.[114] Preliminary results suggested that E1A induces an increase in TFIIICβ mRNA.[114] This would suggest that adenovirus is able to increase the proportion of TFIIIC2 in the transcriptionally active form by stimulating the synthesis of TFIIICβ. However, the ability of E1A to stimulate pol III transcription in vitro[159,160] also suggests that alternative activation pathways may be employed by adenovirus that do not require de novo protein synthesis. For example, E1A has been shown to overcome repression by RB, both in vitro and in transfected cells.[162] It can also release TBP from interaction with Dr1.[163] Since RB and Dr1 are both potent inhibitors of TFIIIB activity,[130,164,165] the ability of E1A to counteract these repressors may provide additional routes that allow adenovirus to stimulate pol III transcription. Adenovirus infection has also been shown to reduce the proportion of Alu genes that are inaccessible to transcription factors due to chromatin-mediated repression.[154] It therefore appears that adenovirus can employ a range of distinct mechanisms in order to ensure that infected cells have a high capacity to transcribe class III genes (Fig. 9.5). The relative importance of these various mechanisms is likely to depend upon the particular cell type and environmental conditions. Adenovirus drives cells into S phase, in order to allow replication of its own

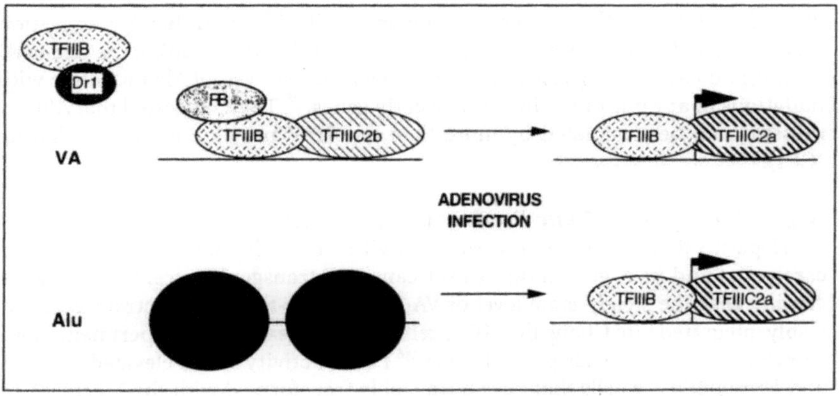

Fig. 9.5. Schematic illustration of how adenovirus infection may influence pol III transcription. Adenovirus converts inactive TFIIIC2b into active TFIIIC2a. It also disrupts nucleosomes that are assembled on Alu genes. The adenoviral E1a protein releases TFIIIB from repression by RB and displaces Dr1 from its interaction with TBP. The net result is a substantial increase in pol III activity.

genes. TFIIIC is the limiting factor during S phase in HeLa cells,[129] and so the primary route employed by adenovirus to stimulate VA expression in infected HeLa cells is to activate TFIIIC.[113,114,151,152]

Pseudorabies Virus

Despite a lack of sequence homology, the pseudorabies virus immediate early (IE) protein has an activity similar to that of E1A, since it can substitute for E1A in inducing the transcription of adenovirus early genes.[150,155] Pseudorabies virus IE protein can also induce the expression of VA genes during coinfection with E1A-deficient adenovirus.[155] Cells that stably express the IE protein transcribe transfected VA_I and tRNA genes at 20-fold higher levels than parental cells.[150] Furthermore, extracts of IE-expressing cells are 10- to 20-fold more active in pol III transcription than control extracts.[150] The mechanism of this effect has yet to be determined, but is likely to be similar to that of E1A.

Herpes Simplex Virus

The steady-state level of small Alu transcripts increases by up to 50-fold following the infection of HeLa cells with herpes simplex virus (HSV).[158,166] HSV also induces Alu expression in fibroblasts and 293 cells.[158] Nuclear run-on experiments have shown that this effect is due to an increase in Alu transcription by pol III.[158,166] A similar increase was not observed with other class III genes.[158,166] The induction of Alu expression required viral protein synthesis and was not prevented by a null mutation in ICP4.[158] This suggests that one or more immediate early proteins are sufficient for activation. Jang and Latchman[166] found that inactivation of the viral ICP27 gene abolishes Alu induction, whereas mutation of the other viral immediate-early genes does not. They also reported that ICP27 stimulates Alu expression when transfected into uninfected cells or when expressed as a stable transformant.[167] These workers proposed that an increase in TFIIIC activity is responsible for Alu induction by ICP27.[167] However, Panning and Smiley[158] found that Alu activation was unaffected by three separate null mutations in ICP27, including the one used by Jang and Latchman. These workers found wild-type levels of Alu induction with mutations in any one of the immediate early genes.[158] They suggested that Alu expression can be stimulated by more than one HSV protein, perhaps including ICP27.[158]

Regulation by the X Gene Product of Hepatitis B Virus

Hepatitis B virus is strongly associated with the development of hepatocellular carcinoma and its X gene induces liver cancer in transgenic mice.[168] The X gene can increase the steady-state level of VA_I RNA by 20-fold when cotransfected or stably integrated into Chang liver (CL) cells.[156,169] Nuclear run-on experiments have shown this to be a transcriptional effect.[156] Pol III activity is also elevated in rat 1A and *Drosophila* S-2 cells that have been stably transformed with the X gene.[170] In each of these cases, extracts prepared from the X-transformed cells transcribe pol III templates more actively than extracts of untransformed parental cells.[156,170] Two or three different X gene products with a common C-terminus can be produced by the use of alternate translation initiation sites within the same reading frame.[169] Although these products vary in their ability to transactivate class II genes, each can stimulate VA_I expression.[169] The C-terminal part of these proteins contains the transactivation domain, whereas the N-terminus contains a regulatory region.[171]

The X gene products are unable to bind DNA directly and instead influence gene expression by protein-protein interactions with other transcription factors and also by stimulating kinase pathways. For example, X activates protein kinase C by raising the level of sn-1,2-diacylglycerol.[172] In vitro binding assays have shown that X protein can interact directly with the conserved core of TBP[173] and with the human RPB5 subunit that is shared between pols I, II and III (homologous to yeast ABC27).[171] Either of these interactions has the potential to explain the highly promiscuous transactivation properties of X. Both the N-terminal regulatory region and the C-terminal transactivation region of X contribute to its interaction with RPB5.[171] Immunoprecipitation assays have confirmed that X associates with RPB5 in transfected HepG2 cells,[171] presumably as part of one or more of the endogenous polymerases. Despite these findings, direct assays of nonspecific polymerization activity in S-2 cells found only a 1.3-fold increase in total pol activity and a 1.5-fold increase in pol III activity in response to X.[170] Although X could be responsible for these slight changes, they seem insufficient to account for the overall level of transactivation of pol III transcription.

Template commitment experiments showed that extracts from X-transformed CL cells contain at least 3- to 6-times more than parental CL extracts of a factor required for stable complex formation on VA_I.[156] Extracts from S-2 cells expressing the X gene displayed elevated TFIIIB activity, whereas TFIIIC activity was unchanged.[170] Addition of PC-B to S-2 extracts stimulated tRNA transcription.[170] Western analysis of unfractionated S-2 or rat 1A extracts showed a 4- to 5-fold increase in TBP levels in response to the expression of the X gene.[170] The levels of eIF-2α were unchanged.[170] An increase in the level of TBP mRNA was also reported.[170] Transfection of a TBP expression vector into S-2 or rat 1A cells increased the expression of tRNA and U6 genes.[117,170] To determine whether the extra TBP acts directly upon class III genes or indirectly by stimulating the production of pol II products, a mutant TBP that is specifically defective in pol III transcription was overexpressed in S-2 cells.[117] This mutant was defective in tRNA and U6 activation, which suggests that the X-induced rise in cellular TBP has a direct effect upon pol III.[117] The data suggest that TBP is the limiting factor for tRNA and U6 transcription in S-2 cells. The X gene appears to activate pol III by raising the levels of this limiting component. Inhibitors of protein kinase C (PKC) were found to block the X-mediated increase in TBP abundance and pol III transcription.[170] This implies that the stimulation of TBP expression by X requires the participation of the PKC pathway. In contrast, the PKC pathway has a negative effect on pol III activity in the absence of the X protein.[170,174]

Regulation by the Tax Protein of HTLV-1

The Tax protein of human T-cell leukemia virus type 1 (HTLV-1) is able to transactivate VA_I by 25-fold when coexpressed in transient transfection assays.[175] Furthermore, recombinant Tax stimulates transcription of VA, tRNA and 5S genes when added to crude extracts or phosphocellulose fractions.[175,176] This effect could be blocked by preincubation of the recombinant protein with antibodies against Tax.[175] Extracts chromatographed over immobilized Tax were depleted of their ability to transcribe VA_I, although transcription from the AdML promoter was unaffected.[175] Pol III transcription in the depleted extract could be restored by the addition of PC-C, but not PC-B.[175] Maximal transactivation by Tax was obtained when PC-C was in excess over PC-B.[176]

Northern analysis revealed higher levels of 5S RNA in an HTLV-1-infected T-lymphocyte line (MT2) when compared to an unrelated T-cell line (CEM) that has not been infected.[176] Transcription of isolated chromatin using exogenous pol III suggested that more 5S and tRNA genes are assembled into active transcription complexes in MT2 cells than in CEM cells.[176] Cytoplasmic extracts of MT2 cells have ~10-fold higher pol III transcriptional activity than CEM cell extracts.[176] Little difference was detected in pol III or B-block binding activities.[176] However, MT2 cell extracts contained ~10-fold higher TATA box-binding activity.[176] Furthermore, the TATA box-binding activity of a PC-B fraction was increased by the presence of Tax.[176] These results lead Gottesfeld et al[176] to propose that Tax protein stimulates pol III transcription by raising the effective concentration of active TFIIIB.

Regulation in Response to Transformation

The steady-state levels of pol III transcripts from B1 and B2 genes are considerably elevated in most transformed cells. The first indication of this was provided by Scott et al,[177] who found that low molecular weight B2 RNA is significantly more abundant in SV40-transformed cell lines than it is in the untransformed parental fibroblasts. A wide range of transformed rodent cell lines has been found to express high levels of small B1 and B2 RNA relative to untransformed tissues or cells.[80,89,112,177-183] This can be the case whether the transforming agent is a DNA tumor virus, an RNA tumor virus, or a chemical carcinogen. The activation is very general, but not universal, there being several examples of transformed lines which do not show the characteristic elevation in B2 transcripts.[80,177] Two fibroblast cell lines transformed by temperature-sensitive mutants of the SV40 large T antigen downregulate B2 RNA at the nonpermissive temperature while reverting to normal morphology and phenotype.[177] This suggests a tight link between B2 activation and transformation. However, an untransformed revertant rat fibroblast line which has lost the integrated polyoma virus early gene continues to produce elevated levels of small B2 RNA.[180] B2 levels vary markedly between different SV40-transformed lines and increased abundance appears to be associated with progression to a more extreme transformed phenotype.[177] The cell lines in which small B2 RNA is most elevated also have the least large T antigen and grow most slowly, indicating that B2 levels are not directly related to the amount of T antigen or the growth rate of the transformed cells.[89,177] The extent of the increase in small B1 and B2 RNA following transformation varies between 5- and 20-fold, depending upon the growth state of the parental cells.[181,182]

Another pol III product has been shown to be induced in rodent tumors. Northern analysis of RNA from breast carcinomas, colonic adenocarcinomas and skin fibrosarcomas showed BC1 transcript which was not detected in RNA from the corresponding untransformed tissues.[184] BC1 expression is normally restricted to neurons.[185] In situ hybridization studies of these tumors confirmed the presence of BC1 RNA in the neoplastic cells, whereas it was absent from the surrounding tissues.[184] Although the fibrosarcomas and adenocarcinomas were induced by local inoculation with cells that had been treated with chemical carcinogens, the breast carcinoma analyzed was a primary tumor induced by ras.[184] Similar studies have shown that BC200 RNA, the primate analog of BC1, is expressed in many, but not all, primary human tumors.[184] Like BC1, BC200 RNA is found exclusively in the malignant cells and not in the adjacent normal tissue.[184] Thus, abnormal activation of pol III expression is a frequent feature of tumors in vivo.

The fact that induction of small B1 and B2 RNA occurs in parallel implies a common mechanism.[80,89,182] In contrast, the steady-state levels of transcripts from a range of other class III genes, including tRNA, 5S and 7SL, show little or no change between murine fibroblast cell lines and their SV40-transformed derivatives.[89,182] Nuclear run-on experiments demonstrated a 5- to 10-fold increase in pol III transcription of B1 and B2 genes in transformed relative to untransformed nuclei.[89,112,182] Transcription of tRNA genes displays a similar increase, whereas 5S transcription shows less of an effect.[89,112,182] Thus, SV40 transformation provides another example of a situation in which changes in the rate of tRNA transcription are not reflected in comparable changes in the steady-state levels of tRNA transcripts. This is likely to be true for many class III genes. It is therefore clear that RNA levels are not a reliable indication of transcriptional activity. However, the abundance of small B2 RNA does reflect the rate of pol III transcription.[89,112,182] This is probably due to the short half-life of these transcripts.[183]

Carey and Singh[103] found that chromatin isolated from the SV40-transformed line SVT-2 gives higher levels of B2 transcription than chromatin from the parental 3T3 cells when supplemented with TFIIIB and pol III. They suggested[103] that transformation may increase the number of accessible B2 genes, thereby driving the formation of additional transcription complexes, or it may cause an increase in TFIIIC activity. However, they observed no increase in pol III transcription in SVT-2 extracts relative to extracts from growing 3T3 cells.[103] In contrast, White et al[89] found that extracts from SV40-transformed SV3T3 cell lines have a substantially greater pol III transcriptional capacity than extracts prepared in parallel from growing 3T3 cells.[89] This applies to all class III genes tested.[89] However, pol II transcription of the HPRT gene is equally efficient in the two types of extract,[89] demonstrating that SV40-transformation has a specific effect upon pol III. Mixing experiments suggested that the general activation of class III transcription is due to an increase in a positive factor rather than the loss of an inhibitor.[89] Add-back experiments implicated TFIIIC as the rate-limiting factor for pol III transcription in both 3T3 and SV3T3 extracts.[89] SV3T3 extracts display a small but consistent and specific increase in B-block-binding activity relative to 3T3 extracts, as measured by gel retardation assays.[89] This may indicate a slight increase in the abundance of TFIIIC2. A more striking effect was that the SV3T3 extracts have a higher proportion of the transcriptionally active TFIIIC2a relative to TFIIIC2b.[89] The TFIIICβ subunit is required to convert TFIIIC2b into TFIIIC2a.[114] We have found recently that TFIIICβ levels are specifically upregulated following SV40 transformation.[186] The resultant increase in activity of TFIIIC2 is likely to contribute substantially to the higher levels of pol III transcription observed in SV40-transformed cells.

It seems likely that TFIIIC is the target of many transforming agents. 293 cells, which are a human embryonic kidney cell line transformed by the left 14% of the adenovirus genome, also have elevated TFIIIC activity and an enhanced capacity for pol III transcription.[149,150,152,156,187] Thus, TFIIIC is targeted as a means to regulate pol III transcription during viral transformation, just as it so often is during the course of viral infection. However, other pol III factors are also likely to become activated following transformation. For example, TBP has been shown to interact directly with both E1A[188] and SV40 large T antigen.[189] TFIIIB may therefore be a direct target for these oncoproteins. TFIIIB is inhibited by the tumor suppressors RB and p53, both of which are frequently inactivated in cancers.[130,162,165,190] Deletion of RB has been shown to cause a 5-fold increase in pol III transcription in primary murine fibroblasts.[162] It is highly likely that the loss of function of RB and/or p53 will contribute substantially to the high level of pol III activity that characterizes many tumors.

Mutations that inactivate RB are found in many human cancers. People who inherit a nonfunctional allele of the Rb gene have a roughly 90% chance of developing retinoblastoma at an early age.[191] Inactivation of the remaining allele by somatic mutation appears to be a universal feature of this tumor and is likely to be the rate-limiting step in its initiation.[192] Somatic inactivation of RB is also found in many other types of human tumor, including osteosarcomas, small cell lung, breast, and bladder carcinomas.[132] In these cases the individual inherits two wild-type copies of Rb, but mutations in both alleles arise during tumorigenesis. Most striking in this category are the small cell lung carcinomas, where Rb mutation is found in nearly all instances.[192] Other tumor types display a lower frequency of RB alteration. For example, RB was found to be altered or missing in one third of bladder carcinomas that were surveyed.[192] Many types of tumor appear to express normal RB, such as melanomas and colon carcinomas.[192] Thus, mutation of RB appears to be an important feature in the pathogenesis of a subset of human tumors.

Deletion and substitution analyses have demonstrated that the region of RB that controls pol III transcription corresponds to the domains that are frequently mutated in tumors.[162,165] Several examples have been characterized of highly localized and subtle mutations that inactivate RB in human cancers. For example, in one small cell lung carcinoma a single base change in a splice acceptor site gave rise to an RB product that lacked the 35 amino acids encoded by exon 21.[192] In another small cell lung carcinoma, a point mutation created a stop codon and a novel splice donor site within exon 22, thereby eliminating 38 amino acids from the product.[192] A third inactivating mutation from a small cell lung carcinoma resulted in a single substitution at residue 706.[193] Each of these three naturally occurring RB mutations has lost its ability to regulate pol III transcription.[162] Although this is clearly a limited survey, it nevertheless demonstrates a correlation between the ability of RB to function as a tumor suppressor and its ability to control pol III.

Although wild-type RB is present in most cancers, its function is generally found to be compromised. Indeed, it has been suggested that the regulatory pathway involving RB may be disrupted in all human malignancies.[132] A survey of seven human cervical carcinoma cell lines found that two bear small inactivating mutations in RB.[194] Whereas neither of these lines were infected by human papillomavirus (HPV), each of the remaining five lines that expressed wild-type RB also contained HPV DNA.[194] HPVs have an etiologic role in most cervical neoplasias.[195] The viral E7 gene product that is expressed in these tumors binds to RB and contributes significantly to the oncogenic capacity of the HPVs. Thus, RB function may be lost in most if not all cervical cancers, either by gene mutation in the minority of HPV-negative cases or by complex formation with E7 protein.[194]

The oncoproteins of several other DNA tumor viruses can also bind RB and neutralize its function.[195] This property is displayed by adenoviral E1A[196,197] and SV40 large T antigen.[198-201] Extensive mutagenesis has shown that the regions of these oncoproteins that are required for binding RB are also necessary for their transforming properties.[197-200] Furthermore, the parts of RB that are needed for interactions with E1A and T antigen are common sites for mutational inactivation.[202] By binding RB, these viral proteins are thought to interfere with its normal cellular functions, thereby mimicking the effects of the Rb mutations that occur in many tumors. Consistent with this, both E1A and T antigen can relieve pol III transcription from repression by RB.[162,186]

The activity of RB can also be switched off through phosphorylation by the cyclin D- and E-dependent kinases.[131,132,203,204] This is part of the normal control mechanism that is used to regulate progress through the cell cycle.[131,132,203,204] Cyclin D-dependent kinases are hyperactive in a variety of cancers and this provides another mechanism whereby RB function is lost.[132,203-205] For example, the gene for cyclin D1 is amplified in at least 15% of primary breast cancers and an even greater proportion of squamous cell carcinomas of the head, neck, esophagus and lung.[203,205] Cyclin D1 RNA and protein is overexpressed in 30-40% of primary breast tumors, indicating that gene amplification is not the only mechanism responsible for increased levels of the product.[205] In some parathyroid adenomas and B cell lymphomas, cyclin D1 is overproduced due to chromosomal translocations.[203,205] In addition to these situations in which cyclins are affected directly, many other cancers have lost the function of p15 and/or p16, which are repressors of the cyclin D-dependent kinases.[132,203,206] For example, the genes for p15 and p16 are deleted in many glioblastomas, esophageal, bladder, lung and pancreatic carcinomas, and are sometimes mutated in familial melanomas.[132,206] Thus, cyclin D-dependent kinase activity is abnormally elevated in a broad spectrum of cancers and this has the effect of switching off RB.

It is therefore clear that RB function is lost in a large proportion of human tumors. This can occur in a variety of ways: gene mutation, association with viral oncoproteins, or hyperphosphorylation. One survey of small cell lung cancers provided a good illustration of the importance of inactivating RB during tumorigenesis.[207] This study tested 55 small cell lung carcinomas and found that 48 lacked normal RB expression but contained wild-type p16: of the remaining seven, six lacked functional p16.[207] Since RB plays a major rôle in repressing class III gene expression, at least in some cell types, it is highly likely that its inactivation is a frequent and substantial contributor to the elevated levels of pol III transcription that characterize so many cancers.

A similar argument could be made for p53, which is mutated in more than half of all human tumors.[208] However, most of these inactivating mutations are clustered in the central domain of p53, which is responsible for DNA binding.[209] It has yet to be determined if this domain is involved in regulating pol III, and whether or not such regulation is compromised by the mutations that arise in tumors.

References

1. Peterson RC, Doering JL, Brown DD. Characterization of two *Xenopus* somatic 5S DNA and one minor oocyte-specific DNA. Cell 1980; 20:131-141.
2. Harper ME, Price J, Korn LJ. Chromosomal mapping of *Xenopus* 5S genes: somatic-type versus oocyte-type. Nucleic Acids Res 1983; 11:2313-2333.
3. Fedoroff NV, Brown DD. The nucleotide sequence of oocyte 5S DNA in *Xenopus laevis*. I. The AT-rich spacer. Cell 1978; 13:701-716.
4. Miller JR, Cartwright EM, Brownlee GG et al. The nucleotide sequence of oocyte 5S DNA in *Xenopus laevis*. II. The GC-rich region. Cell 1978; 13:717-725.
5. Brown DD, Dawid IB. Specific gene amplification in oocytes. Science 1968; 160:272-280.
6. Gall JG. Differential synthesis of the genes for ribosomal RNA during amphibian oogenesis. Proc Natl Acad Sci USA 1968; 60:553-560.
7. Wegnez M, Monier R, Denis H. Sequence heterogeneity of 5S RNA in *Xenopus laevis*. FEBS Lett 1972; 25:13-20.
8. Ford PJ, Southern EM. Different sequences for 5S RNA in kidney cells and ovaries of *Xenopus laevis*. Nature New Biol 1973; 241:7-12.
9. Brown DD, Carroll D, Brown RD. The isolation and characterization of a second oocyte 5S DNA from *Xenopus laevis*. Cell 1977; 12:1045-1056.
10. Wakefield L, Gurdon JB. Cytoplasmic regulation of 5S RNA genes in nuclear-transplant embryos. EMBO J 1983; 2:1613-1619.
11. Newport J, Kirschner M. A major developmental transition in early *Xenopus* embryos: II. Control of the onset of transcription. Cell 1982; 30:687-696.
12. Prioleau M-N, Huet J, Sentenac A et al. Competition between chromatin and transcription complex assembly regulates gene expression during early development. Cell 1994; 77:439-449.
13. Almouzni G, Wolffe AP. Constraints on transcriptional activator function contribute to transcriptional quiescence during early *Xenopus* embryogenesis. EMBO J 1995; 14:1752-1765.
14. Woodland HR, Adamson EP. The synthesis and storage of histone during the oogenesis of *Xenopus laevis*. Dev Biol 1977; 57:118-135.
15. Wormington WM, Brown DD. Onset of 5S RNA gene regulation during *Xenopus* embryogenesis. Dev Biol 1983; 99:248-257.
16. Korn LJ, Gurdon JB. The reactivation of developmentally inert 5S genes in somatic nuclei injected into *Xenopus* oocytes. Nature 1981; 289:461-465.
17. Korn LJ. Transcription of *Xenopus* 5S ribosomal RNA genes. Nature 1982; 295:101-105.
18. Brown DD. The role of stable complexes that repress and activate eukaryotic genes. Cell 1984; 37:359-365.
19. Wolffe AP, Brown DD. Developmental regulation of two 5S ribosomal RNA genes. Science 1988; 241:1626-1632.

20. Millstein LS, Gottesfeld JM. Control of gene expression in eukaryotic cells: lessons from class III genes. Curr Opin Cell Biol 1989; 1:497-502.
21. Wormington WM, Bogenhagen DF, Jordan E et al. A quantitative assay for *Xenopus* 5S RNA gene transcription in vitro. Cell 1981; 24:809-817.
22. Sakonju S, Brown DD. Contact points between a positive transcription factor and the *Xenopus* 5S RNA gene. Cell 1982; 31:395-405.
23. Gottesfeld JM, Bloomer LS. Assembly of transcriptionally active 5S RNA gene chromatin in vitro. Cell 1982; 28:781-791.
24. Brown DD, Schlissel MS. A positive transcription factor controls the differential expression of two 5S RNA genes. Cell 1985; 42:759-767.
25. Andrews MT, Brown DD. Transient activation of oocyte 5S RNA genes in *Xenopus* embryos by raising the level of the trans-acting factor TFIIIA. Cell 1987; 51:445-453.
26. McConkey GA, Bogenhagen DF. TFIIIA binds with equal affinity to somatic and major oocyte 5S RNA genes. Genes Dev 1988; 2:205-214.
27. Blanco J, Millstein L, Razik MA et al. Two TFIIIA activities regulate expression of the *Xenopus* 5S RNA gene families. Genes Dev 1989; 3:1602-1612.
28. Xing YY, Worcel A. The C-terminal domain of transcription factor IIIA interacts differently with different 5S RNA genes. Mol Cell Biol 1989; 9:499-514.
29. Shastry BS, Honda BM, Roeder RG. Altered levels of a 5S gene-specific transcription factor (TFIIIA) during oogenesis and embryonic development of *Xenopus laevis*. J Biol Chem 1984; 259:11373-11382.
30. Taylor W, Jackson IJ, Siegel N et al. The developmental expression of the gene for TFIIIA in *Xenopus laevis*. Nucleic Acids Res 1986; 14:6185-6195.
31. Andrews MT, Loo S, Wilson LR. Coordinate inactivation of class III genes during the Gastrula-Neurula Transition in *Xenopus*. Dev Biol 1991; 146:250-254.
32. Lund E, Dahlberg JE. Control of 4-8S RNA transcription at the midblastula transition in *Xenopus laevis* embryos. Genes Dev 1992; 6:1097-1106.
33. Gilbert DM. Temporal order of replication of *Xenopus laevis* 5S ribosomal RNA genes in somatic cells. Proc Natl Acad Sci USA 1986; 83:2924-2928.
34. Guinta DR, Korn LJ. Differential order of replication of *Xenopus laevis* 5S RNA genes. Mol Cell Biol 1986; 6:2536-2542.
35. Guinta DR, Tso JY, Narayanswami S et al. Early replication and expression of oocyte-type 5S RNA genes in a *Xenopus* somatic cell line carrying a translocation. Proc Natl Acad Sci USA 1986; 83:5150-5154.
36. Wolffe AP, Brown DD. Differential 5S RNA gene expression in vitro. Cell 1987; 51:733-740.
37. Reynolds W. Effect of sequence differences between somatic and oocyte 5S RNA genes on transcriptional efficiency in an oocyte S150 extract. Mol Cell Biol 1988; 8:5056-5058.
38. Reynolds WF, Azer K. Sequence differences upstream of the promoters are involved in the differential expression of the *Xenopus* somatic and oocyte 5S RNA genes. Nucleic Acids Res 1988; 16:3391-3403.
39. Reynolds WF. Sequences preceding the minimal promoter of the *Xenopus* somatic 5S RNA gene increase binding efficiency for transcription factors. Nucleic Acids Res 1989; 17:9381-9394.
40. Keller HJ, Romaniuk PJ, Gottesfeld JM. Interaction of *Xenopus* TFIIIC with the TFIIIA.5S RNA gene complex. J Biol Chem 1992; 267:18190-18198.
41. Peck LJ, Millstein L, Eversole-Cire P et al. Transcriptionally inactive oocyte-type 5S RNA genes of *Xenopus laevis* are complexed with TFIIIA in vitro. Mol Cell Biol 1987; 7:3503-3510.

42. Wolffe AP. Transcription fraction TFIIIC can regulate differential *Xenopus* 5S RNA gene transcription in vitro. EMBO J 1988; 4:1071-1079.
43. Keller HJ, You QM, Romaniuk PJ et al. Additional intragenic promoter elements of the *Xenopus* 5S RNA genes upstream from the TFIIIA-binding site. Mol Cell Biol 1990; 10:5166-5176.
44. Millstein L, Eversole-Cire P, Blanco J et al. Differential transcription of *Xenopus* oocyte and somatic-type 5S genes in a *Xenopus* oocyte extract. J Biol Chem 1987; 262:1-11.
45. Seidel CW, Peck LJ. Kinetic control of 5S RNA gene transcription. J Mol Biol 1992; 227:1009-1018.
46. Lassar AB, Martin PL, Roeder RG. Transcription of class III genes: formation of preinitiation complexes. Science 1983; 222:740-748.
47. Brown DD. Is there a *Xenopus* transcription factor that can substitute for TFIIIA? Re: Two TFIIIA activities regulate expression of the *Xenopus* 5S RNA gene families [letter]. Genes Dev 1991; 5:1737-1738.
48. Bogenhagen DF, Wormington WM, Brown DD. Stable transcription complexes of *Xenopus* 5S RNA genes: a means to maintain the differentiated state. Cell 1982; 28:413-421.
49. Engelke DR, Gottesfeld JM. Chromosomal footprinting of transcriptionally active and inactive oocyte-type 5S RNA genes of *Xenopus laevis*. Nucleic Acids Res 1990; 18:6031-6037.
50. Chipev CC, Wolffe AP. Chromosomal organization of *Xenopus laevis* oocyte and somatic 5S rRNA genes in vivo. Mol Cell Biol 1992; 12:45-55.
51. Schlissel MS, Brown DD. The transcriptional regulation of *Xenopus* 5S RNA genes in chromatin: the roles of active stable transcription complexes and histone H1. Cell 1984; 37:903-913.
52. Darby MK, Andrews TM, Brown DD. Transcription complexes that program *Xenopus* 5S RNA genes are stable in vivo. Proc Natl Acad Sci USA 1988; 85:5516-5520.
53. Gurdon JB, Dingwall C, Laskey RA et al. Developmental inactivity of 5S RNA genes persists when chromosomes are cut between genes. Nature 1982; 299:652-653.
54. Wolffe AP. Dominant and specific repression of *Xenopus* oocyte 5S RNA genes and satellite I DNA by histone H1. EMBO J 1989; 8:527-537.
55. Wolffe AP. Developmental regulation of chromatin structure and function. Trends Cell Biol 1991; 1:61-66.
56. Dimitrov S, Almouzni G, Dasso M et al. Chromatin transitions during early *Xenopus* embryogenesis: Changes in histone H4 acetylation and in linker histone type. Dev Biol 1993; 160:214-227.
57. Nightingale K, Dimitrov S, Reeves R et al. Evidence for a shared structural role for HMG1 and linker histones B4 and H1 in organizing chromatin. EMBO J 1996; 15:548-561.
58. Ura K, Nightingale K, Wolffe AP. Differential association of HMG1 and linker histones B4 and H1 with dinucleosomal DNA: structural transitions and transcriptional repression. EMBO J 1996; 15:4959-4969.
59. Bouvet P, Dimitrov S, Wolffe AP. Specific regulation of *Xenopus* chromosomal 5S rRNA gene transcription in vivo by histone H1. Genes Dev 1994; 8:1147-1159.
60. Kandolf H. The H1A histone variant is an in vivo repressor of oocyte-type 5S gene transcription in *Xenopus laevis* embryos. Proc Natl Acad Sci USA 1994; 91:7257-7261.
61. Andrews DL, Millstein L, Hamkalo BA et al. Competition between *Xenopus* satellite I sequences and pol III genes for stable transcription complex formation. Nucleic Acids Res 1984; 12:7753-7769.

62. Almouzni G, Mechali M, Wolffe AP. Competition between transcription complex assembly and chromatin assembly on replicating DNA. EMBO J 1990; 9:573-582.
63. Almouzni G, Mechali M, Wolffe AP. Transcription complex disruption caused by a transition in chromatin structure. Mol Cell Biol 1991; 11:655-665.
64. Renz M, Day LA. Transition from non-cooperative to cooperative and selective binding of histone H1 to DNA. Biochemistry 1976; 15:3220-3228.
65. Jerzmanowski A, Cole RD. Flanking sequences of *Xenopus* 5S RNA genes determine differential inhibition by H1 histone in vitro. J Biol Chem 1990; 265:10726-10732.
66. Shimamura A, Sapp M, Rodriquez-Campos A et al. Histone H1 represses transcription from minichromosomes assembled in vitro. Mol Cell Biol 1989; 9:5573-5584.
67. Stutz F, Gouilloud E, Clarkson SG. Oocyte and somatic tyrosine tRNA genes in *Xenopus laevis*. Genes Dev 1989; 3:1190-1198.
68. Reynolds WF, Johnson DL. Differential expression of oocyte-type class III genes with fraction TFIIIC from immature or mature oocytes. Mol Cell Biol 1992; 12:946-953.
69. Reynolds WF. The tyrosine phosphatase cdc25 selectively inhibits transcription of the *Xenopus* oocyte-type tRNAtyrC gene. Nucleic Acids Res 1993; 21:4372-4377.
70. Gouilloud E, Clarkson SG. A dispersed tyrosine tRNA gene from *Xenopus laevis* with high transcriptional activity in vitro. J Biol Chem 1986; 261:486-494.
71. Kaplan G, Jelinek WR, Bachvarova R. Repetitive sequence transcripts and U1 RNA in mouse oocytes and eggs. Dev Biol 1985; 109:15-24.
72. Taylor KD, Piko L. Patterns of mRNA prevalence and expression of B1 and B2 transcripts in early mouse embryos. Development 1987; 101:877-892.
73. Bachvarova R. Small B2 RNAs in mouse oocytes, embryos, and somatic tissues. Dev Biol 1988; 130:513-523.
74. Rothstein JL, Johnson D, DeLoia JD et al. Gene expression during preimplantation mouse development. Genes Dev 1992; 6:1190-1201.
75. Nothias J-Y, Miranda M, DePamphilis ML. Uncoupling of transcription and translation during zygotic gene activation in the mouse. EMBO J 1996; 15:5715-5725.
76. Kramerov DA, Tillib SV, Lekakh IV et al. Biosynthesis and cytoplasmic distribution of small poly(A)-containing B2 RNA. Biochim Biophys Acta 1985; 824:85-98.
77. Murphy D, Brickell PM, Latchman DS et al. Transcripts regulated during normal embryonic development and oncogenic transformation share a repetitive element. Cell 1983; 35:865-871.
78. Vasseur M, Condamine H, Duprey P. RNAs containing B2 repeated sequences are transcribed in the early stages of mouse embryogenesis. EMBO J 1985; 4:1749-1753.
79. Ryskov AP, Ivanov PL, Kramerov DA et al. Mouse ubiquitous B2 repeat in polysomal and cytoplasmic poly(A)$^+$ RNAs: unidirectional orientation and 3'-end localization. Nucl Acids Res 1983; 11:6541-6558.
80. Ryskov AP, Ivanov PL, Tokarskaya ON et al. Major transcripts containing B1 and B2 repetitive sequences in cytoplasmic poly(A)$^+$ RNA from mouse tissues. FEBS Letters 1985; 182:73-76.
81. Bernstine EG, Hooper ML, Grandchamp S et al. Alkaline phosphatase activity in mouse teratoma. Proc Natl Acad Sci USA 1973; 70:3899-3903.
82. Hogan BLM, Taylor A, Adamson E. Cell interactions modulate embryonal carcinoma cell differentiation into parietal or visceral endoderm. Nature 1981; 291:235-237.
83. Strickland S, Smith KK, Marotti KR. Hormonal induction of differentiation in teratocarcinoma stem cells: generation of parietal endoderm by retinoic acid and dibutyryl cAMP. Cell 1980; 21:347-355.

84. Bennett KL, Hill RE, Pietras DF et al. Most highly repeated dispersed DNA families in the mouse genome. Mol Cell Biol 1984; 4:1561-1571.
85. Vasseur M, Duprey P, Brulet P et al. One gene and one pseudogene for the cytokeratin endo A. Proc Natl Acad Sci USA 1985; 82:1155-1159.
86. White RJ, Stott D, Rigby PWJ. Regulation of RNA polymerase III transcription in response to F9 embryonal carcinoma stem cell differentiation. Cell 1989; 59:1081-1092.
87. Bladon TS, Fregeau CJ, McBurney MW. Synthesis and processing of small B2 transcripts in mouse embryonal carcinoma cells. Mol Cell Biol 1990; 10:4058-4067.
88. Gokal PK, Cavanaugh AH, Thompson EA. The effects of cycloheximide upon transcription of rRNA, 5S RNA, and tRNA genes. J Biol Chem 1986; 261:2536-2541.
89. White RJ, Stott D, Rigby PWJ. Regulation of RNA polymerase III transcription in response to Simian virus 40 transformation. EMBO J 1990; 9:3713-3721.
90. Meißner W, Ahlers A, Seifart KH. The activity of transcription factor PBP, which binds to the proximal sequence element of mammalian U6 genes, is regulated during differentiation of F9 cells. Mol Cell Biol 1995; 15:5888-5897.
91. Alzuherri HM, White RJ. Unpublished observations.
92. Matthews JL, Zwick MG, Paule MR. Coordinate regulation of ribosomal component synthesis in *Acanthamoeba castellanii*: 5S RNA transcription is down regulated during encystment by alteration of TFIIIA activity. Mol Cell Biol 1995; 15:3327-3335.
93. Hatfield D, Varricchio F, Rice M et al. The aminoacyl-tRNA population of human reticulocytes. J Biol Chem 1982; 257:3183-3188.
94. Sprague KU, Hagenbuchle O, Zuniga MC. The nucleotide sequence of two silk gland alanine tRNAs: implications for fibroin synthesis and for initiator tRNA structure. Cell 1977; 11:561-570.
95. Underwood DC, Knickerbocker H, Gardner G et al. Silkgland-specific $tRNA^{Ala}$ genes are tightly clustered in the silkworm genome. Mol Cell Biol 1988; 8:5504-5512.
96. Young LS, Takahashi N, Sprague KU. Upstream sequences confer distinctive transcriptional properties on genes encoding silkgland-specific $tRNA^{Ala}$. Proc Natl Acad Sci USA 1986; 83:374-378.
97. Sullivan HS, Young LS, White CN et al. Silk gland-specific $tRNA^{Ala}$ genes interact more weakly than constitutive $tRNA^{Ala}$ genes with silkworm TFIIIB and polymerase III fractions. Mol Cell Biol 1994; 14:1806-1814.
98. Young LS, Ahnert N, Sprague KU. Silkworm TFIIIB binds both constitutive and silk gland-specific $tRNA^{Ala}$ promoters but protects only the constitutive promoter from DNase I cleavage. Mol Cell Biol 1996; 16:1256-1266.
99. Clarke EM, Peterson CL, Brainard AV et al. Regulation of the RNA polymerase I and III transcription systems in response to growth conditions. J Biol Chem 1996; 271:22189-22195.
100. Sethy I, Moir RD, Librizzi M et al. In vitro evidence for growth regulation of tRNA gene transcription in yeast. J Biol Chem 1995; 270:28463-28470.
101. Lopez-de-Leon A, Librizzi M, Tuglia K et al. PCF4 encodes an RNA polymerase III transcription factor with homology to TFIIB. Cell 1992; 71:211-220.
102. Huibregtse JM, Engelke DR. Genomic footprinting of a yeast tRNA gene reveals stable complexes over the 5'-flanking region. Mol Cell Biol 1989; 9:3244-3252.
103. Carey MF, Singh K. Enhanced B2 transcription in simian virus 40-transformed cells is mediated through the formation of RNA polymerase III transcription complexes on previously inactive genes. Proc Natl Acad Sci USA 1988; 85:7059-7063.
104. Tower J, Sollner-Webb B. Polymerase III transcription factor B activity is reduced in extracts of growth-restricted cells. Mol Cell Biol 1988; 8:1001-1005.

105. Johnson LF, Abelson HT, Green H et al. Changes in RNA in relation to growth of the fibroblast. I. Amounts of mRNA, rRNA, and tRNA in resting and growing cells. Cell 1974; 1:95-100.
106. Mauck JC, Green H. Regulation of pre-transfer RNA synthesis during transition from resting to growing state. Cell 1974; 3:171-177.
107. Ortwerth BJ, Liu LP. Correlation between a specific isoaccepting lysyl transfer ribonucleic acid and cell division in mammalian tissues. Biochemistry 1973; 12:3978-3984.
108. Johnson LF, Levis R, Abelson HT et al. Changes in RNA in relation to growth of the fibroblast. IV. Alterations in the production and processing of mRNA and rRNA in resting and growing cells. J Cell Biol 1976; 71:933-938.
109. Abelson HT, Johnson LF, Penman S et al. Changes in RNA in relation to growth of the fibroblast: II. The lifetime of mRNA, rRNA, and tRNA in resting and growing cells. Cell 1974; 1:161-165.
110. Schlegel RA, Iversen P, Rechsteiner M. The turnover of tRNAs microinjected into animal cells. Nucleic Acids Res 1978; 5:3715-3729.
111. Edwards DR, Parfett CLJ, Denhardt DT. Transcriptional regulation of two serum-induced RNAs in mouse fibroblasts: equivalence of one species to B2 repetitive elements. Mol Cell Biol 1985; 5:3280-3288.
112. Lania L, Pannuti A, La Mantia G et al. The transcription of B2 repeated sequences is regulated during the transition from quiescent to proliferative state in cultured rodent cells. FEBS Lett 1987; 219:400-404.
113. Hoeffler WK, Kovelman R, Roeder RG. Activation of transcription factor IIIC by the adenovirus E1A protein. Cell 1988; 53:907-920.
114. Sinn E, Wang Z, Kovelman R et al. Cloning and characterization of a TFIIIC2 subunit (TFIIICβ) whose presence correlates with activation of RNA polymerase III-mediated transcription by adenovirus E1A expression and serum factors. Genes Dev 1995; 9:675-685.
115. Garber M, Panchanathan S, Fan RS et al. The phorbol ester, 12-O-tetradecanoylphorbol-13-acetate, induces specific transcription by RNA polymerase III in *Drosophila* Schneider cells. J Biol Chem 1991; 266:20598-20601.
116. Garber M, Vilalta A, Johnson DL. Induction of *Drosophila* RNA polymerase III gene expression by the phorbol ester 12-O-tetradecanoylphorbol-13-acetate (TPA) is mediated by transcription factor IIIB. Mol Cell Biol 1994; 14:339-347.
117. Trivedi A, Vilalta A, Gopalan S et al. TATA-binding protein is limiting for both TATA-containing and TATA-lacking RNA polymerase III promoters in *Drosophila* cells. Mol Cell Biol 1996; 16:6909-6916.
118. Prescott DM, Bender MA. Synthesis of RNA and protein during mitosis in mammalian tissue culture cells. Exp Cell Res 1962; 26:260-268.
119. Terasima T, Tolmach LJ. Growth and nucleic acid synthesis in synchronously dividing populations of HeLa cells. Exp Cell Res 1963; 30:344-362.
120. Davidson D. RNA synthesis in roots of *Vicia faba*. Exp Cell Res 1964; 35:317-325.
121. Fink K, Turnock G. Synthesis of transfer RNA during the synchronous nuclear division cycle in *Physarum polycephalum*. Eur J Biochem 1977; 80:93-96.
122. Gottesfeld JM, Forbes DJ. Mitotic repression of the transcriptional machinery. Trends Biochem Sci 1997; 22:197-202.
123. Hartl P, Gottesfeld J, Forbes DJ. Mitotic repression of transcription in vitro. J Cell Biol 1993; 120:613-624.
124. Wolf VJ, Dang T, Hartl P et al. Role of maturation-promoting factor ($p34^{cdc2}$-cyclin B) in differential expression of the *Xenopus* oocyte and somatic-type 5S RNA genes. Mol Cell Biol 1994; 14:4704-4711.

125. Gottesfeld JM, Wolf VJ, Dang T et al. Mitotic repression of RNA polymerase III transcription in vitro mediated by phosphorylation of a TFIIIB component. Science 1994; 263:81-84.
126. Leresche A, Wolf VJ, Gottesfeld JM. Repression of RNA polymerase II and III transcription during M phase of the cell cycle. Exp Cell Res 1996; 229:282-288.
127. White RJ, Gottlieb TM, Downes CS et al. Mitotic regulation of a TATA-binding-protein-containing complex. Mol Cell Biol 1995; 15:1983-1992.
128. McLees A, White RJ. Unpublished observations.
129. White RJ, Gottlieb TM, Downes CS et al. Cell cycle regulation of RNA polymerase III transcription. Mol Cell Biol 1995; 15:6653-6662.
130. Larminie CGC, Cairns CA, Mital R et al. Mechanistic analysis of RNA polymerase III regulation by the retinoblastoma protein. EMBO J 1997; 16:2061-2071.
131. Sherr CJ. G1 phase progression: cycling on cue. Cell 1994; 79:551-555.
132. Weinberg RA. The retinoblastoma protein and cell cycle control. Cell 1995; 81:323-330.
133. Panning B, Smiley JR. Activation of RNA polymerase III transcription of human Alu repetitive elements by adenovirus type 5: requirement for the E1b 58-kilodalton protein and the products of E4 open reading frames 3 and 6. Mol Cell Biol 1993; 13:3231-3244.
134. Liu W-M, Chu W-M, Choudary PV et al. Cell stress and translational inhibitors transiently increase the abundance of mammalian SINE transcripts. Nucleic Acids Res 1995; 23:1758-1765.
135. Dieci G, Duimio L, Peracchia G et al. Selective inactivation of two components of the multiprotein transcription factor TFIIIB in cycloheximide growth-arrested yeast cells. J Biol Chem 1995; 270:13476-13482.
136. Bohr VA, Smith CA, Okumoto DS et al. DNA repair in an active gene: removal of pyrimidine dimers from the DHFR gene of CHO cells is much more efficient than in the genome overall. Cell 1985; 40:359-369.
137. Vos J-M, Wauthier EL. Differential introduction of DNA damage and repair in mammalian genes transcribed by RNA polymerase I and II. Mol Cell Biol 1991; 11:2245-2252.
138. Dammann R, Pfeifer GP. Lack of gene- and strand-specific DNA repair in RNA polymerase III-transcribed human tRNA genes. Mol Cell Biol 1997; 17:219-229.
139. Fornace AJ, Mitchell JB. Induction of B2 RNA polymerase III transcription by heat shock: enrichment for heat shock induced sequences in rodent cells by hybridization subtraction. Nucleic Acids Res 1986; 14:5793-5811.
140. Fornace AJ, Alamo I, Hollander MC et al. Induction of heat shock protein transcripts and B2 transcripts by various stresses in Chinese hamster cells. Exp Cell Res 1989; 182:61-74.
141. Fradkin LG, Yoshinaga SK, Berk AJ et al. Inhibition of host cell RNA polymerase III-mediated transcription by poliovirus: inactivation of specific transcription factors. Mol Cell Biol 1987; 7:3880-3887.
142. Clark ME, Dasgupta A. A transcriptionally active form of TFIIIC is modified in poliovirus-infected HeLa cells. Mol Cell Biol 1990; 10:5106-5113.
143. Shen Y, Igo M, Yalamanchili P et al. DNA binding domain and subunit interactions of transcription factor IIIC revealed by dissection with poliovirus 3C protease. Mol Cell Biol 1996; 16:4163-4171.
144. Kovelman R, Roeder RG. Purification and characterization of two forms of human transcription factor IIIC. J Biol Chem 1992; 267:24446-24456.
145. Clark ME, Hammerle T, Wimmer E et al. Poliovirus proteinase 3C converts an active form of transcription factor IIIC to an inactive form: a mechanism for inhibition of host cell polymerase III transcription by poliovirus. EMBO J 1991; 10:2941-2947.

146. Clark ME, Lieberman PM, Berk AJ et al. Direct cleavage of human TATA-binding protein by poliovirus protease 3C in vivo and in vitro. Mol Cell Biol 1993; 13:1232-1237.
147. White RJ, Jackson SP. Mechanism of TATA-binding protein recruitment to a TATA-less class III promoter. Cell 1992; 71:1041-1053.
148. Soderlund H, Pettersson U, Vennstom B et al. A new species of virus-coded low molecular weight RNA from cells infected with Adenovirus type 2. Cell 1976; 7:585-593.
149. Berger SL, Folk WR. Differential activation of RNA polymerase III-transcribed genes by the polyomavirus enhancer and the adenovirus E1A gene products. Nucleic Acids Res 1985; 13:1413-1428.
150. Gaynor RB, Feldman LT, Berk AJ. Transcription of class III genes activated by viral immediate early proteins. Science 1985; 230:447-450.
151. Hoeffler WK, Roeder RG. Enhancement of RNA polymerase III transcription by the E1A gene product of adenovirus. Cell 1985; 41:955-963.
152. Yoshinaga S, Dean N, Han M et al. Adenovirus stimulation of transcription by RNA polymerase III: evidence for an E1A-dependent increase in transcription factor IIIC concentration. EMBO J 1986; 5:343-354.
153. Sollerbrant K, Akusjarvi G, Svensson C. Repression of RNA polymerase III transcription by adenovirus E1A. J Virol 1993; 67:4195-4204.
154. Russanova VR, Driscoll CT, Howard BH. Adenovirus type 2 preferentially stimulates polymerase III transcription of *Alu* elements by relieving repression: a potential role for chromatin. Mol Cell Biol 1995; 15:4282-4290.
155. Ahlers SE, Feldman LT. Effects of a temperature-sensitive mutation in the immediate-early gene of pseudorabies virus on class II and class III gene transcription. J Virol 1987; 61:1103-1107.
156. Aufiero B, Schneider RJ. The hepatitis B virus X-gene product trans-activates both RNA polymerase II and III promoters. EMBO J 1990; 9:497-504.
157. Loeken M, Bikel I, Livingston DM et al. Trans-activation of RNA polymerase II and III promoters by SV40 small t antigen. Cell 1988; 55:1171-1177.
158. Panning B, Smiley JR. Activation of RNA polymerase III transcription of human *Alu* elements by herpes simplex virus. Virology 1994; 202:408-417.
159. Patel G, Jones NC. Activation in vitro of RNA polymerase II and III directed transcription by baculovirus produced E1A protein. Nucleic Acids Res 1990; 18:2909-2915.
160. Datta S, Soong CJ, Wang DM et al. A purified adenovirus 289-amino-acid E1A protein activates RNA polymerase III transcription in vitro and alters transcription factor TFIIIC. J Virol 1991; 65:5297-5304.
161. Kovelman R, Roeder RG. Sarkosyl defines three intermediate steps in transcription initiation by RNA polymerase III: application to stimulation of transcription by E1A. Genes Dev 1990; 4:646-658.
162. White RJ, Trouche D, Martin K et al. Repression of RNA polymerase III transcription by the retinoblastoma protein. Nature 1996; 382:88-90.
163. Kraus VB, Inostroza JA, Yeung K et al. Interaction of the Dr1 inhibitory factor with the TATA binding protein is disrupted by adenovirus E1A. Proc Natl Acad Sci USA 1994; 91:6279-6282.
164. White RJ, Khoo BC-E, Inostroza JA et al. The TBP-binding repressor Dr1 differentially regulates RNA polymerases I, II and III. Science 1994; 266:448-450.
165. Chu W-M, Wang Z, Roeder RG et al. RNA polymerase III transcription repressed by Rb through its interactions with TFIIIB and TFIIIC2. J Biol Chem 1997; 272:14755-14761.

166. Jang KL, Latchmann DS. HSV infection induces increased transcription of Alu repeated sequences by RNA polymerase III. FEBS Lett 1989; 258:255-258.
167. Jang KL, Latchman DS. The herpes simplex virus immediate-early protein ICP27 stimulates the transcription of cellular Alu repeated sequences by increasing the activity of transcription factor TFIIIC. Biochem J 1992; 284:667-673.
168. Kim C-M, Koike K, Saito I et al. *HBx* gene of hepatitis B virus induces liver cancer in transgenic mice. Nature 1991; 351:317-320.
169. Kwee L, Lucito R, Aufiero B et al. Alternate translation initiation on hepatitis B virus X mRNA produces multiple polypeptides that differentially transactivate class II and III promoters. J Virol 1992; 66:4382-4389.
170. Wang H-D, Yuh C-H, Dang CV et al. The hepatitis B virus X protein increases the cellular level of TATA-binding protein, which mediates transactivation of RNA polymerase III genes. Mol Cell Biol 1995; 15:6720-6728.
171. Cheong J, Yi M, Lin Y et al. Human RPB5, a subunit shared by eukaryotic nuclear RNA polymerases, binds human hepatitis B virus X protein and may play a role in X transactivation. EMBO J 1995; 14:143-150.
172. Kekule AS, Lauer U, Weiss L et al. Hepatitis B virus transactivator HBx uses a tumor promoter signalling pathway. Nature 1993; 361:742-745.
173. Qadri I, Maguire HF, Siddiqui A. Hepatitis B virus transactivator protein X interacts with the TATA-binding protein. Proc Natl Acad Sci USA 1995; 92:1003-1007.
174. James CBL, Carter TH. Activation of protein kinase C inhibits adenovirus VA gene transcription in vitro. J Gen Virol 1992; 73:3133-3139.
175. Piras G, Dittmer J, Radonovich MF et al. Human T-cell leukemia virus type I Tax protein transactivates RNA polymerase III promoter in vitro and in vivo. J Biol Chem 1996; 271:20501-20506.
176. Gottesfeld JM, Johnson DL, Nyborg JK. Transcriptional activation of RNA polymerase III-dependent genes by the human T-cell leukemia virus type 1 Tax protein. Mol Cell Biol 1996; 16:1777-1785.
177. Scott MRD, Westphal K-H, Rigby PWJ. Activation of mouse genes in transformed cells. Cell 1983; 34:557-567.
178. Kramerov DA, Lekakh IV, Samarina OP et al. The sequences homologous to major interspersed repeats B1 and B2 of mouse genome are present in mRNA and cytoplasmic poly(A)$^+$ RNA. Nucleic Acids Res 1982; 10:7477-7491.
179. Brickell PM, Latchman DS, Murphy D et al. Activation of a *Qa/Tla* class I major histocompatibility antigen gene is a general feature of oncogenesis in the mouse. Nature 1983; 306:756-760.
180. Majello B, La Mantia G, Simeone A et al. Activation of major histocompatibility complex class I mRNA containing an *Alu*-like repeat in polyoma virus-transformed rat cells. Nature 1985; 314:457-459.
181. Singh K, Carey M, Saragosti S et al. Expression of enhanced levels of small RNA polymerase III transcripts encoded by the B2 repeats in simian virus 40-transformed mouse cells. Nature 1985; 314:553-556.
182. Carey MF, Singh K, Botchan M et al. Induction of specific transcription by RNA polymerase III in transformed cells. Mol Cell Biol 1986; 6:3068-3076.
183. Kramerov DA, Tillib SV, Shumyatsky GP et al. The most abundant nascent poly(A)$^+$ RNAs are transcribed by RNA polymerase III in murine tumor cells. Nucleic Acids Res 1990; 18:4499-4506.
184. Chen W, Heierhorst J, Brosius J et al. Expression of neural *BC1* RNA: induction in murine tumors. Eur J Cancer 1997; 33:288-292.
185. DeChiara TM, Brosius J. Neural BC1 RNA: cDNA clones reveal nonrepetitive sequence content. Proc Natl Acad Sci USA 1987; 84:2624-2628.

186. Larminie CGC, White RJ. Unpublished observations.
187. Dean N, Berk AJ. Separation of TFIIIC into two functional components by sequence specific DNA affinity chromatography. Nucleic Acids Res 1987; 15:9895-9907.
188. Lee WS, Kao CC, Bryant GO et al. Adenovirus E1A activation domain binds the basic repeat in the TATA box transcription factor. Cell 1991; 67:365-376.
189. Gruda MC, Zabolotny JM, Xiao JH et al. Transcriptional activation by simian virus 40 large T antigen: interactions with multiple components of the transcription complex. Mol Cell Biol 1993; 13:961-969.
190. Chesnokov I, Chu W-M, Botchan MR et al. p53 inhibits RNA polymerase III-directed transcription in a promoter-dependent manner. Mol Cell Biol 1996; 16:7084-7088.
191. Whyte P. The retinoblastoma protein and its relatives. Seminars in Cancer Biology 1995; 6:83-90.
192. Horowitz JM, Park S-H, Bogenmann E et al. Frequent inactivation of the retinoblastoma anti-oncogene is restricted to a subset of human tumor cells. Proc Natl Acad Sci USA 1990; 87:2775-2779.
193. Kaye FJ, Kratzke RA, Gerster JL et al. A single amino acid substitution results in a retinoblastoma protein defective in phosphorylation and oncoprotein binding. Proc Natl Acad Sci USA 1990; 87:6922-6926.
194. Scheffner M, Munger K, Byrne JC et al. The state of the p53 and retinoblastoma genes in human cervical carcinoma cell lines. Proc Natl Acad Sci USA 1991; 88:5523-5527.
195. Vousden KH. Regulation of the cell cycle by viral oncoproteins. Seminars in Cancer Biology 1995; 6:109-116.
196. Whyte P, Buchkovich KJ, Horowitz JM et al. Association between an oncogene and an anti-oncogene: the adenovirus E1A proteins bind to the retinoblastoma gene product. Nature 1988; 334:124-129.
197. Whyte P, Williamson NM, Harlow E. Cellular targets for transformation by the adenovirus E1A proteins. Cell 1989; 56:67-75.
198. DeCaprio JA, Ludlow JW, Figge J et al. SV40 large tumor antigen forms a specific complex with the product of the retinoblastoma susceptibility gene. Cell 1988; 54:275-283.
199. Moran E. A region of SV40 large T antigen can substitute for a transforming domain of the adenovirus E1A products. Nature 1988; 334:168-170.
200. Ewen ME, Ludlow JW, Marsilio E et al. An N-terminal transformation-governing sequence of SV40 large T antigen contributes to the binding of both $p110^{Rb}$ and a second cellular protein, p120. Cell 1989; 58:257-267.
201. Ludlow JW, DeCaprio JA, Huang C-M et al. SV40 large T antigen binds preferentially to an underphosphorylated member of the retinoblastoma susceptibility gene product family. Cell 1989; 56:57-65.
202. Hu Q, Dyson N, Harlow E. The regions of the retinoblastoma protein needed for binding to adenovirus E1A or SV40 large T antigen are common sites for mutations. EMBO J 1990; 9:1147-1155.
203. Hunter T, Pines J. Cyclins and cancer II: cyclin D and CDK inhibitors come of age. Cell 1994; 79:573-582.
204. Pines J. Cyclins, CDKs and cancer. Seminars in Cancer Biology 1995; 6:63-72.
205. Bates S, Peters G. Cyclin D1 as a cellular proto-oncogene. Seminars in Cancer Biology 1995; 6:73-82.
206. Hirama T, Koeffler HP. Role of the cyclin-dependent kinase inhibitors in the development of cancer. Blood 1995; 86:841-854.

207. Otterson GA, Kratzke RA, Coxon A et al. Absence of $p16^{Ink4}$ protein is restricted to the subset of lung cancer lines that retains wildtype RB. Oncogene 1994; 9:3375-3378.
208. Hollstein M, Sidransky D, Vogelstein B et al. p53 mutations in human cancers. Science 1991; 253:49-53.
209. Haffner R, Oren M. Biochemical properties and biological effects of p53. Curr Opin Genetics Dev 1995; 5:84-90.

CHAPTER 10

Perspective

Our understanding of the details of pol III transcription has increased substantially in recent years. The characterization and cloning of many of the components of the system have been achieved through a combination of biochemical and genetic approaches. The availability of cloned components will greatly facilitate further structural and mechanistic studies. As a result, the pace of pol III research is likely to accelerate.

By far the most progress in this regard has been made using *S. cerevisiae* as a model system. A sustained effort by several laboratories over a period of years has resulted in the purification of the yeast pol III transcription machinery. The majority of polypeptides required for transcription of a class III gene in *S. cerevisiae* have now been cloned. This includes TFIIIA, all three subunits of TFIIIB, all six subunits of TFIIIC and fourteen subunits of pol III itself (qv chapters 3 and 4). TFIIIB activity has now been reconstituted using entirely recombinant subunits.[1-3] It may not be long before a yeast tRNA gene can be transcribed using only cloned components, which will be a landmark achievement. Furthermore, the order of interaction and spatial arrangement of many of these components on template DNA have now been documented. As such, this can be regarded as one of the best characterized eukaryotic transcriptional systems.

To what extent the information obtained using *S. cerevisiae* as a model system can be applied to higher organisms is an important issue that now faces the field. Yeast and human factors are generally incompatible for sustaining tRNA synthesis.[4] However, some functional conservation appears to be shown by TFIIIB, since yeast B" and TFIIIE are reported to substitute for hTFIIIB-α in supporting U6 transcription in a reconstituted human system.[4] The degree of sequence homology between these factors remains to be determined. As yet, only the TBP and BRF subunits of TFIIIB have had their genes cloned from multicellular organisms. TBP is extremely well conserved across evolution.[5] BRF, too, shows significant homology between yeast and metazoa.[6-8] The cloning of genes for mammalian pol III subunits has also revealed substantial conservation in several cases.[9-11] Indeed, four subunits have been shown to be functionally interchangeable between *H. sapiens* and *S. cerevisiae*.[10] There is also evidence for a conserved subcomplex that mediates the interaction between pol III and TFIIIB.[11] In contrast to TFIIIB and pol III, the assembly factors TFIIIC and TFIIIA appear remarkably diverged when vertebrates are compared with yeast. For example, the human B-block-binding polypeptide, TFIIICα, is ~60% larger than its yeast counterpart τ_{138}, and the two polypeptides display only 10% similarity, well below the score attributable to chance.[12,13] The overall similarity between human and rat TFIIICα (75%) is also below average

RNA Polymerase III Transcription, Second Edition, by Robert J. White.

for homologous nuclear genes from these two organisms.[13] Human TFIIICβ shows no significant homology to any of the cloned subunits of yeast TFIIIC.[14] Yeast factors are unable to assemble a stable complex on a human $tRNA^{Meti}$ gene and this gene is not transcribed in yeast cells.[15,16] This is very surprising, given the high conservation of ICR sequences. Indeed, the human factors are less selective and will accurately transcribe yeast tRNA genes.[4,15,17] TFIIIA is also poorly conserved. Although they both contain nine zinc fingers, frog and yeast TFIIIA are only ~20% identical, and the identity drops to ~8% if the consensus residues of the finger motif are excluded.[18] A yeast 5S gene is bound 100- to 1000-fold less tightly by frog TFIIIA than it is by yeast TFIIIA.[19] It therefore appears that the class III assembly factors have been subject to very little evolutionary maintenance. They may even have diverged to such a degree that unrelated genes now perform analogous functions in vertebrates and fungi. This contrasts strikingly with the situation for the basal pol II machinery, in which yeast and human homologues are readily identifiable and generally display 30-70% sequence identity. The subunits of pol III itself also tend to be less well conserved than those of pol II.[11] The use of sequence similarities to isolate homologous genes from different species has proved a very fruitful approach for molecular biology in general. The extreme divergence of the class III factors will often exclude this avenue for the pol III field. This constitutes a severe handicap that has hampered progress to date and, in all probability, will continue to do so. It also means that any conclusions drawn concerning the mechanics of pol III transcription in yeast cannot be assumed to apply in higher organisms. One of the most striking features of the yeast pol III system is the avid DNA binding by TFIIIB once recruited onto a template. Although it remains to be conclusively proved, the indications are that vertebrate TFIIIB does not share this property. If so, this difference may have significant mechanistic and regulatory consequences.

Pol III templates were the first eukaryotic genes to have their promoter regions mapped. The discovery that essential sequence elements are frequently located within the transcribed region came as a major surprise, because it contrasts with the situation that had been characterized in prokaryotes and was later found for the eukaryotic classes I and II. Subsequent studies brought a more gradual realization that upstream sequences also influence most, if not all, pol III transcription. It is now clear that class III genes cannot be rigidly separated from the other groups on the basis of their promoter structure. Although ICRs constitute essential promoter components for the majority of pol III templates, upstream sequences also perform at least a modulatory function in most cases and sometimes contain sequence motifs that are also found in class II genes. The continuum of promoter structure between classes II and III culminates with vertebrate U6 and 7SK genes, which no longer require an ICR, and with class II genes such as c-myc that can be transcribed by pol III under certain circumstances.

There are also significant similarities between the transcription complexes used by pols I, II and III. In each case the initiation complex contains TBP bound to a series of TAFs. One of the TAFs in the class III complex is highly homologous to the TFIIB polypeptide that is essential for the function of the class II complex. Furthermore, overlapping regions of TBP are involved in contacting the class II and class III factors.[20] Although TBP is certainly also part of the class I initiation complex, a class I homologue of TFIIB has not been identified.

The roles of TFIIB and BRF were probably established before the evolutionary divergence of the eukaryotic nuclear RNA polymerases. TFIIB and TFIIIB both associate with DNA near the transcription start site and both interact with RNA polymerase in order to select the position at which transcription initiates.[21-27] Like TFIIIB, TFIIB may have a cryptic DNA-binding potential that becomes activated during transcription complex assembly.[28]

In each system, initiation complex assembly is nucleated by a basal factor binding to a core sequence element. The resultant complex is insufficient to recruit polymerase productively. Instead, all three systems utilize a second factor to provide a molecular bridge between the nucleating factor and the polymerase. For pols I and III, this may be enough to generate a functional initiation complex. However, the class II system requires additional factors to allow initiation. This necessity may be related to the presence of the C terminal domain in the largest subunit of pol II that is absent from pols I and III.

The largest subunits of yeast pols II and III are more closely related to each other than they are to that of pol I.[29] This has led to the suggestion that two nuclear RNA polymerases may have existed at some stage of evolution—one being the precursor of pol I and the other being the precursor of pols II and III.[29] Paradoxically, however, pol III shares more common subunits with pol I than it does with pol II.

The shared components of the different polymerase systems may provide a means for coordinating gene expression. But they may also result in direct competition between the various classes. Cormack and Struhl[20] found that mutations in TBP that specifically prevent it from interacting with class III factors result in increased expression of class II genes in yeast cells. Conversely, pol II-specific mutations in TBP produce a slight increase in pol III transcription. Clearly, such competition demands that any shared components be tightly regulated.

The fact that the class II and class III systems share a number of common promoter elements and factors raises the question as to how polymerase-specificity is achieved. For example, Roberts et al[30] reported that TFIIB and BRF bind equally well to a TBP/TATA complex whether the DNA originates from a class II or a class III promoter. Indeed, the TATA region of the yeast U6 gene directs transcription by both pols II and III in the absence of the downstream U6 elements.[30-33] Raising the level of BRF stimulates pol III transcription at the expense of pol II, indicating a direct competition between the two classes of factors at this site.[30] It is the B-block that converts the yeast U6 gene into a pol III-specific template.[30] Binding of TFIIIC commits the promoter to pol III through preferential recruitment of TFIIIB. Thus, TFIIIC binds BRF but not TFIIB.[34,35] BRF competes against class II factors for limiting amounts of TBP.[20] B" then recognizes a complex containing BRF and TBP, but it does not interact with TFIIB/TBP.[30] The binding of BRF and B" prevents the association of TFIIA or TFIIB with TBP.[30] The combination of BRF and B" directs the specific recruitment of pol III. One element in this specificity involves the C34 subunit of pol III, which binds to BRF but not TFIIB.[34,36] Thus there is a sequential chain of interactions that determine which polymerase transcribes a particular gene.

In general, a great deal can be learned by comparing and contrasting what is already known about the three classes of eukaryotic transcription. The different fields have often adopted different approaches and different attitudes to similar questions. As more and more shared components are discovered, the degree of relevance of one field to another will be more clearly appreciated. Whereas the overall trend during the early years of eukaryotic transcriptional research was for the three polymerase fields to diverge from one another, we are now well into a period of overall convergence.

The mapping and sequencing of the human genome will soon provide detailed information concerning the number and organization of class III genes. A less tractable question that may continue to excite controversy is the function, if any, of the various SINE families that fill mammalian chromosomes. As by far the largest group of class III templates in mammals, these genes have received relatively little attention in terms of their promoter structures, factor requirements and regulation. A question that particularly deserves further study is what is responsible for the general underexpression of SINEs relative to other, much less abundant class III genes.

An enormous amount of work has been invested into the detailed analysis of promoter organization. Although the cis-acting sequences involved in a gene's expression have been precisely defined in many cases, it is worth remembering that the importance of any element can vary considerably according to the conditions of the assay. This point was made convincingly in two studies of silkworm genes,[37,38] and is also clear from the work on *Xenopus* 5S genes.

The cloning of the subunits of *S. cerevisiae* pol III has provided great insight into the composition of this complex enzyme. Most of the credit for this belongs to Sentenac and his colleagues. The next challenge will be to find functions for the different subunits. This would be greatly facilitated if activity could be reconstituted in vitro from the cloned components. It would also be reassuring to learn why mutations in pol III have so little effect upon 5S gene transcription in vivo. The study of mammalian pol III enzyme lags sadly behind.

The cloning and analysis of class III transcription factors is also most advanced in the yeast system. All the subunits of *S. cerevisiae* TFIIIB and TFIIIC are now cloned and sequenced. Detailed structural studies should soon follow. As for the polymerase, a major challenge will then be to reconstitute activity and deduce the functions of the individual components. TFIIIC is one of the largest and most complex transcription factors known. This level of complexity suggests that it performs multiple functions. The best characterized of these is its role as an assembly factor for TFIIIB.[24] This probably also involves activating the DNA-binding capability of TFIIIB. In addition, TFIIIC is involved in keeping class III genes free of repressive chromatin structures.[32] These various essential functions may account for the structural complexity of this factor. In contrast, TFIIIA is an extremely simple protein. Yet TFIIIA performs important roles in RNA storage[39] and transport,[40] as well as its function as a gene-specific transcription factor. This small polypeptide therefore provides a fine example of evolutionary economy. As such, it is surprising that TFIIIA is not more highly conserved between species.

The pol III transcription machinery in higher organisms is much less well defined. Although a consensus has emerged concerning TFIIIC2, the TFIIIC1 activity remains almost completely uncharacterized. The full composition of mammalian TFIIIB is also far from clear.[41] However, substantial progress has been made concerning the PTF/$SNAP_c$ complex. Any understanding of the mechanisms and regulation of pol III transcription in multicellular organisms is likely to remain vague until the molecular composition of the principal factors has been established.

The number of factors involved in pol III transcription is not yet clear, even in *S. cerevisiae*. Quite apart from the basal components, the number of regulatory factors may be enormous. However, one consideration may argue against this. Regulated transcription by pol II has been shown to involve the TAF components of the TFIID complex, which perform a coactivator or mediator function.[5,42,43] TFIIIB appears to have far fewer TAFs than TFIID.[44,45] This may mean that the pol III system is more restricted in its ability to respond to regulatory factors. The far greater diversity of pol II templates may demand a much higher flexibility of regulatory potential. However, it is not clear whether the analogy between TFIIIB and TFIID is sufficiently close to permit this conclusion.

The possibility of a TFIIF-like factor in the class III system deserves some consideration. TFIIF binds to the N-terminal region of TFIIB, including the putative zinc finger.[46] This region is highly conserved between TFIIB and BRF, which suggests that BRF might bind TFIIF or a TFIIF-related factor. However, extracts depleted of TFIIF are unimpaired in pol III transcription.[47] TFIIF facilitates both the initiation and elongation steps of pol II transcription.[48-50] The pol I system also contains a factor, TIF-IC, that stimulates both initiation and elongation.[51] Like TFIIF, TIF-IC associates tightly with its polymerase.[51] It is not yet clear whether TIF-IC bears sequence homology to TFIIF, as its gene(s) remains to be cloned. The fact that the class I and II systems contain a polymerase-associated factor that is involved in both initiation and elongation, raises a clear possibility that a comparable factor may also be part of the class III system. If no such factor exists, it will be interesting to learn why pol III is able to function in its absence. Perhaps one or more of the pol III-specific subunits has subsumed the roles of TFIIF. The small size of class III genes may also obviate the need for elongation factors.

The purification and cloning of factors allows a detailed analysis of how they interact with one another and with template DNA. This is clearly of great importance in understanding the mechanistic basis of transcription initiation. *Xenopus* TFIIIA was the first eukaryotic transcription factor to be cloned[52] and its binding to 5S DNA has now been studied at high resolution. However, the other interactions within a class III transcription complex remain ill-defined in higher organisms. This is particularly the case for type III promoters, where even the order of factor binding has yet to be established unequivocally. A knowledge of how transcription complexes assemble on vertebrate U6 genes is likely to be the key to resolving the issue of polymerase selection by U snRNA genes.

In contrast to this vagueness, the interactions and relative positions of the components in *S. cerevisiae* tRNA and 5S gene complexes have been worked out in considerable detail by Kassavetis, Geiduschek and colleagues. As a result of this outstanding work, these are some of the most closely defined eukaryotic transcription complexes. The way is now open to investigate the enzymatic roles of the individual subunits within these macromolecular assemblies. DNA binding by yeast TFIIIB should prove a particularly fascinating topic to study. It is remarkable that a factor can bind DNA in a largely sequence-independent fashion with such high affinity that it can resist a variety of chaotropic reagents. The fact that this binding needs to be activated by other components of the transcription complex is probably necessary to ensure that TFIIIB does not become attached to inappropriate sites in the genome. This must be of great importance in yeast, where TFIIIB is so difficult to displace once bound to DNA.

A promoter-deficient U6 gene can be expressed in yeast cells if TFIIIC or TFIIIB is recruited by means of a heterologous Gal4 DNA-binding domain.[53] This effect is obtained if Gal4 is fused to the τ_{138} or τ_{131} subunits of TFIIIC or to BRF, but not if it is fused to TBP or any of 10 different pol III subunits that were tested.[53] Expression is possible with the Gal4 recognition site inserted at various locations within or upstream of the gene and in either orientation.[53] Transcripts initiate near +1 in all cases.[53] It is assumed that the τ_{138} and τ_{131} fusion proteins are assembled into complete TFIIIC molecules which are directed to the Gal4 binding sites.[53] This seems reasonable, since each of these chimeric subunits can sustain cell growth in the absence of the corresponding wild-type polypeptide.[53] After the initial Gal4-mediated recruitment, the factors may reposition to their natural sites on the promoter. These surprising results have several implications. One is that polymerase recruitment is not sufficient for transcription, since no expression was observed when Gal4-pol III constructs were used.[53] The integrity of the A-block was needed for the function of Gal4-BRF.[53] This suggests that Gal4-BRF does not bypass TFIIIC, but may help to recruit it to the A-block. TFIIIC, in turn, might then serve to position TFIIIB correctly, which would explain how initiation at +1 is possible with a Gal4 site at -180.[53] A clear implication of these unexpected results is that initiation does not require a strictly ordered cascade of interactions.

An intriguing question that remains to be answered fully is how pol III can transcribe a template when so many polypeptides are positioned in its path. It seems that TFIIIC and TFIIIA are retained in the transcription complex by protein-protein interactions. Thus, either factor on its own is displaced by the polymerase, whereas they are retained in a complete initiation complex.[54-58] It is likely that the upstream binding site of TFIIIB allows it to anchor the other factors while the ICR is being transcribed. Nevertheless, a powerful imagination is needed to comprehend the molecular gymnastics that allow one large multisubunit complex to pass another on a gene. One should also bear in mind that vertebrate TFIIIB appears not to interact tightly with DNA.

After the slow process of preinitiation complex assembly, a class III gene generally remains stably committed to transcription, with pol III the only recycling factor. Even this recycling step has been optimized so that the polymerase can rapidly reinitiate without being released into a free pool.[59] The small class III templates which are bent by transcription factors bound to internal sites may be ideally suited to this facilitated recycling procedure. The system appears to be geared towards maximum transcript production once the preinitiation complex has formed. Pol I transcription is also highly efficient, and it has been suggested that the tandem arrangement of rRNA genes facilitates the direct transfer of pol I from the terminator of one template to the promoter of the next without dissociation.[60] A very different situation exists for pol II, where most components of the transcription complex dissociate following initiation; only TFIID remains at the promoter, and the other factors must reassemble for each cycle.[61] Although the release and reassociation of the class II factors makes reinitiation slow and inefficient, it provides considerable scope for regulating each cycle of transcription. The facilitated recycling pathway used by pol III bypasses almost all of the steps that are required for the first round of transcription; this greatly increases the efficiency of RNA synthesis but reduces the potential for fine regulation.

Integration of the *S. cerevisiae* retrotransposon Ty3 is directed specifically to the 5'-flanking regions of tRNA, 5S and U6 genes.[62] The site of integration occurs within 1-3 bp of the transcription start site, with Ty3 and the class III gene orientated head-to-head.[62] Such targeting requires functional promoter elements, implying that it involves an interaction between class III transcription complexes and the Ty3 integration machinery.[62-64] Position-specific integration can be reconstituted in vitro using Ty virus-like particles.[64] Promoter-directed targeting in vitro requires fractions containing TFIIIC and TFIIIB, whereas pol III is dispensable.[64] Indeed, inclusion of pol III decreased the frequency of integration.[64] This raises the possibility that the Ty3 integration machinery and pol III may compete for interaction with the initiation complex. Indeed, reducing pol III transcription with a temperature-sensitive mutation in C160 can increase the pheromone-inducible expression of Ty3 genes by up to 60-fold.[65] The upstream localization of Ty3 insertion events and the requirement for TFIIIB suggest that the feature recognized by the integration complex is TFIIIB or a TFIIIB-induced bend in DNA.

Ty1, Ty2 and Ty4 are also frequently found close to tRNA genes, although the precision of targeting is much lower.[66] Whereas Ty3 is only found adjacent to pol III templates, the other Ty elements can also integrate elsewhere.[66] Although Ty3 inserts within a few bp of the initiation site, Ty1 integrates anywhere within 100-700 bp upstream of a class III gene.[66] Indeed, the region immediately upstream of the initiation site is resistant to Ty1 integration, with the nearest insertions occurring at -70.[66] In plasmid-based in vivo assays, tRNA, 5S and U6 genes increased the frequency of Ty1 integration, sometimes by several hundred-fold.[66] This effect required an active pol III promoter.[66] Since Ty1 and Ty3 are not thought to have arisen from a common progenitor, it seems that the similarity in target selection has arisen through convergent evolution. A further example of this may be provided by the DRE retrotransposon element of *Dictyostelium*, which is only found in proximity to tRNA genes, usually within 50 bp of the start site.[67]

Most of the work done on pol III transcription has made use of naked DNA as template. Yet the genetic material transcribed in vivo is chromatin. Chromatin is, of course, much more difficult to manipulate than simple plasmid DNA. However, an appreciation of the biology of the system requires that data obtained using the more tractable systems be related ultimately back to how genes and factors behave in chromatin. This has become particularly clear during studies of gene regulation. Chromatin has been shown to play a major role in silencing the inappropriate expression of oocyte 5S genes in *Xenopus* and of SINEs in mammals. The *Xenopus* 5S genes provide one of the most precisely characterized chromatin systems, primarily due to the sustained efforts of Wolffe and colleagues.

The regulation of pol III transcription is the one area of the field in which the yeast system lags behind. Indeed, very little is known about the control of class III gene expression in yeast. This is a great pity because the wealth of structural and mechanistic information, the availability of cloned reagents, the opportunity for genetic analysis and the relative simplicity of the system should greatly facilitate the analysis of regulatory phenomena. Many studies have shown that pol III transcription in higher organisms is strongly regulated in response to growth and differentiation, viral infection and tumorigenesis. From a biomedical point of view, this is probably the most important aspect of the pol III field. And yet the work on regulation in vertebrates has most often, of necessity, examined activities that are vague and ill-defined at the molecular level. Many of the conclusions reached will need to be re-examined as the factors concerned become better characterized and as new factors are discovered. The cloning of genes for class III transcription factors and the raising of specific antibodies will allow regulatory studies to be conducted with a much higher level of confidence in the future.

A recent advance that may be of considerable significance is that the tumor suppressors RB and p53 are capable of repressing pol III transcription.[68-72] Indeed, endogenous RB is responsible for reducing the synthesis of tRNA and 5S rRNA by 5-fold in primary mouse embryonic fibroblasts.[68] If this is true of many cell types, then RB may constitute one of the most important regulators of mammalian pol III activity. It has been suggested that the inhibition of pol III transcription could provide a mechanism for suppressing cell growth.[73-75] This must be the case, to some extent, since pol III is clearly an essential contributor to cellular biosynthesis. However, it remains to be determined whether the activity of pol III is ever limiting for the growth of cells under physiological conditions. In *S. cerevisiae*, a 2-fold reduction in the level of initiator tRNA results in a 3-fold increase in doubling time.[16] If the same were true of a mammalian cell, then the 5-fold depression in tRNA synthesis that is imposed by RB must surely have a substantial impact upon the rate of growth. During tumor development, when RB function is compromised, the release of pol III from this major constraint may be an important step towards neoplasia.

An important question that has been largely neglected is the extent to which the activities of the three nuclear RNA polymerases are coordinated with respect to each other and how this is achieved. Such regulation could potentially be of considerable significance, since a lack of balance might place a huge unnecessary burden on the metabolic economy of the cell.[76] Ribosome synthesis requires the equimolar accumulation of ~80 ribosomal proteins and 28S, 18S, 5.8S and 5S rRNAs. One might anticipate that the genes encoding these products would be controlled coordinately. To a considerable extent this is the case, and the production of ribosomal components is well-balanced. In particular, the rates of ribosome biogenesis and transcription by pols I and III are closely coupled with cellular growth. However, the coordination of these activities is far from absolute. The levels of 5S rRNA tend to be controlled less precisely than large rRNA and ribosomal proteins, with an excess of up to 20% detected in many cells.[77] When yeast are starved of nutrients, pol I transcription declines rapidly, whereas 5S and then tRNA synthesis follow more slowly.[78] Thus, although pols I and III show similar responses to growth conditions, pol I transcription appears to be the most sensitive. A far more extreme example of this is seen during the differentiation of rat L6 myoblasts, where a rapidly proliferating population of cells changes into a nondividing syncytial population. The rate of pol I transcription and ribosome accumulation is 5-fold reduced in the differentiated myotubes, but synthesis of 5S rRNA and ribosomal proteins is unchanged, leading to considerable overproduction.[77] The excess of these components is rapidly degraded, thereby preventing the accumulation of a static pool of unassembled ribosomal precursors.[77] The apparently unnecessary synthesis and destruction of such large populations of macromolecules would seem to be an extraordinary example of cellular profligacy.

The obvious way to coordinate transcription by pols I, II and III would be to target shared components. These three RNA polymerases have five common subunits, which could, in theory, be used to achieve coregulation. However, no evidence for such control has been reported to date. The synthesis of 5S rRNA and large rRNA decrease almost coordinately during the encystment of *Acanthamoeba castellanii.*[79] The downregulation of large rRNA synthesis is due to the specific inactivation of the pol I enzyme.[80] This is accompanied by a change in the electrophoretic mobility of the *Acanthamoeba* homologue of the AC40 subunit that is shared between pols I and III.[80] Inactivation of AC40 would appear to provide an ideal mechanism for coregulating pols I and III during *Acanthamoeba* encystment. However, the downregulation of rRNA synthesis precedes any change in the activity of the pol III enzyme, and 5S regulation is instead achieved through the specific inactivation of TFIIIA.[79] Thus, coordinate regulation appears to be achieved through apparently unrelated mechanisms.

The other protein that is shared by the three transcription systems is TBP. By targeting TBP, the repressor protein Dr1 has been shown to coregulate pols II and III in vitro and in vivo.[81,82] However, pol I transcription is immune to this repressor.[81,82] Thus, even factors that regulate a shared component may have differential effects. It may be that Dr1 is able to shift the balance of nuclear transcription in favor of pol I. This could be of value under conditions in which the production of large rRNA lags behind the synthesis of other transcripts. It is likely that additional regulatory proteins serve to coordinate the activities of other combinations of polymerases. For example, RB has been shown to repress pols I and III without having a general effect upon pol II transcription.[68,70,71,83,84] Topoisomerase I inhibits pols I and II, but not pol III.[85-87] Factors such as these may provide a regulatory network that interlinks the activities of the three nuclear RNA polymerases. Such control must have been extremely important in the evolution of the tripartite eukaryotic transcription system.

References

1. Kassavetis GA, Nguyen ST, Kobayashi R et al. Cloning, expression, and function of *TFC5*, the gene encoding the B" component of the *Saccharomyces cerevisiae* RNA polymerase III transcription factor TFIIIB. Proc Natl Acad Sci USA 1995; 92:9786-9790.
2. Roberts S, Miller SJ, Lane WS et al. Cloning and functional characterization of the gene encoding the TFIIIB90 subunit of RNA polymerase III transcription factor TFIIIB. J Biol Chem 1996; 271:14903-14909.
3. Ruth J, Conesa C, Dieci G et al. A suppressor of mutations in the class III transcription system encodes a component of yeast TFIIIB. EMBO J 1996; 15:1941-1949.
4. Teichmann M, Dieci G, Huet J et al. Functional interchangeability of TFIIIB components from yeast and human cells in vitro. EMBO J 1997; 16:4708-4716.
5. Hernandez N. TBP, a universal eukaryotic transcription factor? Genes Dev 1993; 7:1291-1308.
6. Wang Z, Roeder RG. Structure and function of a human transcription factor TFIIIB subunit that is evolutionarily conserved and contains both TFIIB- and high-mobility-group protein 2-related domains. Proc Natl Acad Sci USA 1995; 92:7026-7030.
7. Mital R, Kobayashi R, Hernandez N. RNA polymerase III transcription from the human U6 and adenovirus type 2 VAI promoters has different requirements for human BRF, a subunit of human TFIIIB. Mol Cell Biol 1996; 16:7031-7042.
8. Larminie CGC, White RJ. Identification of a putative BRF homologue in the genome of *Caenorhabditis elegans.* DNA Sequence 1998; in press.
9. Mann C, Micouin J-Y, Chiannilkulchai N et al. *RPC53* encodes a subunit of *Saccharomyces cerevisiae* RNA polymerase C (III) whose inactivation leads to a predominantly G_1 arrest. Mol Cell Biol 1992; 12:4314-4326.

10. Shpakovski GV, Acker J, Wintzerith M et al. Four subunits that are shared by the three classes of RNA polymerase are functionally interchangeable between *Homo sapiens* and *Saccharomyces cerevisiae.* Mol Cell Biol 1995; 15:4702-4710.
11. Wang Z, Roeder RG. Three human RNA polymerase III-specific subunits form a subcomplex with a selective function in specific transcription initiation. Genes Dev 1997; 11:1315-1326.
12. L'Etoile ND, Fahnestock-ML, Shen Y et al. Human transcription factor IIIC box B binding subunit. Proc Natl Acad Sci USA 1994; 91:1652-1656.
13. Lagna G, Kovelman R, Sukegawa J et al. Cloning and characterization of an evolutionarily divergent DNA-binding subunit of mammalian TFIIIC. Mol Cell Biol 1994; 14:3053-3064.
14. Sinn E, Wang Z, Kovelman R et al. Cloning and characterization of a TFIIIC2 subunit (TFIIICβ) whose presence correlates with activation of RNA polymerase III-mediated transcription by adenovirus E1A expression and serum factors. Genes Dev 1995; 9:675-685.
15. Drabkin HJ, RajBhandary UL. Attempted expression of a human initiator tRNA gene in *Saccharomyces cerevisiae.* J Biol Chem 1985; 260:5596-5602.
16. Francis MA, Rajbhandary UL. Expression and function of a human initiator tRNA gene in the yeast *Saccharomyces cerevisiae.* Mol Cell Biol 1990; 10:4486-4494.
17. Standring DN, Venegas A, Rutter WJ. Yeast $tRNA^{Leu3}$ gene transcribed and spliced in a HeLa cell extract. Proc Natl Acad Sci USA 1981; 78:5963-5967.
18. Archambault J, Milne CA, Schappert KT et al. The deduced sequence of the transcription factor TFIIIA from *Saccharomyces cerevisiae* reveals extensive divergence from *Xenopus* TFIIIA. J Biol Chem 1992; 267:3282-3288.
19. Wang CK, Weil PA. Purification and characterization of *Saccharomyces cerevisiae* transcription factor IIIA. J Biol Chem 1989; 264:1092-1099.
20. Cormack BP, Struhl K. Regional codon randomization: defining a TATA-binding protein surface required for RNA polymerase III transcription. Science 1993; 262:244-248.
21. Braun BR, Riggs DL, Kassavetis GA et al. Multiple states of protein-DNA interaction in the assembly of transcription complexes on *Saccharomyces cerevisiae* 5S ribosomal RNA genes. Proc Natl Acad Sci USA 1989; 86:2530-2534.
22. Buratowski S, Hahn S, Sharp PA et al. Five intermediate complexes in transcription initiation by RNA polymerase II. Cell 1989; 56:549-561.
23. Kassavetis GA, Riggs DL, Negri R et al. Transcription factor IIIB generates extended DNA interactions in RNA polymerase III transcription complexes on tRNA genes. Mol Cell Biol 1989; 9:2551-2566.
24. Kassavetis GA, Braun BR, Nguyen LH et al. *S. cerevisiae* TFIIIB is the transcription initiation factor proper of RNA polymerase III, while TFIIIA and TFIIIC are assembly factors. Cell 1990; 60:235-245.
25. Moncollin V, Fischer L, Cavallini B et al. Class II (B) general transcription factor (TFIIB) that binds to the template-committed preinitiation complex is different from general transcription factor BTF3. Proc Natl Acad Sci USA 1992; 89:397-401.
26. Pinto I, Ware DE, Hampsey M. The yeast *SUA7* gene encodes a homolog of human transcription factor TFIIB and is required for normal start site selection in vivo. Cell 1992; 68:977-988.
27. Tschochner H, Sayre MH, Flanagan PM et al. Yeast RNA polymerase II initiation factor e: isolation and identification as the functional counterpart of human transcription factor IIB. Proc Natl Acad Sci USA 1992; 89:11292-11296.
28. Malik S, Lee DK, Roeder RG. Potential RNA polymerase II-induced interactions of transcription factor TFIIB. Mol Cell Biol 1993; 13:6253-6259.

29. Memet S, Saurin W, Sentenac A. RNA polymerases B and C are more closely related to each other than to RNA polymerase A. J Biol Chem 1988; 263: 10048-10051.
30. Roberts S, Colbert T, Hahn S. TFIIIC determines RNA polymerase III specificity at the TATA-containing yeast U6 promoter. Genes Dev 1995; 9:832-842.
31. Margottin F, Dujardin G, Gerard M et al. Participation of the TATA factor in transcription of the yeast U6 gene by RNA polymerase C. Science 1991; 251:424-426.
32. Burnol A-F, Margottin F, Huet J et al. TFIIIC relieves repression of U6 snRNA transcription by chromatin. Nature 1993; 362:475-477.
33. Joazeiro CAP, Kassavetis GA, Geiduschek EP. Identical components of yeast transcription factor IIIB are required and sufficient for transcription of TATA box-containing and TATA-less genes. Mol Cell Biol 1994; 14:2798-2808.
34. Khoo B, Brophy B, Jackson SP. Conserved functional domains of the RNA polymerase III general transcription factor BRF. Genes Dev 1994; 8:2879-2890.
35. Chaussivert N, Conesa C, Shaaban S et al. Complex interactions between yeast TFIIIB and TFIIIC. J Biol Chem 1995; 270:15353-15358.
36. Werner M, Chaussivert N, Willis IM et al. Interaction between a complex of RNA polymerase III subunits and the 70-kDa component of transcription factor IIIB. J Biol Chem 1993; 268:20721-20724.
37. Wilson ET, Larson D, Young LS et al. A large region controls tRNA gene transcription. J Mol Biol 1985; 183:153-163.
38. Wilson ET, Condliffe DP, Sprague KU. Transcriptional properties of BmX, a moderately repetitive silkworm gene that is an RNA polymerase III template. Mol Cell Biol 1988; 8:624-631.
39. Picard B, Wegnez M. Isolation of a 7S particle from *Xenopus laevis* oocytes: a 5S RNA-protein complex. Proc Natl Acad Sci USA 1979; 76:241-245.
40. Guddat U, Bakken AH, Pieler T. Protein-mediated nuclear export of RNA: 5S rRNA containing small RNPs in *Xenopus* oocytes. Cell 1990; 60:619-628.
41. Rigby PWJ. Three in one and one in three: it all depends on TBP. Cell 1993; 72:7-10.
42. Gill G, Tjian R. Eukaryotic coactivators associated with the TATA box binding protein. Curr Opin Genet Dev 1992; 2:236-242.
43. Tansey WP, Herr W. TAFs: guilt by association? Cell 1997; 88:729-732.
44. Kassavetis GA, Joazeiro CAP, Pisano M et al. The role of the TATA-binding protein in the assembly and function of the multisubunit yeast RNA polymerase III transcription factor, TFIIIB. Cell 1992; 71:1055-1064.
45. Chiang C-M, Ge H, Wang Z et al. Unique TATA-binding protein-containing complexes and cofactors involved in transcription by RNA polymerases II and III. EMBO J 1993; 12:2749-2762.
46. Ha I, Roberts S, Maldonado E et al. Multiple functional domains of human transcription factor IIB: distinct interactions with two general transcription factors and RNA polymerase II. Genes Dev 1993; 7:1021-1032.
47. Burton ZF, Killeen M, Sopta M et al. RAP30/74: a general initiation factor that binds to RNA polymerase II. Mol Cell Biol 1988; 8:1602-1613.
48. Orphanides G, Lagrange T, Reinberg D. The general transcription factors of RNA polymerase II. Genes Dev 1996; 10:2657-2683.
49. Reines D, Conaway JW, Conaway RC. The RNA polymerase II general elongation factors. Trends Biochem Sci 1996; 21:351-355.
50. Roeder RG. The role of general initiation factors in transcription by RNA polymerase II. Trends Biochem Sci 1996; 21:327-334.
51. Schnapp G, Schnapp A, Rosenbauer H et al. TIF-IC, a factor involved in both transcription initiation and elongation of RNA polymerase I. EMBO J 1994; 13:4028-4035.

52. Ginsberg AM, King BO, Roeder RG. *Xenopus* 5S gene transcription factor, TFIIIA: characterization of a cDNA clone and measurement of RNA levels throughout development. Cell 1984; 39:479-489.
53. Marsolier M-C, Chaussivert N, Lefebvre O et al. Directing transcription of an RNA polymerase III gene via GAL4 sites. Proc Natl Acad Sci USA 1994; 91:11938-11942.
54. Bogenhagen DF, Wormington WM, Brown DD. Stable transcription complexes of *Xenopus* 5S RNA genes: a means to maintain the differentiated state. Cell 1982; 28:413-421.
55. Jahn D, Wingender E, Seifart KH. Transcription complexes for various class III genes differ in parameters of formation and stability towards salt. J Mol Biol 1987; 193:303-313.
56. Wolffe AP, Jordan E, Brown DD. A bacteriophage RNA polymerase transcribes through a *Xenopus* 5S RNA gene transcription complex without disrupting it. Cell 1986; 44:381-389.
57. Campbell FE, Setzer DR. Displacement of *Xenopus* transcription factor IIIA from a 5S rRNA gene by a transcribing RNA polymerase. Mol Cell Biol 1991; 11:3978-3986.
58. Bardeleben C, Kassavetis GA, Geiduschek EP. Encounters of *Saccharomyces cerevisiae* RNA polymerase III with its transcription factors during RNA chain elongation. J Mol Biol 1994; 235:1193-1205.
59. Dieci G, Sentenac A. Facilitated recycling pathway for RNA polymerase III. Cell 1996; 84:245-252.
60. Baker SM, Platt T. Pol I transcription: which comes first, the end or the beginning? Cell 1986; 47:839-840.
61. Zawel L, Kumar KP, Reinberg D. Recycling of the general transcription factors during RNA polymerase II transcription. Genes Dev 1995; 9:1479-1490.
62. Chalker DL, Sandmeyer SB. Ty3 integrates within the region of RNA polymerase III transcription initiation. Genes Dev 1992; 6:117-128.
63. Chalker DL, Sandmeyer SB. Sites of RNA polymerase III transcription initiation and *Ty3* integration at the U6 gene are positioned by the TATA box. Proc Natl Acad Sci USA 1993; 90:4927-4931.
64. Kirchner J, Connolly CM, Sandmeyer SB. Requirement of RNA polymerase III transcription factors for in vitro position-specific integration of a retroviruslike element. Science 1995; 267:1488-1491.
65. Hull MW, Erickson J, Johnston M et al. tRNA genes as transcriptional repressor elements. Mol Cell Biol 1994; 14:1266-1277.
66. Devine SE, Boeke JD. Integration of the yeast retrotransposon Ty1 is targeted to regions upstream of genes transcribed by RNA polymerase III. Genes Dev 1996; 10:620-633.
67. Marschalek R, Brechner T, Amon-Bohm E et al. Transfer RNA genes: landmarks for integration of mobile genetic elements in Dictyostelium discoideum. Science 1989; 244:1493-1496.
68. White RJ, Trouche D, Martin K et al. Repression of RNA polymerase III transcription by the retinoblastoma protein. Nature 1996; 382:88-90.
69. Chesnokov I, Chu W-M, Botchan MR et al. p53 inhibits RNA polymerase III-directed transcription in a promoter-dependent manner. Mol Cell Biol 1996; 16:7084-7088.
70. Chu W-M, Wang Z, Roeder RG et al. RNA polymerase III transcription repressed by Rb through its interactions with TFIIIB and TFIIIC2. J Biol Chem 1997; 272:14755-14761.
71. Larminie CGC, Cairns CA, Mital R et al. Mechanistic analysis of RNA polymerase III regulation by the retinoblastoma protein. EMBO J 1997; 16:2061-2071.

72. Cairns CA, White RJ. Unpublished observations.
73. Nasmyth K. Another role rolls in. Nature 1996; 382:28-29.
74. Larminie CGC, Alzuherri HM, Cairns CA et al. Transcription by RNA polymerases I and III: a potential link between cell growth, protein synthesis and the retinoblastoma protein. J Mol Med 1998; In press.
75. White RJ. Regulation of RNA polymerases I and III by the retinoblastoma protein: a mechanism for growth control? Trends Biochem Sci 1997; 22:77-80.
76. White RJ. Coordination of nuclear RNA polymerase activity. J NIH Research 1995; 7:48-49.
77. Zahradka P, Larson DE, Sells BH. Regulation of ribosome biogenesis in differentiated rat myotubes. Mol Cell Biochem 1991; 104:189-194.
78. Clarke EM, Peterson CL, Brainard AV et al. Regulation of the RNA polymerase I and III transcription systems in response to growth conditions. J Biol Chem 1996; 271:22189-22195.
79. Matthews JL, Zwick MG, Paule MR. Coordinate regulation of ribosomal component synthesis in *Acanthamoeba castellanii*: 5S RNA transcription is down regulated during encystment by alteration of TFIIIA activity. Mol Cell Biol 1995; 15:3327-3335.
80. Bateman E, Paule MR. Regulation of eukaryotic ribosomal RNA synthesis by RNA polymerase modification. Cell 1986; 47:445-450.
81. White RJ, Khoo BC-E, Inostroza JA et al. The TBP-binding repressor Dr1 differentially regulates RNA polymerases I, II and III. Science 1994; 266:448-450.
82. Kim S, Na JG, Hampsey M et al. The Dr1/DRAP1 heterodimer is a global repressor of transcription in vivo. Proc Natl Acad Sci USA 1997; 94:820-825.
83. Cavanaugh AH, Hempel WM, Taylor LJ et al. Activity of RNA polymerase I transcription factor UBF blocked by *Rb* gene product. Nature 1995; 374:177-180.
84. Voit R, Schafer K, Grummt I. Mechanism of repression of RNA polymerase I transcription by the retinoblastoma protein. Mol Cell Biol 1997; 17:4230-4237.
85. Brill SJ, DiNardo S, Voelkel-Meiman K et al. Need for DNA topoisomerase activity as a swivel for DNA replication for transcription of ribosomal RNA. Nature 1987; 326:414-416.
86. Choder M. A general topoisomerase I-dependent transcriptional repression in the stationary phase in yeast. Genes Dev 1991; 5:2315-2326.
87. Merino A, Madden KR, Lane WS et al. DNA topoisomerase I is involved in both repression and activation of transcription. Nature 1993; 365:227-232.

Index

A

α-amanitin 38, 40, 57, 64
A-block 23-29, 31, 32, 36-40, 84, 94, 95, 97, 103, 133, 134, 138-140, 142-146, 149, 150, 165, 180-183, 202, 258
ABC10α 59, 67
ABC10β 67, 70
ABC14.5 66, 67
ABC23 66-68, 107, 204
ABC27 60, 61, 66, 67, 139, 163, 164, 236
ABF1 70
AC19 59, 65-67, 70, 204
AC40 58, 59, 65-67, 70, 163, 164, 261
Acanthamoeba 224, 261, 265
Acetylation 178, 182, 187
Adenovirus 1, 4, 11, 21, 31-33, 42, 55, 71, 78, 82, 112, 233-236, 240
Alu 6-11, 23, 27, 34, 35, 78, 175, 182-185, 196-198, 202, 231, 233-236
AP1 35, 202
Arabidopsis 40, 79, 81, 191
ATF 33, 34, 42, 202

B

B' 83, 84, 87, 89, 100, 135, 136, 138, 149, 150
B" 83, 84, 88, 89, 91, 93, 96, 100, 135, 136, 138, 139, 145, 150, 200, 231, 253, 255
B-block 23, 27, 28, 31, 32, 34, 36, 38, 40, 93, 94, 96-99, 133, 139, 140, 143, 146, 148, 150, 165
B-TFIID 82, 92
B1 7, 9-11, 23, 27, 34, 77, 78, 100, 202, 203, 221, 222, 228, 231, 237, 238
B2 7, 8, 11, 23, 27, 31, 77, 78, 175, 183, 194, 198, 203, 221, 222, 226, 231, 232, 237, 238
BC1 6, 8, 34, 35, 238
BC200 6, 238
BmX 7, 31, 199
BN51 68, 69
Bombyx 7, 26, 28, 78, 107, 199, 200, 224
BRF1 84, 85, 87, 95

C

C-block 23, 24, 26, 27, 103, 104, 141, 142, 144, 145, 150, 176
c-myc 11, 41, 42, 95, 254
C. albicans 86, 87
C11 60
C25 60, 61, 70
C31 60, 61, 64, 68-70, 163, 164
C32 58, 60, 70, 141
C34 60, 61, 64, 68-70, 138, 139, 141, 163, 164, 255
C37 59, 163, 164
C39 60, 69, 141
C53 60, 61, 66, 68-70, 139, 148, 163
C62 60, 68, 141
C82 60, 61, 64, 68-70, 139, 148, 163, 164
C128 59-61, 65, 70, 139, 163, 164, 166
C160 59-64, 68, 70, 139, 163-165, 259
CACCC 41
Caenorhabditis 89
Calpains 198
Candida 86
Casein kinase II 66, 100, 204
cdc2 203, 204, 228
cdc25 204, 221
CeBRF 89-91
Cell cycle 69, 176, 197, 204, 215, 217, 228-230, 240
Chromatin 35, 149, 173-184, 199, 213-220, 226-228, 234-238, 256, 259
CL cells 236
CpG 35, 183-185
Creatine kinase gene 11, 42
CREB 202
Cyclin 85, 194, 195, 197, 203, 204, 227-229, 240
Cycloheximide 69, 225, 230

D

Differentiation 222-224, 259, 260
Dr1 191-193, 196, 234, 235, 261
DRAP1 191, 192
DRE 259
Drosophila 3-5, 26, 27, 29, 31, 42, 60, 62, 65, 67, 77, 79, 85, 138, 141, 144, 148, 166, 179, 180, 205, 227, 236
  D. melanogaster 3, 67
DSE 37, 38, 41, 42, 111, 147, 201

E

E-box 39, 202
E1A 233-235, 239, 240
E2E 11, 42, 82
EBER 4, 6, 23, 27, 32-34, 78, 194, 198, 202
EC cells 222, 223
Embryonal carcinoma 222

F

F9 cells 222, 223
Founder elements 9, 35

G

G0 phase 230
G1 phase 69, 194, 195, 229, 230
G2 phase 229, 230
Growth 4, 7, 10, 61, 66, 70, 84, 94, 107-109, 168, 191, 193, 194, 196, 197, 204, 205, 222, 225, 226, 230, 231, 238, 258-260

H

H1 *See also* Histone. 1, 5, 7, 40, 62, 89, 108, 109, 173-180, 183, 184, 217-220
hBRF 91, 92, 110, 141, 146, 148, 195, 196, 229
Heat shock 232
HeLa cells 32, 34, 41, 90, 98, 107, 109, 183, 184, 191, 202, 205, 220, 227-229, 232-235
Hepatitis B virus 236
Herpes simplex virus 235
Histone 173-185, 191, 192, 213, 217-220, 228
  B4 218
  H1 174, 175, 178-180, 217-220
  H2A 173, 176, 177, 179
  H2B 173, 176, 179
  H3 173, 177, 182
  H4 173, 182
  H5 176, 184
HLH 95
HMG 91, 94, 95, 133, 178, 218
HTLV-1 237

I

ICP27 235, 236
ICR 23-28, 31, 32, 34, 40, 93, 94, 103, 104, 133, 141, 142, 144, 146, 166, 175, 177-181, 184, 215, 254, 258
ID gene 10

K

Kluveromyces lactis 86, 87

L

L1 11
La protein 101
Lupus antigen 100

M

Maize 40, 79
MAP kinase 203, 228
Methylation 35, 142, 183-185, 202
Mitosis 203, 229, 230
MRP 1, 5, 7, 40, 41, 201
MSP 40

N

N3RdU 84, 94, 134, 136, 144, 164
Nucleosome 173-184, 217-219, 228, 235

O

OAX 7, 31, 179
Oct-1 109-112, 146-148, 201-203
Oct-2 111, 112, 146, 202, 203
Octamer 34, 37, 40, 41, 111, 129, 147, 173, 176-178, 183, 184, 201, 202
Okadaic acid 228
Oocyte 1-3, 7, 8, 11, 12, 24-27, 30, 31, 37, 41-43, 102, 105, 107, 141-144, 175, 178-180, 183, 184, 198, 200, 201, 204, 213-221, 227, 228, 259

P

p43 107
p53 136, 196-198, 203, 239, 241, 260
Parietal endoderm 222
PBP 109-111, 146
PCF4 84
PE cells 222, 223
PKC 204, 205, 237
Poliovirus 232, 233
Polyoma virus 237
POU 111, 112, 147
PP2A 205, 229
Protein kinase C 66, 204, 236, 237
PSE 34, 37-42, 80, 109-111, 146-148, 224
Pseudorabies virus 235
PTF 109-112, 146-148, 256

PTFα 109, 146, 147
PTFβ 109, 110, 146, 147
PTFδ 109-111, 146
PTFγ 109-111, 146
PZF1 107

R

Rana catesbeiana 108
RB 85, 193-196, 198, 227, 229, 230, 234, 235, 239-241, 260, 261
RET1 65
Retinoblastoma 193, 239
Retrotransposition 6, 8
Retrotransposon 11, 259
RNase P 5
RNP 5-7, 105-107
Ro 6
RPB5 236
RPC19 65
RPC31 68, 69
RPC34 69
RPC40 65, 70
RPC53 68
RPC82 68
RPC160 62, 64, 68, 70
RPO31 62

S

S phase 69, 194-196, 215, 220, 226, 229, 230, 234, 235
S-2 cells 79, 205, 227, 236, 237
SABP 183, 202
Saccharomyces 1, 28, 37, 57, 58, 65, 68, 69, 81, 101, 108, 151, 179, 191
  S. cerevisiae 1, 3-5, 26, 29, 34, 36, 40, 59, 63, 66, 67, 69, 70, 85-87, 89, 90, 91, 94, 95, 101, 107, 108, 134, 137, 141, 145, 179, 182, 199, 202, 205, 253, 256, 257, 259, 260
  S. pombe 3, 4, 34, 36, 67, 69
SANT domain 88
Satellite 1 215
Schizosaccharomyces 1, 5
Serum 10, 83, 222, 226, 227, 230
5S genes 1, 2, 25, 26, 29, 61, 77-79, 84, 90, 91, 100, 102-104, 107, 131, 140-144, 146, 149-151, 164, 174-180, 182-184, 198, 204, 213-220, 226, 228, 237, 256, 259
  Oocyte 5S genes 175, 178-180, 184, 213-220, 228, 259
  Somatic 5S gene 2, 24-26, 141-143, 151, 173, 176-180, 201, 213-217, 219, 220, 227, 228
Silk gland 199, 200, 224, 225
Silkworm 28-31, 57, 77, 100, 107, 139, 144, 199, 200, 224, 256
SINE 1, 7-11, 31, 182-184, 231, 232, 256, 259
7SK 1, 5, 7, 23, 34, 41-43, 92, 96, 109-111, 146, 196, 201, 202, 231, 232, 254
7SL 1, 5, 6, 9, 10, 23, 34, 35, 40, 96, 197, 202, 231-233, 238
SNAP43 110, 111, 146
SNAP45 110, 111, 146
SNAP50 146, 147
SNAPc 110-112, 146-148, 256
snRNA 4, 31, 36-41, 62, 109-111, 222, 257
Sp1 26, 33, 37, 40, 202, 203
Staf 31, 37, 41, 147, 200, 201
Stationary phase 225, 226
SV3T3 cells 226
SV40 149, 175, 180, 181, 202, 237-240

T

3T3 cells 183, 226, 238
3T6 cells 226
T antigen 202, 237, 238-240
TAF 82, 83, 89-92, 195, 197, 200, 203, 223, 224, 228, 229, 257
Tagetitoxin 12, 38, 40, 42, 57
TAP1 198, 199
TATA 26, 28-30, 32-34, 36-43, 77-82, 84, 89, 90, 131, 135-139, 146-148, 150, 151, 181, 182, 193, 200, 237, 255
Tax 237
TBP 69, 77-92, 110, 131, 135-139, 141, 146, 148, 150, 182, 191-193, 195-198, 200, 204, 223, 227-229, 231, 233-237, 239, 253-255, 258, 261
TDS4 84
Telomerase 7
TFC1 94, 95
TFC2 107
TFC3 94, 95
TFC4 95, 96
TFC5 88
TFIIA 89, 148, 200, 255
TFIIB 85-91, 139, 192, 193, 254, 255, 257
TFIIIA 12, 42, 43, 66, 77, 92, 102-108, 131, 133, 141-146, 149, 150, 165, 176, 178, 179, 214-218, 220, 224, 228, 231, 232, 253-258, 261
  S-TFIIIA 102

TFIIIB 60, 61, 69, 77, 82-84, 87-93, 96, 97, 100, 101, 110, 131, 133-141, 144-151, 163, 165, 179-182, 192, 195, 196, 198, 200, 201, 203-205, 215, 216, 221, 223-239, 253-259
  0.38M-TFIIIB 90-92, 146
  0.48M-TFIIIB 90, 92, 146
  hTFIIIB-α 92, 93, 146, 253
  hTFIIIB-β 92, 93
  TFIIIB90 91
TFIIIC 60, 77, 83, 84, 86-89, 93-96, 98-101, 131, 133-146, 148-150, 165, 179-182, 192, 195, 196, 198-202, 204, 205, 215, 216, 219, 220, 221, 223-236, 238, 239, 253-256, 258, 259
  τ131 88, 93-96, 134-136, 138, 145, 146, 149, 150, 204, 225, 258
  τ138 88, 93-95, 99, 133, 145, 204, 253, 258
  τ55 93, 94, 133, 134
  τ60 93
  τ91 93, 94, 133, 145
  τ95 93-95, 133, 134, 144, 145, 204
  TFIIIC0 97, 140
  TFIIIC1 96, 97, 99, 100, 131, 140, 143, 146, 148, 149, 256
  TFIIIC2 96-100, 131, 140, 143, 144, 146, 148, 149, 196, 227, 230, 232-234, 239, 256
  TFIIIC2a 97-99, 148, 227, 230, 232-235, 239
  TFIIIC2b 97-99, 227, 230, 232-235, 239
  TFIIICα 98, 99, 148, 149, 227, 233, 234, 253
  TFIIICβ 97-99, 148, 227, 233, 234, 239, 254
TFIIID 100, 139, 144
TFIIIE 89, 93, 100, 138, 231, 253
TFIIIR 199, 200
TPA 227
TPR 95, 96, 135, 138
Transformation 10, 136, 197, 237-239
tRNA 1-6, 8, 9, 23, 27-34, 36, 38, 40, 60-62, 64, 65, 68-70, 77-79, 83-85, 88-90, 92, 93, 95-97, 100, 101, 107, 109, 131-141, 144-146, 148-151, 163-166, 168, 175, 180, 181, 183, 184, 192, 194, 197-202, 204, 205, 213, 215, 216, 218-222, 224-231, 233, 235-238, 253, 254, 257, 259, 260
Trypanosome 4
TUF 70
Ty1 259
Ty3 259

U

U1 37, 39, 101, 109, 110, 146, 180, 200, 218
U2 4, 37, 39, 40, 101, 109, 110, 146, 180, 201, 203, 218
U3 4
U4 4, 5
U6 1, 4, 5, 23, 36-42, 62, 78-80, 88, 92-94, 96, 97, 100, 109-111, 133, 139, 146-148, 150, 181, 182, 194, 196-198, 200-202
URP2 199
USE 40
USF 43, 202

V

VAI 4, 31, 32, 57, 69, 82, 90-92, 96, 97, 99, 101, 110, 131, 140, 149, 150, 168, 184, 194, 196, 197, 202, 204, 205, 223, 224, 227, 232-237
VAII 4
Vault 6, 32, 33
VBP 202
vRNA 6, 32, 77

W

WD40 99

X

Xenopus 1, 5, 7, 11, 12, 23-27, 29-31, 37-39, 41-43, 77, 78, 85, 90, 96, 98, 99, 101-103, 105-108, 111, 140-143, 148, 149, 151, 165, 166, 168, 174, 175, 178-180, 182, 184, 198, 200-204, 213-216, 218-220, 227, 228, 256, 257, 259
  X. borealis 108, 173, 176, 177, 179, 184
  X. laevis 1, 3, 7, 26, 90, 108, 176-178, 200, 201

Y

Y RNA 6, 41
  Y1 6
  Y3 6, 148
  Y4 6, 41, 201
  Y5 6
YB3 96, 99
yBRF 69, 84-91, 96, 135-139, 145, 150, 200
YY1 202

Z

Zinc 60, 64-68, 85-87, 90, 91, 102-105, 107-109, 142, 143, 200, 201, 254, 257

MIX
Papier aus verantwortungsvollen Quellen
Paper from responsible sources
FSC® C105338

If you have any concerns about our products,
you can contact us on
ProductSafety@springernature.com

In case Publisher is established outside the EU,
the EU authorized representative is:
Springer Nature Customer Service Center GmbH
Europaplatz 3, 69115 Heidelberg, Germany

Printed by Libri Plureos GmbH
in Hamburg, Germany